AF573969

OXFORD LECTURE SERIES IN MATHEMATICS AND ITS APPLICATIONS

For a full list of titles, please visit http://www.oup.co.uk/academic/science/maths/series/OLSMA/

10. P.L. Lions: *Mathematical topics in fluid mechanics, Vol. 2: Compressible models*
11. W.T. Tutte: *Graph theory as I have known it*
12. Andrea Braides and Anneliese Defranceschi: *Homogenization of multiple integrals*
13. Thierry Cazenave and Alain Haraux: *An introduction to semilinear evolution equations*
14. J.Y. Chemin: *Perfect incompressible fluids*
15. Giuseppe Buttazzo, Mariano Giaquinta and Stefan Hildebrandt: *One-dimensional variational problems: an introduction*
16. Alexander I. Bobenko and Ruedi Seiler: *Discrete integrable geometry and physics*
17. Doina Cioranescu and Patrizia Donato: *An introduction to homogenization*
18. E.J. Janse van Rensburg: *The statistical mechanics of interacting walks, polygons, animals and vesicles*
19. S. Kuksin: *Hamiltonian partial differential equations*
20. Alberto Bressan: *Hyperbolic systems of conservation laws: the one-dimensional Cauchy problem*
21. B. Perthame: *Kinetic formulation of conservation laws*
22. A. Braides: *Gamma-convergence for beginners*
23. Robert Leese and Stephen Hurley: *Methods and Algorithms for Radio Channel Assignment*
24. Charles Semple and Mike Steel: *Phylogenetics*
25. Luigi Ambrosio and Paolo Tilli: *Topics on Analysis in Metric Spaces*
26. Eduard Feireisl: *Dynamics of Viscous Compressible Fluids*
27. Antonín Novotný and Ivan Straškraba: *Introduction to the Mathematical Theory of Compressible Flow*
28. Pavol Hell and Jarik Nesetril: *Graphs and Homomorphisms*
29. Pavel Etingof and Frederic Latour: *The dynamical Yang-Baxter equation, representation theory, and quantum integrable systems*
30. Jorge Ramirez Alfonsin: *The Diophantine Frobenius Problem*
31. Rolf Niedermeier: *Invitation to Fixed Parameter Algorithms*
32. Jean-Yves Chemin, Benoit Desjardins, Isabelle Gallagher and Emmanuel Grenier: *Mathematical Geophysics: An introduction to rotating fluids and the Navier-Stokes equations*
33. Juan Luis Vázquez: *Smoothing and Decay Estimates for Nonlinear Diffusion Equations*
34. Geoffrey Grimmett and Colin McDiarmid: *Combinatorics, Complexity and Chance*
35. Alessio Corti: *Flips for 3-folds and 4-folds*
36. Kirsch and Grinberg: *The Factorization Method for Inverse Problems*
37. Ghergu and Rădulescu, *Singular Elliptic Problems: Bifurcation and Asymptotic Analysis*

Oxford Lecture Series in
Mathematics and its Applications

Singular Elliptic Problems: Bifurcation and Asymptotic Analysis

Marius Ghergu
Vicenţiu D. Rădulescu

CLARENDON PRESS • OXFORD
2008

OXFORD
UNIVERSITY PRESS

Oxford University Press, Inc., publishes works that further
Oxford University's objective of excellence
in research, scholarship, and education.

Oxford New York
Auckland Cape Town Dar es Salaam Hong Kong Karachi
Kuala Lumpur Madrid Melbourne Mexico City Nairobi
New Delhi Shanghai Taipei Toronto

With offices in
Argentina Austria Brazil Chile Czech Republic France Greece
Guatemala Hungary Italy Japan Poland Portugal Singapore
South Korea Switzerland Thailand Turkey Ukraine Vietnam

Published by Oxford University Press, Inc.
198 Madison Avenue, New York, New York 10016

www.oup.com

Library of Congress Cataloging-in-Publication Data

Ghergu, Marius.
Singular elliptic problems : bifurcation and asymptotic analysis /
Marius Ghergu, Vicenţiu D. Rădulescu.
p. cm. — (Oxford lecture series in mathematics and its applications; 37)
Includes bibliographical references and index.
ISBN 978-0-19-533472-2
1. Differential equations, Elliptic—Asymptotic theory.
2. Differential equations, Nonlinear. 3. Bifurcation theory.
I. Radulescu, V. II. Title.
QA377.G47 2008
515′.3533—dc22 2007060129

1 3 5 7 9 8 6 4 2

Printed in the United States of America
on acid-free paper

To our families,
for their patience and continuous support over the years

PREFACE

> The most incomprehensible thing about the world is that it is comprehensible.
>
> Albert Einstein (1879–1955)

The development of nonlinear analysis during the last few decades has been profoundly influenced by attempts to understand various phenomena from mathematical physics. One of the beauties of the subject is the immense breadth of mathematics that has been applied in this pursuit.

There is an enormous body of literature in nonlinear elliptic partial differential equations that stretches back half a century. However, we shall make almost no reference to this literature, and shall rely almost entirely upon personal results. These lecture notes are primarily intended to fill, in a substantial way, the absence of a book dealing with the qualitative analysis of some basic *singular* stationary processes arising in nonlinear sciences. This volume aims to offer an introduction to this subject, and also to present some research problems. The models that we analyze represent a compromise between the description of physical phenomena and analytical requirements; accordingly, our presentation is characterized by a strict interplay between mathematics and nonlinear sciences.

The book is an outgrowth of our original research on the subject during the last few years, and much of the development is motivated by problems arising in applications. However, most of the proofs have been completely reworked and we are especially careful to explain where each chapter is going, why it matters, and what background material is required. Although the theory that we describe could have been carried out on differentiable manifolds even from the beginning, we have chosen to develop it on domains on the Euclidean space. However, the techniques we develop can be extended to Laplace–Beltrami operators on Riemannian manifolds.

The major thrust of this book is the qualitative analysis of some classes of nonlinear stationary problems involving different types of singularities. Be aware, this is definitely a research book. We are mainly concerned with the following types of problems. We first study singular solutions of the logistic equation, with a basic model that is described by the semilinear elliptic equation $\Delta u = u^p$, where $p > 1$. The research program around this equation flourished after the pioneering papers by Bieberbach and Rademacher, continued with the deep contributions of Loewner and Nirenberg in Riemannian geometry, and creating recently (because of the works by Dynkin and Le Gall) a nonlinear analogue of the classical relation between Brownian motion and potential theory. Equations of this type arise in astrophysics, genetics, meteorology, theory of atomic spectra, and the Yamabe problem in geometry. A first consequence of such types of nonlinearities is the possibility of the existence of a "large solution"—that is, a solution blowing up at the boundary. When the large solution is unique, it is a maximal solution and dominates any solution. In connection with the previously mentioned applications, the existence of the large solution in a ball was used by Iscoe to establish the compact support property of super-Brownian motion, demonstrating the importance of the relationship between properties of superdiffusion and the equation. Next, we are concerned with Lane–Emden–Fowler equations and Gierer–Meinhardt systems with singular nonlinearity. The model problem in such cases is described by equations like $-\Delta u = u^{-\alpha}$, where α is a positive real number. To our best knowledge, the first study in this direction is from Fulks and Maybee, who proved existence and uniqueness results by using a fixed point argument; moreover, they showed that solutions of the associated parabolic problem tend to the unique solution of the corresponding elliptic equation. Different approaches are the result of Coclite and Palmieri, respectively Crandall, Rabinowitz, and Tartar, who approximated the singular equation with regular problems, where the standard monotonicity techniques do work. Singular problems of this type arise in the context of chemical heterogeneous catalysts and chemical catalyst kinetics, in the theory of heat conduction in electrically conducting materials, singular minimal surfaces, as well as in the study of non-Newtonian fluids, boundary layer phenomena for viscous fluids, glacial advance, transport of coal slurries down conveyor belts, and in several other geophysical and industrial contents. In both cases, because of the meaning of the unknowns (concentrations, populations, etc.), the positive solutions are relevant in most situations.

We intend to give a systematic treatment of the basic mathematical theory and constructive methods for these classes of nonlinear elliptic equations, as well as their applications to various processes arising in mathematical physics. Our approach leads not only to the basic results of existence, uniqueness, and multiplicity of solutions, but also to several qualitative properties, including bifurcation, asymptotic analysis, blow-up and so forth. Moreover, because the book is concerned primarily with classical solutions, the monotone iteration processes we apply for various classes of nonlinear singular problems are adaptable to numerical solutions of the corresponding discrete processes. To place the text in better perspective, each chapter is concluded with a section on historical notes that includes references to all important and relatively new results. In addition to cited works, the list of references contains many other works related to the material developed in this volume.

The organization of the book is briefly summarized as follows. The first chapter deals with preliminary material, such as the method of sub- and supersolution, several variants of the maximum principle (Stampacchia, Vázquez, Pucci, and Serrin), and various existence and uniqueness results for nonlinear elliptic boundary value problems.

Part II is composed of two chapters, which are concerned with singular solutions of logistic-type equations or systems. There are studied both equations with blow-up boundary solutions and entire solutions blowing up at infinity for elliptic systems. In all these cases, the major role played by the Keller–Osserman condition is discussed.

In the third part of this book we are concerned with elliptic problems involving singular nonlinearities, either in isotropic or in anisotropic media. Chapter 4 deals with sublinear elliptic problems that are affected by singular perturbations. We distinguish between equations on bounded domains or on the whole space and we are also concerned with a related bifurcation problem. Chapters 5 and 6 are devoted to the study of a bifurcation problem in the case of linear growth for the nonlinearity. Two different situations are distinguished and a complete discussion is developed in both circumstances. The superlinear case is studied in Chapter 7, by means of variational arguments, whereas Chapter 8 is concerned with stability properties of solutions. Chapter 9 is devoted to the study of the "competition" between various terms in a singular Lane–Emden–Fowler equation with convection and variable (possible, singular) potential. In the last chapter of these lecture notes, the qualitative analysis of solutions is extended to the case of singular Gierer–Meinhardt systems. We refer to the works of J.M. Ball [11,12], V. Barbu [19], L. Beznea and N. Boboc [22], H. Brezis [30], and P.G. Ciarlet [46,47] for related results and various applications to concrete phenomena.

Four appendices illustrate some basic mathematical tools applied in this book: elements of spectral theory for differential operators, the implicit function theorem, Ekeland's variational principle, and the mountain pass theorem. These auxiliary chapters deal with some analytical methods used in this volume, but also include some complements.

Each problem we develop in this book has its own difficulties. That is why we intend to develop some standard and appropriate methods that are useful and that can be extended to other problems. However, we do our best to restrict the prerequisites to the essential knowledge. We define as few concepts as possible and give only basic theorems that are useful for our topic. The only prerequisite for this volume is a standard graduate course in partial differential equations, drawing especially from linear elliptic equations to elementary variational methods, with a special emphasis on the maximum principle (weak and strong variants). This volume may be used for self-study by advanced graduate students and engineers, and as a valuable reference for researchers in pure and applied mathematics and physics.

Our vision throughout this volume is closely inspired by the following words of Henri Poincaré on the role of partial differential equations in the development of other fields of mathematics and in applications: *Nevertheless, each time I can, I aim the absolute rigor for two reasons. In the first place, it is always hard for a geometer to consider a problem without resolving it completely. In the second place, these equations that I will study are susceptible, not only to physical applications, but also to analytical applications. It is using the existence theory of the Dirichlet problem that Riemann founded his magnificent theory of Abelian functions. Since then, other geometers have made important applications of the same principle to the most fundamental parts of pure analysis. Is it still permitted to content oneself with a demi-rigor? And who will say that the other problems of mathematical physics will not, one day, be called to play in analysis a considerable role, as has been the case of the most elementary of them?* (Henri Poincaré [164])

May 2007

ACKNOWLEDGMENTS

We acknowledge, with unreserved gratitude, the crucial role of Professor Philippe G. Ciarlet, who encouraged writing a book on this subject, but also for his constant support over the years.

We are greatly indebted to Sir John M. Ball for accepting our work in his prestigious series at Oxford University Press. We are also grateful to Dr. Michael Penn, the Oxford book program editor, and to Paul Hobson, the production editor, for their efficient and enthusiastic help, as well as for numerous suggestions related to this volume.

We extend our warm thanks to Professors Catherine Bandle, Viorel Barbu, Lucian Beznea, Bernard Brighi, Olivier Goubet, Patrizia Pucci, and Michel Willem for their useful comments and remarks during the preparation of this work.

We thank Dr. Nicolae Constantinescu for the professional drawing of figures contained in this book.

Both authors acknowledge the support of grants 2-CEx06-11-18/2006 and CNCSIS PNII-79/2007. They have been also supported by grants CNCSIS AT-191/2007 (M. Ghergu) and CNCSIS A-589/2007 (V. Rădulescu).

CONTENTS

PART I

PRELIMINARIES

1

BASIC METHODS

> If I have seen further than others,
> it is by standing upon the
> shoulders of giants.
>
> Sir Isaac Newton (1642–1727),
> *Letter to Robert Hooke*, 1675

The main purpose of nonlinear analysis is to study and describe various material systems arising in mathematical physics and other applied sciences. The language of nonlinear analysis is very close to that of functional analysis and partial differential equations.

This chapter deals with some general features about the qualitative analysis of nonlinear elliptic partial differential equations. The main issues we raise in what follows concern the existence and uniqueness of the solution. The existence of solutions is essential for a prescribed model to make sense. The uniqueness of the solution is a natural requirement for many problems. In some cases, uniqueness becomes a difficult question, but related interesting questions concern multiplicity of solutions, as well as the existence of maximal or minimal solutions. That is why monotonicity properties, described by means of comparison principles, play a central role in this chapter and throughout the book.

1.1 A fixed point result

This type of result applies successfully to classical ordinary differential equations (ODE) boundary value problems. Given a second order equation $y''(t) = f(t, y(t), y'(t))$ subject to the Dirichlet boundary condition, we look for a solution as a *fixed point* of a compact map F (i.e., $F(y) = y$) defined on a suitable closed subset of a normed vector space. There are many results in this direction, but we restrict our attention to the following basic property, which is due to Leray and Schauder [131].

Theorem 1.1.1 (Leray–Schauder) *Let C be a convex set in a normed vector space E and let $U \subset C$ be an open set that contains the origin. Then each compact map $F : \overline{U} \to E$ has at least one of the following properties:*

(i) *F has a fixed point.*

(ii) *There exist $x \in \partial U$ and $0 < \lambda < 1$ such that $x = \lambda F(x)$.*

For the proof we refer the reader to Granas and Dugundji [98], O'Regan [152], and O'Regan and Precup [153]. We next give an application of this result to a singular ODE.

Consider the Sturm–Liouville problem

$$\begin{cases} y''(t) + f(t, y(t)) = 0 & 0 < t < 1, \\ y(0) = a, \; y(1) = b, \end{cases} \tag{1.1}$$

where $f : (0,1) \times \mathbb{R} \to \mathbb{R}$ is a continuous function such that
$(F1)$ for all $r > 0$ there exists $\psi_r \in L^1_{\text{loc}}(0,1)$ satisfying

$$|f(t,y)| \le \psi_r(t) \qquad \text{for all } (t,y) \in (0,1) \times [-r, r], \tag{1.2}$$

and

$$\int_0^1 t(1-t)\psi_r(t)dt < \infty. \tag{1.3}$$

We also consider the problem

$$\begin{cases} y''(t) + \lambda f(t, y(t)) = 0 & 0 < t < 1, \\ y(0) = y(1) = 0\,. \end{cases} \tag{1.4}$$

Theorem 1.1.2 *Assume that f fulfills $(F1)$ and there exists $M > 0$ such that for all $0 < \lambda < 1$ and for any solution $y \in C^2(0,1) \cap C[0,1]$ of (1.4), we have $|y|_\infty < M$.*

Then problem (1.1) has a solution $y \in C^2(0,1) \cap C[0,1]$ such that $|y|_\infty < M$.

Proof We first remark that it suffices to argue if $a = b = 0$. Indeed, if this is not the case, we define $z(t) = y(t) - a(1-t) - bt$, for any $0 \le t \le 1$. Thus, z verifies

$$\begin{cases} z''(t) + g(t, z(t)) = 0 & 0 < t < 1, \\ z(0) = z(1) = 0, \end{cases}$$

where $g(t,x) = f(t, x + a(1-t) + bt)$, $(t,x) \in (0,1) \times \mathbb{R}$. We also observe that g satisfies the hypothesis $(F1)$. Thus, we can assume $a = b = 0$.

By virtue of (1.2) and (1.3), problem (1.1) is equivalent to

$$y(t) = (1-t)\int_0^t sf(s,y(s))ds + \lambda t \int_t^1 (1-s)f(s,y(s))ds. \tag{1.5}$$

Set $U := \{y \in C[0,1] : |y|_\infty < M\}$ and define the map $F : \overline{U} \to C[0,1]$ by

$$Fy(t) = a(1-t) + bt + (1-t)\int_0^t sf(s,y(s))ds + t\int_t^1 (1-s)f(s,y(s))ds.$$

Then F is a compact operator. We claim that property (ii) in Theorem 1.1.1 cannot be true, which implies immediately that F has a fixed point. Indeed, if there exist $y \in \partial U$ and $0 < \lambda < 1$ such that $y = \lambda F y$, then y is a solution of (1.4). Thus, by our hypothesis, $|y|_\infty < M$—that is, $y \notin \partial U$. This is clearly a contradiction. So, by Theorem 1.1.1, we deduce that F has a fixed point $y \in \overline{U}$. In view of the equivalent formulation given in (1.5), this means that y is a solution of (1.1). This ends our proof. □

1.2 The method of sub- and supersolution

Let $\Omega \subset \mathbb{R}^N (N \geq 1)$ be a bounded domain with a smooth boundary $\partial\Omega$ (for instance, we can assume that $\partial\Omega$ is of class C^3) and $\Phi = \Phi(x,t,\xi) : \overline{\Omega}\times\mathbb{R}\times\mathbb{R}^N \to \mathbb{R}$ be a Hölder continuous function with exponent $\gamma \in (0,1)$ and continuously differentiable with respect to the variables t, ξ and such that

$(A1)$ for any $\Omega_0 \subset\subset \Omega$ and any $0 < a < b < \infty$, there exists a constant $C > 0$ such that

$$|\Phi(x,t,\xi)| \leq C(1 + |\xi|^2) \text{ for all } x \in \overline{\Omega}_0,\ t \in [a,b],\ \xi \in \mathbb{R}^N.$$

Consider the nonlinear Dirichlet problem

$$\begin{cases} -\Delta u = \Phi(x,u,\nabla u) & \text{in } \Omega, \\ u = 0 & \text{on } \partial\Omega. \end{cases} \tag{1.6}$$

By classical solution of (1.6) we mean a function $u \in C^{2,\gamma}(\Omega) \cap C(\overline{\Omega})$ that satisfies (1.6).

Definition 1.2.1 *A function $\underline{u} \in C^2(\Omega) \cap C(\overline{\Omega})$ is said to be a subsolution of the problem (1.6) if $\underline{u} \leq 0$ on $\partial\Omega$ and*

$$-\Delta \underline{u} \leq \Phi(x,\underline{u},\nabla \underline{u}) \quad \textit{in } \Omega. \tag{1.7}$$

Similarly, $\overline{u} \in C^2(\Omega) \cap C(\overline{\Omega})$ is a supersolution of (1.6) if $\overline{u} \geq 0$ on $\partial\Omega$ and

$$-\Delta \overline{u} \geq \Phi(x,\overline{u},\nabla \overline{u}) \quad \textit{in } \Omega.$$

The method of sub- and supersolution in its classical form was developed by H. Amann [5], [6] in the framework of nonlinear elliptic partial differential equations. Roughly speaking, this method establishes the existence of a solution, provided suitable sub- and supersolutions do exist. This result is a crucial property for showing the existence of solutions to wide classes of nonlinear elliptic problems.

Theorem 1.2.2 *Let $\underline{u}$ and $\overline{u}$ be a sub- and a supersolution of problem (1.6) such that $\underline{u} \leq \overline{u}$ in Ω. Then the following properties hold true:*

(i) *There exists a solution u of (1.6) that satisfies $\underline{u} \leq u \leq \overline{u}$.*
(ii) *There exists a minimal and a maximal solution $\underline{U}$ and $\overline{U}$ of problem (1.6) with respect to the interval $[\underline{u}, \overline{u}]$.*

Let us assume that Φ is defined only on the set $\Omega \times (0,\infty) \times \mathbb{R}^N$, so that Φ may by singular in $\partial\Omega \times \{0\} \times \mathbb{R}^N$. In this case we are looking for a solution of the problem

$$\begin{cases} -\Delta u = \Phi(x,u,\nabla u) & \text{in } \Omega, \\ u > 0 & \text{in } \Omega, \\ u = 0 & \text{on } \partial\Omega. \end{cases} \tag{1.8}$$

A subsolution $\underline{u}$ of (1.8) is a function $\underline{u} \in C^2(\Omega) \cap C(\overline{\Omega})$, which is positive in Ω, $\underline{u} = 0$ on $\partial\Omega$, and fulfills (1.7). Accordingly, $\overline{u} \in C^2(\Omega) \cap C(\overline{\Omega})$ is a supersolution of (1.8) if $\overline{u}$ is positive in Ω, $\underline{u} = 0$ on $\partial\Omega$, and $-\Delta\overline{u} \geq \Phi(x, \overline{u}, \nabla\overline{u})$ in Ω.

Theorem 1.2.3 *If $\underline{u}$ and $\overline{u}$ are respectively sub- and supersolutions of (1.8) such that $\underline{u} \leq \overline{u}$ in Ω, then (1.8) has at least one classical solution.*

Proof The proof relies on the domain approximation method. In this way we avoid the possible singularities of $\Phi(x, u, \nabla u)$ at the boundary. Let $(\Omega_k)_{k\geq1}$ be a sequence of subdomains of Ω having smooth boundaries and such that

$$\Omega_1 \subset\subset \Omega_2 \subset\subset \cdots \subset\subset \Omega_k \subset\subset \Omega_{k+1} \subset\subset \cdots,$$

and

$$\Omega = \bigcup_{k\geq1} \Omega_k.$$

For each $k \geq 1$, consider the problem

$$\begin{cases} -\Delta u = \Phi(x, u, \nabla u) & \text{in } \Omega_k, \\ u > 0 & \text{in } \Omega_k, \\ u = \underline{u} & \text{on } \partial\Omega_k. \end{cases} \tag{1.9}$$

Obviously, the restriction of $\underline{u}$ and $\overline{u}$ on Ω_k are sub- and supersolutions of problem (1.9). Thus, by Theorem 1.2.2, there exists a minimal solution $u_k \in C^{2,\gamma}(\Omega_k) \cap C(\overline{\Omega}_k)$ of (1.9) such that $\underline{u} \leq u_k \leq \overline{u}$ in Ω_k. We extend u_k to the whole $\overline{\Omega}$ by taking $u_k = \underline{u}$ in $\overline{\Omega} \setminus \overline{\Omega}_k$. Hence, $(u_k)_{k\geq1}$ is a sequence of continuous functions such that

$$\underline{u} \leq u_1 \leq u_2 \leq \cdots \leq u_k \leq u_{k+1} \leq \cdots \leq \overline{u} \quad \text{in } \overline{\Omega}, \tag{1.10}$$

$$-\Delta u_k = \Phi(x, u_k, \nabla u_k) \quad \text{in } \Omega_k, \quad \text{for all } k \geq 1. \tag{1.11}$$

For all $x \in \overline{\Omega}$, define $u(x) := \lim_{k\to\infty} u_k(x)$. To obtain $u \in C^{2,\gamma}(\Omega) \cap C(\overline{\Omega})$ and u is a solution of (1.8), we need the following result.

Lemma 1.2.4 *For each $j \geq 1$ there exists a corresponding constant $C_j > 0$ such that*

$$\|u_k\|_{C^{2,\gamma}(\overline{\Omega}_j)} \leq C_j \quad \textit{for all } k \geq j+1. \tag{1.12}$$

Proof Let $j \geq 1$ be fixed and take two subdomains Q_1 and Q_2 such that

$$\Omega_j \subset\subset Q_1 \subset\subset Q_2 \subset\subset \Omega_{j+1}.$$

Then, for all $k \geq j+1$ we have

$$-\Delta u_k = \Phi(x, u_k, \nabla u_k) \quad \text{in } \overline{\Omega}_{j+1}.$$

If we denote $\phi_k := \Phi(x, u_k, \nabla u_k)$, $k \geq j+1$, the previous equality reads

$$-\Delta u_k = \phi_k \quad \text{in } \overline{\Omega}_{j+1}.$$

Because $\underline{u} \leq u_k \leq \overline{u}$ in $\overline{\Omega}_{j+1}$, for all $k \geq j+1$, we deduce that $(u_k)_{k\geq j+1}$ is uniformly bounded in $\overline{\Omega}_{j+1}$. On the other hand, by the interior gradient estimate [123, Theorem 3.1, p. 266], there exists $C_1 > 0$ independent of k such that

$$\max_{x\in\overline{Q}_2} |\nabla u_k(x)| \leq C_1 \max_{x\in\overline{\Omega}_{j+1}} u_k(x) \leq C_1 \|u\|_\infty.$$

Hence, the sequence $(\nabla u_k)_{k\geq j+1}$ is uniformly bounded in $\overline{Q}_2$. Therefore, the sequence $(\phi_k)_{k\geq j+1}$ is bounded in $\overline{Q}_2$. Next, by the interior L^p estimate [95, Theorem 9.11, p. 235], we conclude that for any $p > 1$ there exists $C_2 > 0$ that does not depend on k such that

$$\begin{aligned}\|u_k\|_{W^{2,p}(Q_1)} &\leq C_2(\|\phi_k\|_{L^p(Q_2)} + \|u_k\|_{L^p(Q_2)}) \\ &\leq C_2|Q_2|^{1/p}(\max_{x\in\overline{Q}_2} |\phi_k(x)| + \max_{x\in\overline{Q}_2} |u_k(x)|).\end{aligned}$$

Thus, the sequence $(u_k)_{k\geq j+1}$ is bounded in $W^{2,p}(Q_1)$. Letting $p = N/(1-\gamma)$, it follows by Sobolev–Morrey's inequality that the sequence $(\nabla u_k)_{k\geq j+1}$ is bounded in $C^\gamma(\overline{Q}_1)$. Finally, by Hölder interior estimates [95, Theorem 6.2, p. 90], there exists a constant $C_3 > 0$ independent of k such that

$$\|u_k\|_{C^{2,\gamma}(\overline{\Omega}_j)} \leq C_3(\|\phi_k\|_{C^\gamma(\overline{Q}_1)} + \max_{x\in\overline{Q}_1} |u_k(x)|).$$

This proves the inequality (1.12). □

By Lemma 1.2.4, the sequence $(u_k)_{k\geq 1}$ is bounded in $C^{2,\gamma}(\overline{\Omega}_j)$, for all $j \geq 1$. Because the embedding $C^{2,\gamma}(\overline{\Omega}_j) \hookrightarrow C^2(\overline{\Omega}_j)$ is compact, there exists a subsequence of $(u_k)_{k\geq 1}$ that converges in $C^2(\overline{\Omega}_j)$. This yields $u \in C^2(\overline{\Omega}_j)$, for all $j \geq 1$—that is, $u \in C^2(\Omega)$. Moreover, we have

$$-\Delta u_k = \Phi(x, u_k, \nabla u_k) \quad \text{in } \Omega_j, \quad \text{for all } k \geq j+1\,.$$

Thus, taking the limit of the subsequence converging in $C^2(\overline{\Omega}_j)$, we conclude that u satisfies $-\Delta u = \Phi(x, u, \nabla u)$ in the whole Ω. Furthermore, from $\underline{u} \leq u \leq \overline{u}$ in Ω we obtain $\lim_{x\to x_0} u(x) = 0$, for all $x_0 \in \partial\Omega$, so $u \in C(\overline{\Omega})$. It remains to prove $u \in C^{2,\gamma}(\Omega)$, which follows easily from the interior regularity theory of elliptic equations. This finishes the proof of Theorem 1.2.3. □

Consider the problem

$$\begin{cases} -\Delta u = \Phi(x, u) & \text{in } \Omega, \\ u > 0 & \text{in } \Omega, \\ u = 0 & \text{on } \partial\Omega. \end{cases} \tag{1.13}$$

Assume that $\Phi : \overline{\Omega} \times (0, \infty) \to (0, \infty)$ is a Hölder continuous function with exponent γ $(0 < \gamma < 1)$ on each compact subset of $\overline{\Omega} \times (0, \infty)$ and it satisfies the following assumptions:

$(A2)$ $\limsup_{t\to\infty} \frac{\Phi(x,t)}{t} < \lambda_1 := \lambda_1(-\Delta,\Omega)$ uniformly for $x \in \Omega$.

$(A3)$ $\lim_{t\searrow 0} \frac{\Phi(x,t)}{t} = \infty$ uniformly for $x \in \Omega$.

Note that Φ may be singular at the origin with respect to the second variable. The existence of a classical solution to (1.13) is given by Theorem 1.2.5.

Theorem 1.2.5 *Assume Φ satisfies hypotheses $(A1)$, $(A2)$, and $(A3)$. Then problem (1.13) has at least one positive solution $u \in C^{2,\gamma}(\Omega) \cap C(\overline{\Omega})$.*

Proof For any positive integer k, consider the approximated problem

$$\begin{cases} -\Delta u = \Phi(x,u) & \text{in } \Omega, \\ u > 0 & \text{in } \Omega, \\ u = \dfrac{1}{k} & \text{on } \partial\Omega. \end{cases} \tag{1.14}$$

Let φ_1 be the normalized positive eigenfunction corresponding to the first eigenvalue λ_1 of the problem

$$\begin{cases} -\Delta u = \lambda u & \text{in } \Omega, \\ u = 0 & \text{on } \partial\Omega. \end{cases} \tag{1.15}$$

Using $(A3)$, there exists $c > 0$, which is small enough and $k_0 \geq 1$, which is large enough, such that

$$v_k := c\varphi_1 + \frac{1}{k}$$

is a subsolution of (1.14) for all $k \geq k_0$. Without loss of generality, we assume that $k_0 = 1$. To provide a supersolution, let us fix $0 < \lambda < \lambda_1$ such that

$$\lim_{t\to\infty} \frac{\Phi(x,t)}{t} < \lambda < \lambda_1, \quad \text{uniformly for } x \in \Omega.$$

Then there exists $\xi \in C^2(\overline{\Omega})$ such that

$$\begin{cases} -\Delta\xi > \lambda\xi & \text{in } \Omega, \\ \xi > 0 & \text{on } \partial\Omega. \end{cases}$$

We now choose $M > 0$ large enough such that $\zeta := M\xi$ is a supersolution of (1.14) for all $k \geq 1$ and $v_k \leq \zeta$ in Ω. There exists $u_1 \in C^2(\overline{\Omega})$ a solution of (1.14) with $k = 1$ such that $v_1 \leq u_1 \leq \zeta$ in Ω. Now, u_1 is a supersolution of (1.14) with $k = 2$ and $v_2 \leq u_1$ in Ω. Hence, there exists $u_2 \in C^2(\overline{\Omega})$ a solution of (1.14) (with $k = 2$) such that $v_2 \leq u_2 \leq u_1$ in Ω.

Repeating the previous process, we obtain a sequence $(u_k)_{k\geq 1}$ such that

(i) u_k is a solution of (1.14);

(ii) $0 < c\varphi_1 \leq v_{k+1} \leq u_{k+1} \leq u_k$ in Ω.

We set $u(x) := \lim_{k\to\infty} u_k(x)$, for all $x \in \overline{\Omega}$. From (ii) we have

$$0 < c\varphi_1 \le u \le u_k \quad \text{in } \Omega, \quad \text{for all } k \ge 1,$$

which yields $u = 0$ on $\partial\Omega$. Standard elliptic regularity arguments imply that $u \in C^{2,\gamma}(\Omega) \cap C(\overline{\Omega})$ is a solution of (1.13). □

1.3 Comparison principles

1.3.1 *Weak and strong maximum principle*

Consider the linear differential operator

$$\mathcal{L}u := -\sum_{1\le i,j\le N} a_{ij}(x)\frac{\partial^2 u}{\partial x_i \partial x_j} + \sum_{i=1^N} b_i(x)\frac{\partial u}{\partial x_i} + c(x)u, \quad u \in C^2(\Omega) \cap C(\overline{\Omega})$$

where Ω is a bounded domain in $\mathbb{R}^N$, $N \ge 1$, and a_{ij}, b_i, $c \in C(\overline{\Omega})$, for all $1 \le i, j \le N$.

Definition 1.3.1 *We say that $\mathcal{L}$ is uniformly elliptic if there exists $c_1, c_2 > 0$ such that*

$$c_1 |\xi|^2 \le \sum_{1\le i,j\le N} a_{ij}(x)\xi_{ij} \le c_2 |\xi|^2, \tag{1.16}$$

for all $\xi = (\xi_{ij})_{1\le i,j\le N} \in \mathbb{R}^N$.

Theorem 1.3.2 (Weak maximum principle) *Assume that $c \ge 0$ in Ω and let $u \in C^2(\Omega) \cap C(\overline{\Omega})$ be such that*

$$\begin{cases} \mathcal{L}u \ge 0 & \textit{in } \Omega, \\ u \ge 0 & \textit{on } \partial\Omega. \end{cases}$$

Then $u \ge 0$ in Ω.

In particular, the weak maximum principle states that if $u \in C^2(\Omega) \cap C(\overline{\Omega})$ is a superharmonic function (that is, $-\Delta u \ge 0$ in Ω) such that $u \ge 0$ on $\partial\Omega$, then $u \ge 0$ in Ω. Accordingly, if $u \in C^2(\Omega) \cap C(\overline{\Omega})$ is a subharmonic function (that is, $-\Delta u \le 0$ in Ω) such that $u \le 0$ on $\partial\Omega$, then $u \le 0$ in Ω. We point out that, quoting J. L. Doob [69], "F. Riesz inaugurated the systematic study of superharmonic and subharmonic functions" (see Riesz [177]).

A refined result that provides better estimates in a more general case is the following theorem.

Theorem 1.3.3 *Assume that $c \ge 0$ in Ω, $f \in C(\Omega)$, and let $u \in C^2(\Omega) \cap C(\overline{\Omega})$ be such that $\mathcal{L}u \ge f$ in Ω. Then there exists $C > 0$ depending only on diam Ω and the coefficients of $\mathcal{L}$ such that*

$$\sup_{\Omega} u \le \sup_{\partial\Omega} u^+ + C\|f\|_\infty.$$

The strong maximum principle is mainly needed to exclude the existence of nontrivial interior maximum points. The domain Ω is required to be smooth in the sense of the definition stated next.

Definition 1.3.4 *We say that a domain Ω satisfies the interior sphere condition at a point $x_0 \in \partial\Omega$ if there exists an open ball $B \subset \Omega$ such that $\overline{B} \cap \Omega = \{x_0\}$.*

Accordingly, if $B \subset \mathbb{R}^N \setminus \Omega$, then we obtain the exterior sphere condition.

Theorem 1.3.5 (Strong maximum principle) *Assume that Ω is a bounded domain satisfying the interior sphere condition and let $\mathcal{L}$ be an elliptic operator as noted earlier with $c \geq 0$. Let $u \in C^2(\Omega) \cap C(\overline{\Omega})$ be such that*

$$\begin{cases} \mathcal{L}u \geq 0 & \text{in } \Omega, \\ u \geq 0 & \text{on } \partial\Omega. \end{cases}$$

Then the following properties hold true:

(i) *If u achieves its minimum in Ω, then u is constant.*

(ii) *If $x_0 \in \partial\Omega$ is such that $u > u(x_0)$ in Ω and Ω satisfies the interior sphere condition at x_0, then the outer normal derivative of u at x_0, if it exists, satisfies $\frac{\partial u}{\partial n}(x_0) < 0$.*

We refer to the monograph by Gilbarg and Trudinger [95] for the proofs of the results contained in this section.

1.3.2 *Maximum principle for weakly differentiable functions*

Let

$$\mathcal{L}u := -\sum_{1 \leq i,j \leq N} \frac{\partial}{\partial x_j}\left(a_{ij}\frac{\partial u}{\partial x_i}\right) + \sum_{i=1^N} b_i(x)\frac{\partial u}{\partial x_i} + c(x)u, \quad u \in H^1(\Omega) \quad (1.17)$$

be a linear uniformly elliptic operator (in the sense that the matrix (a_{ij}) satisfies condition (1.16) in Ω) in divergence form. We assume for simplicity that a_{ij}, b_i, $c \in C(\overline{\Omega})$.

Let $a : H^1(\Omega) \times H^1(\Omega) \to \mathbb{R}$ be the associated bilinear form of $\mathcal{L}$, defined by

$$a(u,v) := \int_\Omega \sum_{i,j=1}^N a_{ij}\frac{\partial u}{\partial x_i}\frac{\partial v}{\partial x_j} + \int_\Omega \left(\sum_{i=1}^N b_i\frac{\partial u}{\partial x_i} + cu\right) dx,$$

for all $u, v \in H^1(\Omega)$.

Definition 1.3.6 *We say that $u \in H^1(\Omega)$ satisfies $\mathcal{L}u \geq 0$ in Ω if $a(u,v) \geq 0$ for all $v \in C_0^\infty(\Omega)$ with $v \geq 0$.*

The following is an analogous result to Theorem 1.3.2 for weakly differentiable functions.

Theorem 1.3.7 *Assume that $c \geq 0$ in Ω and let $u \in H^1(\Omega)$ be such that*

$$\begin{cases} \mathcal{L}u \geq 0 & \text{in } \Omega, \\ u \geq 0 & \text{on } \partial\Omega. \end{cases}$$

Then $u \geq 0$ in Ω.

Corollary 1.3.8 *Let $k \geq 0$ and $u \in H^1(\Omega)$ be such that*

$$\begin{cases} -\Delta u + ku \geq 0 & \text{in } \Omega, \\ u \geq 0 & \text{on } \partial\Omega. \end{cases}$$

Then either $u \equiv 0$ or there exists $\gamma > 0$ such that $u(x) \geq \gamma \operatorname{dist}(x, \partial\Omega)$, for any $x \in \Omega$.

Proof Assume that $u \not\equiv 0$. Then, there exists $m > 0$ and a ball $B \subset\subset \Omega$ with the property $u \geq m$ in $\overline{B}$. Let $v \in C^2(\overline{\Omega} \setminus B)$ be a classical solution of

$$\begin{cases} -\Delta v + kv = 0 & \text{in } \Omega \setminus \overline{B}, \\ u = m & \text{on } \partial B, \\ u = 0 & \text{on } \partial\Omega. \end{cases}$$

By the strong maximum principle we find $\gamma > 0$ such that $v(x) \geq \gamma \operatorname{dist}(x, \partial\Omega)$, for all $x \in \Omega \setminus B$. Also note that

$$\begin{cases} -\Delta(u - v) + k(u - v) \geq 0 & \text{in } \Omega \setminus B, \\ u - v \geq 0 & \text{on } \partial(\Omega \setminus B). \end{cases}$$

By Theorem 1.3.7 we obtain $u \geq v$ in Ω, hence the conclusion. □

1.3.3 *Stampacchia's maximum principle*

Another comparison result for weakly differentiable functions is Stampacchia's maximum principle.

Definition 1.3.9 *The bilinear form a associated with the differential operator $\mathcal{L}$ defined in (1.17) is called coercive if there exists $\alpha > 0$ such that*

$$a(u, u) \geq \alpha \|u\|^2_{H^1(\Omega)} \qquad \text{for all } u \in H^1(\Omega).$$

Theorem 1.3.10 (Stampacchia's maximum principle) *Assume that the bilinear form a associated with the linear operator $\mathcal{L}$ is coercive. Let $u \in H^1(\Omega)$ be such that $a(u, v) \geq 0$ for all $v \in H^1(\Omega)$, $v \geq 0$. Then $u \geq 0$ in Ω.*

Proof Let $u^- = \max\{-u, 0\}$. Then $u^- \in H^1(\Omega)$ and $\nabla u^- = -\nabla u$ on the set where $u < 0$ and $\nabla u^- = 0$ on the set where $u \geq 0$. Because $\mathcal{L}u \geq 0$, it follows that $a(u, u^-) \geq 0$. On the other hand, $a(u, u^-) = -a(u^-, u^-)$ and, from the coercivity of a, we deduce that

$$\alpha \|u^-\|^2_{H^1(\Omega)} \leq a(u^-, u^-) = -a(u, u^-) \leq 0,$$

which implies $u^- \equiv 0$. Hence, $u \geq 0$ in Ω. □

More generally, Stampacchia's maximum principle states that if the bilinear form a is coercive and $u, v \in H^1(\Omega)$ are such that $\mathcal{L}u \geq 0$ and $\mathcal{L}v \geq 0$ in Ω, then $w = \min\{u, v\}$ also verifies $\mathcal{L}w \geq 0$ in Ω. We refer to the book of Stampacchia [183] for further details.

1.3.4 *Vázquez's maximum principle*

We recall that the standard version of the maximum principle asserts that if $u \in C^2(\Omega) \cap C(\overline{\Omega})$ is a superharmonic function such that $u \geq 0$ on $\partial\Omega$, then the following alternative holds: Either $u > 0$ in Ω or $u \equiv 0$ in Ω. Stampacchia extended this result for *linear* perturbations of the Laplace operator and showed that the same conclusion holds if $u \geq 0$ on $\partial\Omega$ and $-\Delta u + au \geq 0$ in Ω, provided $-\Delta + aI$ is coercive in $H^1(\Omega)$, where $a \in L^\infty(\Omega)$. A natural question is to determine whether a similar result holds true when the Laplace operator is affected by a suitable *nonlinear* perturbation. In [190], J. L. Vázquez established an important extension of the maximum principle for semilinear and quasilinear problems, in close relationship with the behavior of the nonlinearity around the origin.

Theorem 1.3.11 (Vázquez's maximum principle) *Let Ω be a bounded domain in $\mathbb{R}^N$, $N \geq 3$, and let $u \in C^2(\Omega) \cap C(\overline{\Omega})$ be such that $u \geq 0$ in Ω and*

$$\Delta u \leq f(u) \qquad \text{in } \Omega, \tag{1.18}$$

where $f : [0, \infty) \to \mathbb{R}$ is a continuous and increasing function such that $f(0) = 0$ and f satisfies the integral condition

$$\int_0^1 \frac{dt}{\sqrt{F(t)}} = \infty\,, \qquad \text{with } F(t) := \int_0^t f(s)ds. \tag{1.19}$$

Then the following alternative holds: Either $u > 0$ in Ω or $u \equiv 0$ in Ω.

Proof We start with the following auxiliary result.

Lemma 1.3.12 *Consider the Sturm–Liouville problem*

$$\begin{cases} v'' = K_1 v' + K_2 f(v) & \text{in } (0, r_1), \\ v(0) = 0, \quad v(r_1) = v_1, \end{cases} \tag{1.20}$$

where K_1, K_2, r_1, and v_1 are positive real numbers. If $f(0) = 0$ and f is increasing, then problem (1.20) has a unique solution. If, moreover, f satisfies condition (1.19), then the unique solution v of (1.20) satisfies $v'(0) > 0$ and $0 < v < v_1$ in $(0, r_1)$.

Proof We first observe that $\underline{v} := 0$ is a subsolution and $\overline{v} := C$ is a supersolution of (1.20), provided C is large enough. Thus, problem (1.20) has at least a solution. To argue that this solution is unique, let us assume that V_1 and V_2 are solutions of (1.20) and set $v := V_1 - V_2$. Then

$$\begin{cases} v'' = K_1 v' + K_2 \left(f(V_1) - f(V_2) \right) & \text{in } (0, r_1) \\ v(0) = v(r_1) = 0. \end{cases} \tag{1.21}$$

We claim that $V_1 \leq V_2$ in $(0, r_1)$. Indeed, if not, there exists $x_0 \in (0, r_1)$ such that $v(x_0) = \sup_{x \in (0,r_1)} v(x) > 0$. Hence, $v'(x_0) = 0$ and $v''(x_0) \leq 0$. Using (1.21), we deduce that

$$0 \geq v''(x_0) = K_1 v'(x_0) + K_2 \left(f(V_1(x_0)) - f(V_2(x_0)) \right) > 0 \,,$$

which is a contradiction. Thus, $V_1 \leq V_2$ in $(0, r_1)$ and, after changing the roles of V_1 and V_2, we deduce that $V_1 = V_2$, which shows that problem (1.20) has a unique solution.

Let us now assume that f satisfies condition (1.19) and let v be the unique solution of (1.20). Set

$$r_0 := \sup\{0 \leq r \leq r_1 : v(r) = 0\} \,.$$

Then $0 \leq r_0 < r_1$ and $v(r_0) = 0$. We show in what follows that $r_0 = 0$, which is enough to conclude the proof of the lemma. Arguing by contradiction, we assume that $r_0 > 0$. Then $v'(r_0) = 0$ and $v' \geq 0$ in (r_0, r_1). Because v cannot have local maxima in (r_0, r_1), it follows that the mapping $v : [r_0, r_1] \to [0, v_1]$ is a bijection. Multiplying (1.20) by v' and integrating, we find

$$w' = 2K_1 w + 2K_2 (F(v))' \,, \tag{1.22}$$

where $w := v'^2$. After multiplication in (1.22) by $e^{-2K_1 r}$ and integration on $[r_0, r]$ we deduce that

$$e^{-2K_1 r} w(r) - e^{-2K_1 r_0} w(r_0) = 2K_2 \int_{r_0}^{r} e^{-2K_1 s} \left(F(v(s)) \right)' ds.$$

Because $w(r_0) = 0$, this relation yields

$$\begin{aligned} e^{-2K_1 r} w(r) &= 2K_2 \int_{r_0}^{r} e^{-2K_1 s} \left(F(v(s)) \right)' ds \\ &\leq 2K_2 \int_{r_0}^{r} e^{-2K_1 r_0} \left(F(v(s)) \right)' ds \\ &= 2K_2 \, e^{-2K_1 r_0} \left(F(v(r)) - F(v(r_0)) \right). \end{aligned}$$

Hence,

$$\frac{v'(r)}{\sqrt{F(v(r))}} \leq \sqrt{2K_2} e^{K_1 (r_1 - r_0)} (r_1 - r_0) \leq C \quad \text{for any } 0 < r < r_1. \tag{1.23}$$

Because $v : [r_0, r_1] \to [0, v_1]$ is a bijection, after integration in (1.23) we obtain

$$\int_0^{v_1} \frac{dt}{\sqrt{F(t)}} \leq C(r_1 - r_0) < \infty,$$

which contradicts our assumption (1.19). □

We are now in position to prove Vázquez's maximum principle. We first denote by $v(r; K_1, K_2, r_1, v_1)$ the unique solution of problem (1.20).

Denote $\Omega_0 := \{x \in \Omega : u(x) = 0\}$, where $u \geq 0$ solves the differential inequality (1.18). Let $x_1 \in \Omega \setminus \Omega_0$ be such that $\operatorname{dist}(x_1, \Omega_0) < \operatorname{dist}(x_1, \partial\Omega)$. Assuming that Ω_0 is a proper subset of Ω, set $R := \operatorname{dist}(x_1, \Omega_0)$ and fix $x_0 \in \Omega_0 \cap \overline{B}(x_1, R)$. Then $u(x_0) = 0$.

Next, we apply Lemma 1.3.12 for $r_1 = R/2$, $v_1 = \inf\{u(x) : |x - x_1| = R/2\}$, $K_1 = 2(N-1)/R$, and $K_2 = 1$. Define $\overline{u}(x) := v(R - |x - x_1| : K_1, K_2, r_1, v_1)$, where v is given by Lemma 1.3.12. Then $\overline{u}(R) = v(0) = 0$ and $\overline{u}(R/2) = v(R/2)$. Using the equation fulfilled by v we obtain

$$\Delta\overline{u} = \overline{u}'' - \frac{N-1}{r}\overline{u}' \geq v'' - \frac{2(N-1)}{r}v' = f(v) \tag{1.24}$$

in $\omega := \{x \in \Omega : R/2 < |x - x_1| < R\}$.

We claim that $\overline{u} \leq u$ in ω. We first observe that the equality holds on $\partial\omega$ and we show that for any $\varepsilon > 0$,

$$\overline{u}(x) \leq u(x) + \varepsilon\,(1 + |x|^2)^{-1/2} \qquad \text{in } \omega. \tag{1.25}$$

Assuming by contradiction that (1.25) is not true, there exists $x_M \in \omega$ such that the mapping $\overline{u}(x) - u(x) - \varepsilon\,(1 + |x|^2)^{-1/2}$ attains its maximum in x_M and that this value is positive. Therefore

$$\begin{aligned} 0 &\geq \Delta\left(\overline{u}(x) - u(x) - \varepsilon\,(1 + |x|^2)^{-1/2}\right)_{|x=x_M} \\ &= \Delta\overline{u}(x_M) - \Delta u(x_M) + \varepsilon\,(N-3)(1 + |x_M|^2)^{-3/2} + 3\varepsilon\,(1 + |x_M|^2)^{-5/2} \\ &\geq \Delta\overline{u}(x_M) - f(u(x_M)) + 3\varepsilon\,(1 + |x_M|^2)^{-5/2}. \end{aligned}$$

So, by relation (1.24),

$$0 \geq 3\varepsilon\,(1 + |x_M|^2)^{-5/2} > 0.$$

This contradiction shows that relation (1.25) is true. Passing, now, at the limit as $\varepsilon \to 0$ we conclude that $\overline{u} \leq u$ in ω. Therefore $\overline{u}(x_0) = u(x_0) = 0$, where $x_0 \in \partial\omega$. Because $u \geq 0$ in Ω, we deduce that x_0 is an interior minimum point of u in Ω. Thus, $\nabla u(x_0) = 0$ and hence $\partial u(x_0)/\partial n = 0$. On the other hand, we compute the normal derivative of u at $x_0 \in \partial\omega$. We have

$$\begin{aligned} \frac{\partial u}{\partial n}(x_0) &= \lim_{t \searrow 0} \frac{u(x_0 - t(x - x_0))}{t} \\ &\geq \lim_{t \searrow 0} \frac{\overline{u}(x_0 - t(x - x_0))}{t} = \lim_{t \searrow 0} \frac{v(tR)}{t} = Rv'(0) > 0\,. \end{aligned}$$

This contradiction shows that our assumption that Ω_0 is a proper subset of Ω is false, hence either $\Omega_0 = \Omega$ (and, in this case, $u \equiv 0$ in Ω) or $\Omega_0 = \emptyset$ (which corresponds to $u > 0$ in Ω). This completes the proof of Vázquez's maximum principle. □

We refer to Vázquez [190] for a corresponding result in the quasilinear case.

1.3.5 *Pucci and Serrin's maximum principle*

The growth assumption (1.19) in the statement of Theorem 1.3.11 is a Keller–Osserman–type condition around the origin, which establishes that, to guarantee the maximum principle for the nonlinear operator $-\Delta u + f(u)$, it suffices that f has at least a linear decay rate near the origin. A straightforward computation shows that all functions $f(u) = u^p$ (where $p \geq 1$) satisfy relation (1.19). The particular case $p = 1$ corresponds to the linear setting, as described in Stampacchia's maximum principle.

It has been argued by Pucci and Serrin [169–172] that the maximum principle may fail if assumption (1.19) is not fulfilled. Moreover, Pucci and Serrin have extended the maximum principle to nonhomogeneous differential operators. They have considered the inequality problem

$$\begin{cases} \operatorname{div}\{A(|\nabla u|)\nabla u\} \leq f(u) & \text{in } \Omega, \\ u \geq 0 & \text{in } \Omega, \end{cases} \tag{1.26}$$

where Ω is a bounded in $\mathbb{R}^N$. The main assumptions on A and f are the following:

($A1$) $A : (0, \infty) \to (0, \infty)$ is a continuous function.
($A2$) The mapping $\Phi(t) := tA(t)$ increases in $(0, \infty)$ and $\lim_{t\searrow 0} \Phi(t) = 0$.
($F1$) $f : [0, \infty) \to [0, \infty)$ is continuous and $f(0) = 0$.
($F2$) f is nondecreasing on some interval $[0, \delta]$, $\delta > 0$.

Some important particular cases are the following:

(i) If $A(t) \equiv 1$, we obtain the Laplace operator.
(ii) If $A(t) = t^{p-2}$ $(p > 1)$, we obtain the p–Laplace operator.
(iii) If $A(t) = (1 + t^2)^{-1/2}$, we obtain the mean curvature operator.
(iv) If $A(t) = (1 + t^2)^{-p/2}$ $(p \leq 1)$, we obtain the generalized mean curvature operator.
(v) If $A(t) = t^{p-2} + t^{q-2}$ $(1 < p < q)$, we obtain an important operator occurring in quantum physics (see, for example, [20]).

Definition 1.3.13 *A function $u \in C^1(\Omega)$ is a solution to the problem (1.26) if for any $\phi \in C_0^\infty(\Omega)$, $\phi \geq 0$,*

$$\int_\Omega [A(|\nabla u|)\nabla u \cdot \nabla\phi + f(u)\phi]\, dx \geq 0.$$

Definition 1.3.14 *We say that the strong maximum principle holds for problem (1.26) if u is an arbitrary solution of (1.26) with $u(x_0) = 0$ for some $x_0 \in \Omega$, then $u \equiv 0$ in Ω.*

An important role in the statement of the maximum principle of Pucci and Serrin is played by the mapping

$$H(t) := t\Phi(t) - \int_0^t \Phi(s)ds\,.$$

Theorem 1.3.15 (Pucci–Serrin maximum principle) *Assume conditions* $(A1)$, $(A2)$, $(F1)$, *and* $(F2)$ *are fulfilled. Then the strong maximum principle is valid for the inequality problem (1.26) provided that either* $f(t) \equiv 0$ *for* $t \in [0, \mu)$, $\mu > 0$, *or* $f(t) > 0$ *for* $t \in (0, \delta)$ *with* $\delta > 0$ *and*

$$\int_0^1 \frac{dt}{H^{-1}(F(t))} = \infty, \qquad \text{where } F(t) := \int_0^t f(s)ds. \tag{1.27}$$

Conversely, if $f(t) > 0$ *for* $t \in (0, \delta)$, *then condition (1.27) is also necessary for the validity of the strong maximum principle.*

We observe that assumption (1.27) reduces to the Keller–Osserman condition (1.19) in the case of the Laplace operator.

A related phenomenon concerns the existence of *dead cores.* More precisely, an elliptic equation or inequality is said to have a *dead core solution* u in some domain $\Omega \subset \mathbb{R}^N$ provided that there exists an open subset ω strictly contained in Ω (called the *dead core* of u) such that

$$u \equiv 0 \text{ in } \Omega \qquad \text{and} \qquad u > 0 \text{ in } \Omega \setminus \overline{\omega}.$$

Such singular phenomena arise in chemical models, when the solution represents the density of a reactant. The case when the solution vanishes corresponds to some region (*dead core*) where no reactant is present (see [18], [171]). The boundary of such subdomains is then called the *free boundary.* Two such situations are chemical reactions and nonlinear diffusion models in population dynamics. In these situations the dead core region is the subdomain where the concentration of reactant is zero or the population dies. More details on free boundary problems can be found in the book by Diaz [66] in which existence and regularity of such problems is studied.

By Theorem 1.3.15, problem (1.26) can have a dead core only if condition (1.27) fails—that is, if $f > 0$ in $(0, \infty)$ and

$$\int_0^1 \frac{dt}{H^{-1}(F(t))} < \infty\,. \tag{1.28}$$

Actually, condition (1.28) is not only necessary, but also sufficient for the existence of dead core solutions. The following result is from Pucci and Serrin [171], [172].

Theorem 1.3.16 *Assume hypotheses* $(A1)$ *and* $(A2)$ *are fulfilled. Let* $f : \mathbb{R} \to \mathbb{R}$ *be a continuous nondecreasing function such that* $f(0) = 0$ *and satisfies condition (1.28). Suppose* $\Phi(\infty) = H(\infty) = \infty$. *Let* u *be a solution of the inequality problem (1.26), with* $0 \le u(x) \le m$ *on* $\partial\Omega$ *for some constant* $m > 0$.

Then the following properties are valid:

(i) $0 \le u < m$ *in* Ω.

(ii) *Assume that*

$$R_0 := \int_0^\infty \frac{dt}{H^{-1}(F(t)/N)} < \infty,$$

and let B_R *be a ball of radius* $R \geq R_0$ *compactly contained in* Ω. *Then* u *has a dead core in* Ω *for all* $m > 0$.

(iii) *If* B *is any ball compactly contained in* Ω, *then* $u \equiv 0$ *in* B *provided that* $m > 0$ *is sufficiently small.*

The previous remarks show that the equation $\Delta u = |u|^{p-1}u$ allows a dead core only if $0 < p < 1$. We will illustrate this phenomenon in one of the following sections of this chapter.

1.3.6 *A comparison principle in the presence of singular nonlinearities*

The result in this section concerns problem (1.13). By the maximum principle we deduce that if $t \longmapsto \Phi(x,t)$ is decreasing at each $x \in \Omega$, then problem (1.13) has at most one solution. The following theorem is a comparison principle that holds in a more general setting.

Theorem 1.3.17 *Let* $\Phi : \Omega \times (0,\infty) \to \mathbb{R}$ *be a continuous function such that*

$(H1)$ *the mapping* $(0,\infty) \ni t \longmapsto \Phi(x,t)/t$ *is strictly decreasing at each* $x \in \Omega$.

Assume that there exist $v, w \in C^2(\Omega) \cap C(\overline{\Omega})$ *such that*

(i) $\Delta w + \Phi(x,w) \leq 0 \leq \Delta v + \Phi(x,v)$ *in* Ω;

(ii) $v, w > 0$ *in* Ω *and* $v \leq w$ *on* $\partial\Omega$;

(iii) $\Delta v \in L^1(\Omega)$ *or* $\Delta w \in L^1(\Omega)$.

Then $v \leq w$ *in* Ω.

Proof We follow an idea of Brezis and Kamin [32] (see also [87], [182]).

Suppose by contradiction that $v \leq w$ is not true in Ω. Then, we can find ε_0, $\delta_0 > 0$ and a ball $B \subset\subset \Omega$ such that

$$v - w \geq \varepsilon_0 \quad \text{in } B, \tag{1.29}$$

$$\int_B vw\left(\frac{\Phi(x,w)}{w} - \frac{\Phi(x,v)}{v}\right)dx \geq \delta_0. \tag{1.30}$$

Let us assume that $\Delta w \in L^1(\Omega)$ and set

$$M = \max\{1,\ \|\Delta w\|_{L^1(\Omega)}\},\ \varepsilon = \min\left\{1,\ \varepsilon_0,\ \frac{\delta_0}{4M}\right\}.$$

Consider $\theta \in C^1(\mathbb{R})$ a nondecreasing function such that $0 \leq \theta \leq 1$, $\theta(t) = 0$, if $t \leq 1/2$ and $\theta(t) = 1$ for all $t \geq 1$. Define

$$\theta_\varepsilon(t) = \theta\left(\frac{t}{\varepsilon}\right), \quad t \in \mathbb{R}.$$

Because $w \geq v$ on $\partial\Omega$, we can find a smooth subdomain $\Omega^* \subset\subset \Omega$ such that

$$B \subset \Omega^* \quad \text{and} \quad v - w < \frac{\varepsilon}{2} \quad \text{in } \Omega \setminus \Omega^*.$$

Using hypotheses (i) and (ii) we deduce

$$\int_{\Omega^*} (w\Delta v - v\Delta w)\theta_\varepsilon(v-w)dx \geq \int_{\Omega^*} vw\left(\frac{\Phi(x,w)}{w} - \frac{\Phi(x,v)}{v}\right)\theta_\varepsilon(v-w)dx. \tag{1.31}$$

By relation (1.30), we have

$$\begin{aligned}
\int_{\Omega^*} & vw\left(\frac{\Phi(x,w)}{w} - \frac{\Phi(x,v)}{v}\right)\theta_\varepsilon(v-w)dx \\
&\geq \int_B vw\left(\frac{\Phi(x,w)}{w} - \frac{\Phi(x,v)}{v}\right)\theta_\varepsilon(v-w)dx \\
&= \int_B vw\left(\frac{\Phi(x,w)}{w} - \frac{\Phi(x,v)}{v}\right)dx \\
&\geq \delta_0.
\end{aligned}$$

To raise a contradiction, we need only to prove that the left-hand side in (1.31) is smaller than δ_0. For this purpose, define

$$\Theta_\varepsilon(t) := \int_0^t s\theta'_\varepsilon(s)ds, \quad t \in \mathbb{R}.$$

It is easy to see that

$$\Theta_\varepsilon(t) = 0, \ \text{ if } t < \frac{\varepsilon}{2} \quad \text{and} \quad 0 \leq \Theta_\varepsilon(t) \leq 2\varepsilon, \ \text{ for all } t \in \mathbb{R}. \tag{1.32}$$

Now, using Green's first formula, we evaluate the left side of (1.31):

$$\begin{aligned}
\int_{\Omega^*} & (w\Delta v - v\Delta w)\theta_\varepsilon(v-w)dx \\
&= \int_{\partial\Omega^*} w\theta_\varepsilon(v-w)\frac{\partial v}{\partial n}d\sigma(x) - \int_{\Omega^*} (\nabla w \cdot \nabla v)\theta_\varepsilon(v-w)dx \\
&\quad - \int_{\Omega^*} w\theta'_\varepsilon(v-w)\nabla v \cdot \nabla(v-w)dx - \int_{\partial\Omega^*} v\theta_\varepsilon(v-w)\frac{\partial w}{\partial n}d\sigma(x) \\
&\quad + \int_{\Omega^*} (\nabla w \cdot \nabla v)\theta_\varepsilon(v-w)dx + \int_{\Omega^*} v\theta'_\varepsilon(v-w)\nabla w \cdot \nabla(v-w)dx \\
&= \int_{\Omega^*} \theta'_\varepsilon(v-w)(v\nabla w - w\nabla v) \cdot \nabla(v-w)dx.
\end{aligned}$$

The previous relation can be rewritten as

$$\int_{\Omega^*}(w\Delta v - v\Delta w)\theta_\varepsilon(v-w)dx = \int_{\Omega^*} w\theta'_\varepsilon(v-w)\nabla(w-v)\cdot\nabla(v-w)dx$$
$$+\int_{\Omega^*}(v-w)\theta'_\varepsilon(v-w)\nabla w\cdot\nabla(v-w)dx.$$

Because $\int_{\Omega^*} w\theta'_\varepsilon(v-w)\nabla(w-v)\cdot\nabla(v-w)dx \le 0$, the last equality yields

$$\int_{\Omega^*}(w\Delta v - v\Delta w)\theta_\varepsilon(v-w)dx \le \int_{\Omega^*}(v-w)\theta'_\varepsilon(v-w)\nabla w\cdot\nabla(v-w)dx.$$

Therefore,

$$\int_{\Omega^*}(w\Delta v - v\Delta w)\theta_\varepsilon(v-w)dx \le \int_{\Omega^*}\nabla w\cdot\nabla(\Theta_\varepsilon(v-w))dx.$$

Again by Green's first formula, and by (1.32), we have

$$\begin{aligned}\int_{\Omega^*}(w\Delta v - v\Delta w)\theta_\varepsilon(v-w)dx &\le \int_{\partial\Omega^*}\Theta_\varepsilon(v-w)\frac{\partial w}{\partial n}d\sigma(x)\\ &\quad -\int_{\Omega^*}\Theta_\varepsilon(v-w)\Delta w dx\\ &\le -\int_{\Omega^*}\Theta_\varepsilon(v-w)\Delta w dx \le 2\varepsilon\int_{\Omega^*}|\Delta w|dx\\ &\le 2\varepsilon M < \frac{\delta_0}{2}.\end{aligned}$$

Thus, we have obtained a contradiction. Hence $v \le w$ in Ω, which completes the proof. □

Theorem 1.3.17 states that if v and w are respectively sub- and supersolutions of problem (1.13) such that either $\Delta v \in L^1(\Omega)$ or $\Delta w \in L^1(\Omega)$, then we always have $v \le w$ in Ω. As a consequence, problem (1.13) has at most one solution, $u \in C^2(\Omega)\cap C(\overline{\Omega})$, such that $\Delta u \in L^1(\Omega)$.

The proof of Theorem 1.3.17 was given in [87] (see also [182] for a particular case), following an idea that goes back to the method of Brezis and Kamin [32].

Theorems 1.2.5 and 1.3.17 are suitable for semilinear elliptic problems involving singular nonlinearities. In both these results, the nonlinearity Φ is defined only on $(0,\infty)$ with respect to its second variable, so Φ may be singular at the origin. Some simple examples of functions that fulfill the hypotheses in Theorems 1.2.5 and 1.3.17 are

(i) $\Phi(x,t) = t^{-\alpha}$, $\alpha > 0$;

(ii) $\Phi(x,t) = a(x)t^{-\alpha} + t^p$, where $a \in C^\gamma(\overline{\Omega})$, $a > 0$ in $\overline{\Omega}$, $0 < p < 1$, and $\alpha > 0$.

1.4 Existence properties and related maximum principles

In this section we establish a general property that asserts the existence of a unique positive subharmonic function satisfying a nonlinear boundary value problem with nontrivial Dirichlet boundary value data. We illustrate this result with two maximum principles corresponding to nonlinear perturbations of the Laplace operator.

Theorem 1.4.1 *Let $\Omega \subset \mathbb{R}^N$ be a bounded smooth domain. Assume $0 \not\equiv p \in C^{0,\gamma}(\overline{\Omega})$ is nonnegative, and $f \in C^1[0,\infty)$ is positive and nondecreasing on $(0,\infty)$ such that $f(0)=0$. If $0 \not\equiv \phi \in C^{0,\gamma}(\partial\Omega)$ is nonnegative, then the boundary value problem*

$$\begin{cases} \Delta u = p(x)f(u) & \text{in } \Omega, \\ u \geq 0 & \text{in } \Omega, \\ u = \phi & \text{on } \partial\Omega \end{cases} \tag{1.33}$$

has a unique classical solution, which is positive in Ω.

Proof We first observe that $\underline{u}=0$ is a subsolution of (1.33), whereas $\overline{u} \equiv n$ is a supersolution of (1.33) if n is large enough. Hence problem (1.33) has at least one nonnegative solution u. Then, taking into account the regularity of p and f, a standard bootstrap argument based on Schauder and Hölder regularity shows that $u \in C^2(\overline{\Omega})$.

To prove the uniqueness, let u_1 and u_2 be two solutions of (1.33); we show that $u_1 \geq u_2$ in Ω. Denote $\omega := \{x \in \Omega : u_1(x) < u_2(x)\}$ and suppose that $\omega \neq \emptyset$. Then the function $v = u_1 - u_2$ satisfies

$$\Delta v = p(x)(f(u_1) - f(u_2)) \quad \text{in } \omega,$$

and $v = 0$ on $\partial\omega$. Because f is nondecreasing and $p \geq 0$, it follows that v is a superharmonic function in ω that vanishes on $\partial\omega$. Thus, by the maximum principle, either $v=0$ or $v>0$ in ω, which yields a contradiction. Thus $u_1 \leq u_2$ in Ω. Similarly we get $u_2 \leq u_1$ in Ω, thus $u_1 = u_2$.

We state in what follows two proofs for the positivity of u. The first one relies essentially on Vázquez's maximum principle, whereas the second proof offers a direct approach.

First proof: Let w be the unique nonnegative classical solution of the problem

$$\begin{cases} \Delta u = \|p\|_\infty f(w) & \text{in } \Omega, \\ w = \phi & \text{on } \partial\Omega. \end{cases}$$

To conclude that $u>0$ in Ω it is enough to show that $u \geq w > 0$ in Ω. Because $f \in C^1[0,\infty)$, we have

$$\lim_{t\searrow 0} \frac{t^2}{F(t)} = \lim_{t\searrow 0} \frac{2t}{f(t)} = \frac{2}{f'(0)} > 0, \tag{1.34}$$

which implies immediately that $\int_0^1 dt/\sqrt{F(t)} = \infty$. By Vázquez's maximum principle (Theorem 1.3.11) we conclude that $w>0$ in Ω.

We now prove that $u \geq w$ in Ω. To this aim, fix $\varepsilon > 0$. We claim that

$$w(x) \leq u(x) + \varepsilon(1 + |x|^2)^{-1/2} \quad \text{for any } x \in \Omega. \tag{1.35}$$

Assume the contrary. Because $u = w$ on $\partial\Omega$, we deduce that

$$\max_{x \in \overline{\Omega}} \{w(x) - u(x) - \varepsilon(1 + |x|^2)^{-1/2}\}$$

is achieved in Ω. At this point we have

$$\begin{aligned} 0 \geq & \Delta\left(w(x) - u(x) - \varepsilon(1 + |x|^2)^{-1/2}\right) \\ = & \|p\|_\infty f(w(x)) - p(x) f(u(x)) - \varepsilon\Delta(1 + |x|^2)^{-1/2} \\ \geq & p(x)\left(f(w(x)) - f(u(x))\right) + \varepsilon(N - 3)(1 + |x|^2)^{-3/2} + 3\varepsilon(1 + |x|^2)^{-5/2} > 0, \end{aligned}$$

which is a contradiction. Because $\varepsilon > 0$ is chosen arbitrarily, inequality (1.35) implies $u \geq w$ in Ω.

SECOND PROOF: Because ϕ is not identically zero, there exists $x_0 \in \Omega$ such that $u(x_0) > 0$. To conclude that $u > 0$ in Ω, it is sufficient to prove that $u > 0$ in $B(x_0, \overline{r})$, where $\overline{r} = \operatorname{dist}(x_0, \partial\Omega)$. Without loss of generality we can assume $x_0 = 0$. By the continuity of u, there exists $\underline{r} \in (0, \overline{r})$ such that $u(x) > 0$ for all x with $|x| \leq \underline{r}$. So, $\rho := \min_{|x| = \underline{r}} u(x) > 0$. We define

$$M := \max_{\overline{\Omega}} p, \quad \eta := \int_\rho^{\rho+1} \frac{dt}{f(t)} \quad \text{and} \quad \nu(\varepsilon) := \int_\varepsilon^{\rho+1} \frac{dt}{f(t)} \quad \text{for } 0 < \varepsilon < \rho.$$

It remains to show that $u > 0$ in $A(\underline{r}, \overline{r})$, where

$$A(\underline{r}, \overline{r}) := \{x \in \mathbb{R}^N : \underline{r} < |x| < \overline{r}\}.$$

For this purpose, we need the following auxiliary result.

Lemma 1.4.2 *For $\varepsilon > 0$ small enough, the problem*

$$\begin{cases} -\Delta v = M & \text{in } A(\underline{r}, \overline{r}), \\ v = \eta & \text{as } |x| = \underline{r}, \\ v = \nu(\varepsilon) & \text{as } |x| = \overline{r} \end{cases} \tag{1.36}$$

has a unique solution, which is increasing in $A(\underline{r}, \overline{r})$.

Proof By the maximum principle, problem (1.36) has a unique solution. Moreover, v is radially symmetric in $A(\underline{r}, \overline{r})$—namely, $v(x) = v(r)$, $r = |x|$. The function v satisfies

$$v''(r) + \frac{N-1}{r} v'(r) = -M, \quad \underline{r} < r < \overline{r}.$$

Integrating this relation twice we find

$$v(r) = -\frac{M}{2N}r^2 - \frac{C_1}{N-2}r^{2-N} + C_2, \quad \underline{r} < r < \overline{r},$$

where C_1 and C_2 are real constants. The boundary conditions $v(\underline{r}) = \eta$ and $v(\overline{r}) = \nu(\varepsilon)$ imply

$$C_1 = \left(\nu(\varepsilon) - \eta + \frac{M}{2N}(\overline{r}^2 - \underline{r}^2)\right) \frac{N-2}{\underline{r}^{2-N} - \overline{r}^{2-N}}.$$

From (1.34) we deduce that $\nu(\varepsilon) \to \infty$ as $\varepsilon \to 0$. Thus, taking $\varepsilon > 0$ sufficiently small, C_1 becomes large enough to ensure that $v'(r) > 0$ for all $r \in (\underline{r}, \overline{r})$. This concludes the proof. □

Set $\varepsilon > 0$ sufficiently small such that the conclusion of Lemma 1.4.2 holds. Let $\underline{u}$ be the function defined implicitly as follows:

$$\int_{\underline{u}(x)+\varepsilon}^{\rho+1} \frac{dt}{f(t)} = v(x) \quad \text{for all } x \in A(\underline{r}, \overline{r}). \tag{1.37}$$

It is easy to check that

$$\begin{cases} \Delta \underline{u} \geq M f(\underline{u} + \varepsilon) \geq p(x) f(\underline{u}) & \text{in } A(\underline{r}, \overline{r}), \\ \underline{u}(x) = \rho - \varepsilon < u(x) & \text{as } |x| = \underline{r}, \\ \underline{u}(x) = 0 \leq u(x) & \text{as } |x| = \overline{r}. \end{cases}$$

Using the maximum principle, as in the proof of (1.35), we deduce that $\underline{u} \leq u$ in $A(\underline{r}, \overline{r})$. By (1.37) and Lemma 1.4.2 we deduce that $\underline{u}$ decreases in $A(\underline{r}, \overline{r})$. Thus, $\underline{u} > 0$ in $A(\underline{r}, \overline{r})$. This completes the proof. □

Under the same assumptions on p and f as in the statement of Theorem 1.4.1 we have the following corollary.

Corollary 1.4.3 (Strong maximum principle) *Let $\Omega \subset \mathbb{R}^N$ be a domain. If u is a nonnegative classical solution of the equation $\Delta u = p(x) f(u)$ in Ω, then the following alternative holds: Either $u \equiv 0$ in Ω or u is positive in Ω.*

Proof Assume that $u \not\equiv 0$ in Ω. Hence, there exists $x_0 \in \Omega$ such that $u(x_0) > 0$. We claim that $u > 0$ in Ω. To this aim, let us fix $x_1 \in \Omega$ and let $\omega \subset\subset \Omega$ be a smooth bounded domain such that $x_1 \in \omega$ and $x_0 \in \partial\omega$. By Theorem 1.4.1, there exists a unique $v \in C^2(\overline{\Omega})$ such that

$$\begin{cases} \Delta v = p_0 f(v) & \text{in } \omega, \\ v > 0 & \text{in } \omega, \\ v = u \not\equiv 0 & \text{on } \partial\omega, \end{cases} \tag{1.38}$$

where $p_0 := 1 + \sup_\omega p > 0$. Because f is nondecreasing, we easily deduce $u \geq v > 0$ in ω. Thus, $u(x_1) > 0$, which yields $u > 0$ in Ω. □

Corollary 1.4.4 *Let $\Omega \subset \mathbb{R}^N$ be a smooth bounded domain. If $u \in C^2(\Omega) \cap C(\overline{\Omega})$ satisfies*

$$\begin{cases} \Delta u + \lambda u = p(x) f(u) & \text{in } \Omega, \\ u \geq 0 & \text{in } \Omega, \\ u \not\equiv 0 & \text{on } \partial\Omega, \end{cases}$$

then $u > 0$ in Ω.

Proof Let $\phi \in C^{0,\gamma}(\partial\Omega)$ be a nonnegative function such that $\phi \not\equiv 0$ and $\phi \leq u$ on $\partial\Omega$. By Theorem 1.4.1, there exists a unique $u \in C^2(\Omega) \cap C(\overline{\Omega})$ such that

$$\begin{cases} \Delta v = |\lambda| v + \|p\|_\infty f(v) & \text{in } \Omega, \\ v \geq 0 & \text{in } \Omega, \\ v = \phi & \text{on } \partial\Omega. \end{cases} \tag{1.39}$$

Moreover, $v > 0$ in Ω. Note that u is a supersolution for (1.39) and a standard argument shows that $u \geq v > 0$ in Ω. □

1.4.1 *Dead core solutions of sublinear logistic equations*

The positiveness of the solution in Theorem 1.4.1 follows essentially by the assumption $f \in C^1[0, \infty)$. We show in what follows that if f is not differentiable at the origin, then problem (1.33) has a unique solution that is not necessarily positive in Ω. However, in this case, the positiveness of the solution may depend on c and on the geometry of Ω. Indeed, let us consider the problem

$$\begin{cases} \Delta u = \sqrt{u} & \text{in } \Omega, \\ u \geq 0 & \text{in } \Omega, \\ u = c & \text{on } \partial\Omega, \end{cases} \tag{1.40}$$

where $c > 0$ is a constant. The existence of a solution follows after observing that $\underline{u} = 0$ and $\overline{u} = c$ are respectively sub- and supersolutions for our problem, whereas the uniqueness follows from Theorem 1.3.17.

The following example illustrates that in certain situations, the unique solution of problem (1.40) may vanish.

Example 1.1 Let $\Omega = B(0,1) \subset \mathbb{R}^N$, $w(x) = a|x|^4$ with $c \leq a \leq (4N+8)^{-2}$. We have

$$\begin{cases} \Delta w = (4N+8) a |x|^2 \leq \sqrt{a} |x|^2 = \sqrt{w} & \text{in } \Omega, \\ w = a \geq c & \text{on } \partial\Omega. \end{cases}$$

This means that w is a supersolution of (1.40). Hence, problem (1.40) has a solution u such that $0 \leq u \leq w$ in Ω. Because $w(0) = 0$ then, necessarily, $u(0) = 0$.

The next example shows that in some cases, depending on c and on $\operatorname{diam} \Omega$, the unique solution of (1.40) is positive.

Example 1.2 Suppose that Ω can be included in a ball $B(x_0, R)$ with $R \leq R_c := 2\sqrt[4]{c(N+2)^2}$. Define $w(x) := a|x - x_0|^4$, where a is chosen so that

$$\frac{c}{R^4} \geq a \geq \frac{1}{(4N+8)^2}.$$

Then w satisfies

$$\begin{cases} \Delta w = (4N+8)a|x-x_0|^2 \geq \sqrt{a}|x-x_0|^2 = \sqrt{w} & \text{in } \Omega, \\ w = a|x-x_0|^4 \leq c & \text{on } \partial\Omega, \end{cases}$$

which shows that w is a subsolution of (1.40). We conclude that $u(x) \geq w(x) > 0$, for any $x \in \Omega\backslash\{x_0\}$.

If $\operatorname{diam}\Omega < 2R \leq 2R_c$, then there exists two points, x_0 and x_1, such that Ω can be included in each of the balls $B(x_0, R)$ and $B(x_1, R)$. Using the previous conclusion we have

$$u(x) \geq a \max\{|x-x_0|^4, |x-x_1|^4\} \geq a\left|\frac{x_1 - x_0}{2}\right|^4 > 0.$$

Choosing $a = c/R^4$, $|x_1 - x_0| = 2R - \operatorname{diam}\Omega$, and $R = R_c$, we find

$$u(x) \geq \frac{c}{R^4}\left(\frac{2R - \operatorname{diam}\Omega}{2}\right)^4 = c\left(1 - \frac{\operatorname{diam}\Omega}{2R}\right)^4 > 0 \quad \text{in } \Omega.$$

Hence, u is positive solution of (1.40).

1.4.2 *Singular solutions of the logistic equation*

We now illustrate how Theorem 1.4.1 can be applied in the qualitative analysis of a class of singular solutions for the logistic equation. We are first concerned with the nonlinear problem

$$\begin{cases} \Delta u = u^p & \text{in } B_1, \\ u > 0 & \text{in } B_1, \\ u = \infty & \text{on } \partial B_1, \end{cases} \tag{1.41}$$

where $p > 1$ and $B_1 \subset \mathbb{R}^N$ is the unit ball centered at the origin. The notation $u = \infty$ on ∂B_1 signifies that $\lim_{|x|\nearrow 1} u(x) = \infty$. A function $u \in C^2(B_1)$ satisfying (1.41) is called *blow-up boundary (explosive)* solution of problem (1.41).

Theorem 1.4.5 *Assume that $p > 1$. Then problem (1.41) has a unique solution u. Moreover, u is infinitely differentiable in B_1, u is radially symmetric, and*

$$\lim_{r\nearrow 1}(1-r)^{2/(p-1)}u(r) = \left[\frac{2(p+1)}{(p-1)^2}\right]^{1/(p-1)}.$$

Proof Using Theorem 1.4.1 we deduce that for any positive integer n, there exists a unique function $u_n \in C^2(\overline{B_1})$ such that

$$\begin{cases} \Delta u_n = u_n^p & \text{in } B_1, \\ u_n > 0 & \text{in } B_1, \\ u_n = n & \text{on } \partial B_1. \end{cases} \tag{1.42}$$

By uniqueness we deduce that u_n has radial symmetry. Standard elliptic regularity implies that $u_n \in C^\infty(B_1)$. Next, by the maximum principle, $u_n < u_{n+1}$ in B_1, for any $n \geq 1$. We now observe that $\overline{u}(x) := C(1-r^2)^{-2/(p-1)}$ is a supersolution of problem (1.42), provided C is large enough. This shows that for any $n \geq 1$, $u_n \leq \overline{u}$, hence (u_n) is locally bounded in B_1. Thus, for any $x \in B_1$, there exists $u(x) := \lim_{n\to\infty} u_n(x) \in \mathbb{R}$, which is a solution of (1.42). Schauder and Hölder regularity arguments imply that $u \in C^\infty(B_1)$.

To deduce the blow-up rate of u near the boundary, we observe that the differential equation fulfilled by the solution is

$$u'' + \frac{N-1}{r}u' = u^p \qquad \text{in } (0,1).$$

Therefore

$$\left(r^{N-1}u'\right)' = r^{N-1}u^p \qquad \text{in } (0,1). \tag{1.43}$$

After multiplication by $r^{N-1}u'$ in (1.43) and integration, we find

$$u'^2(r) = 2\left[\frac{u^{p+1}(r)}{p+1} - \frac{2N-2}{(p+1)r^{2N}}\int_0^r t^{2N-1}u^{p+1}(t)dt\right]. \tag{1.44}$$

At this stage, it is enough to argue that

$$B(r) = o\left(A(r)\right) \qquad \text{as } r \nearrow 1, \tag{1.45}$$

where

$$A(r) = u^{p+1}(r) \qquad B(r) := r^{-2N}\int_0^r t^{2N-1}u^{p+1}(t)dt.$$

Indeed, relations (1.44) and (1.45) yield

$$1 - \frac{\sqrt{p+1}\,u'(r)}{\sqrt{2u^{p+1}(r)}} = 1 - \left(1 - \frac{B(r)}{A(r)}\right)^{1/2} \sim \frac{B(r)}{2A(r)} = o(1) \quad \text{as } r \nearrow 1.$$

Integrating this relation between r and 1 we deduce that

$$(1-r) - \int_{u(r)}^{\infty} \frac{\sqrt{p+1}}{\sqrt{2t^{p+1}}}dt = o(1) \quad \text{as } r \nearrow 1,$$

which implies the desired asymptotic estimate.

To prove (1.45), let us fix $\varepsilon \in (0,1/2)$. Thus, for any $r \in (1-\varepsilon, 1)$,

$$\begin{aligned}\frac{B(r)}{A(r)} &= \frac{1}{r}\int_0^r \left(\frac{t}{s}\right)^{2N-1}\left(\frac{u(t)}{u(r)}\right)^{p+1} dt\\ &= \frac{1}{r}\int_0^{1-\varepsilon} \left(\frac{t}{s}\right)^{2N-1}\left(\frac{u(t)}{u(r)}\right)^{p+1} dt + \frac{1}{r}\int_{1-\varepsilon}^r \left(\frac{t}{s}\right)^{2N-1}\left(\frac{u(t)}{u(r)}\right)^{p+1} dt\\ &\le 2\,\frac{u^{p+1}(1-\varepsilon)}{u^{p+1}(r)} + \varepsilon\,.\end{aligned}$$

Letting first $r \nearrow 1$ and then $\varepsilon \searrow 0$, we obtain (1.45). In fact, all solutions of (1.41) have the same blow-up rate near the boundary, as described by (1.45). This implies, by the maximum principle, that problem (1.41) has a unique solution. We refer to [71] for details and many related results. □

With similar arguments as in the proof of Theorem 1.4.5, we can show that the nonlinear problem

$$\begin{cases} \Delta u = u^p & \text{if } 1 < |x| < 2,\\ u > 0 & \text{if } 1 < |x| < 2,\\ u = \infty & \text{if } |x| = 1,\\ u = 0 & \text{if } |x| = 2 \end{cases}$$

has a unique solution, which is of class C^∞ and which has the same blow-up rate near $|x| = 1$, as in the statement of Theorem 1.4.5. We leave it to the reader to determine the details of the proof.

1.5 Brezis–Oswald theorem

Consider the problem

$$\begin{cases} -\Delta u = f(x,u) & \text{in } \Omega,\\ u \ge 0,\ u \not\equiv 0 & \text{in } \Omega,\\ u = 0 & \text{on } \partial\Omega, \end{cases} \tag{1.46}$$

where $\Omega \subset \mathbb{R}^N$ is a smooth bounded domain and $f : \Omega \times [0,\infty) \to \mathbb{R}$ satisfies

($f1$) f is a Hölder continuous function and the mapping $(0,\infty) \ni t \longmapsto \frac{f(t)}{t}$ is decreasing;

($f2$) for all $t \ge 0$, $x \longmapsto f(x,t)$ belongs to $L^\infty(\Omega)$;

($f3$) there exists a positive constant $c > 0$ such that $f(x,t) \le c(1+t)$ for all $(x,t) \in \Omega \times [0,\infty)$.

By a solution of problem (1.46), we mean a function $u \in C^2(\overline{\Omega})$ that satisfies (1.46). In view of ($f1$), for all $x \in \Omega$ there exist

$$a_0(x) := \lim_{t\searrow 0} \frac{f(t)}{t} \quad \text{and} \quad a_\infty(x) := \lim_{t\to\infty} \frac{f(t)}{t}.$$

We have

$$-\infty < a_0(x) \leq \infty \quad \text{and} \quad -\infty \leq a_\infty(x) < \infty \quad \text{in } \Omega.$$

It follows that any solution $u \in H_0^1(\Omega) \cap L^\infty(\Omega)$ of (1.46) verifies

$$-f(x, \|u\|_\infty) \leq f(x, u(x)) \leq c(\|u\|_\infty + 1) \quad \text{in } \Omega.$$

This yields $f(x,u) \in L^\infty(\Omega)$. Therefore, any solution u of (1.46) belongs to $W^{2,p}(\Omega)$ for every $1 < p < \infty$. Denote by λ_0 and λ_∞ the first eigenvalue of the operators $-\Delta - a_0(x)$ and $-\Delta - a_\infty(x)$ respectively, with a zero Dirichlet boundary condition.

Because a_0 and a_∞ may be $\pm\infty$, the precise meaning of this first eigenvalue must be understood in the following sense: For any measurable function $a(x)$, we define

$$\lambda_1(-\Delta - a(x)) := \inf_{u \in H_0^1(\Omega), u \neq 0} \left\{ \int_\Omega |\nabla u|^2 dx - \int_{\{u \neq 0\}} a(x) u dx \right\}.$$

Remark 1.5.1 *Any solution u of (1.46) satisfies $u > 0$ in Ω and $\frac{\partial u}{\partial n} < 0$ on $\partial\Omega$.*

Indeed, from $(f1)$ we have

$$\frac{f(x,u)}{u} \geq \frac{f(x, \|u\|_\infty)}{\|u\|_\infty} \quad \text{in } \Omega.$$

Hence $f(x,u) \geq -Mu$ in Ω for some positive constant M. Thus, u satisfies $-\Delta u + Mu \geq 0$ in Ω. By the maximum principle (Theorem 1.3.5) we obtain the conclusion.

Brezis and Oswald [37] established the following general result.

Theorem 1.5.2 *Assume that f satisfies $(f1) - (f3)$. Then, problem (1.46) has solutions if and only if $\lambda_0 < 0 < \lambda_\infty$. Moreover, in this case problem (1.46) has a unique solution.*

Proof We divide the proof of Theorem 1.5.2 into three steps.

Step 1: Necessary condition. Let $u \in H_0^1(\Omega) \cap L^\infty(\Omega)$ be a solution of (1.46). Multiplying by u in (1.46) and then integrating by parts we obtain

$$\int_\Omega |\nabla u|^2 dx = \int_\Omega f(x,u) u dx < \int_\Omega a_0(x) u dx.$$

This implies that $\lambda_0 < 0$. Set

$$b(x) := \frac{f(x, \|u\|_\infty + 1)}{\|u\|_\infty + 1} \in L^\infty(\Omega)$$

and let (μ, ψ) be the first eigenvalue and eigenfunction of the linear operator $-\Delta - b(x)$. That is,

$$\begin{cases} -\Delta\psi - b(x)\psi = \mu\psi & \text{in } \Omega, \\ \psi > 0 & \text{in } \Omega, \\ \psi = 0 & \text{on } \partial\Omega. \end{cases}$$

Because $a_\infty \leq b$ in Ω, it follows that $\mu \leq \lambda_\infty$. Multiplying by ψ in (1.46) and integrating over Ω we find

$$\int_\Omega (b(x) + \mu)\psi u dx = \int_\Omega f(x, u)\psi dx.$$

Note that $f(x, u) > b(x)u$ by virtue of $(f1)$, which also yields $\mu \int_\Omega u\psi dx > 0$. Thus, $\lambda_\infty \geq \mu > 0$.

Step 2: Sufficient condition. Consider the energy functional

$$E : H_0^1(\Omega) \to (-\infty, \infty],$$

defined by

$$E(u) = \frac{1}{2}\int_\Omega |\nabla u|^2 dx - \int_\Omega F(x, u)dx \quad \text{for all } u \in H_0^1(\Omega),$$

where $F(x, t) = \int_0^t f(x, s)ds$ and $f(x, s)$ is extended to be $f(x, 0)$ for $s \leq 0$. Remark that $E \in C^1$ because

$$F(x, t) \leq c(t^2 + |t|) \quad \text{for all } (x, t) \in \Omega \times \mathbb{R}, \tag{1.47}$$

by virtue of $(f3)$.

In the statement of the next auxiliary result, we need the notion of *lower semicontinuity.* We recall that if S is a topological space, then a functional $E : S \to \mathbb{R} \cup \{+\infty\}$ is *lower semicontinuous* if and only if for any $a \in \mathbb{R}$, the sublevel set $S_a := \{u \in S : E(u) \leq a\}$ is closed. If S is a metric space, this condition is equivalent to $E(u) \leq \liminf_{n\to\infty} E(u_n)$ whenever $u_n \to u$.

Lemma 1.5.3 *The following properties hold true:*

(i) $E(u) \to \infty$ *as* $\|u\|_{H_0^1} \to \infty$.
(ii) E *is lower semicontinuous.*
(iii) $E(w) < 0$ *for some* $w \in H_0^1(\Omega)$.

Proof (i) Assume by contradiction that there exists $(u_n)_{n\geq 1} \subset H_0^1(\Omega)$ such that $\|u_n\|_{H_0^1} \to \infty$ and $E(u_n) \leq C$. From the definition of E and (1.47) we obtain

$$\frac{1}{2}\int_\Omega |\nabla u_n|^2 dx \leq M \int_\Omega (u_n^2 + 1)dx. \tag{1.48}$$

Set $t_n = \|u_n\|_2$ and $v_n = u_n/t_n$. From our assumption and (1.48) we have

$$t_n \to \infty, \quad \|v_n\|_2 = 1 \quad \text{and} \quad (v_n)_{n\geq 1} \text{ bounded in } H_0^1(\Omega).$$

Therefore, one can assume that

$$\begin{aligned} v_n &\rightharpoonup v \quad \text{weakly in } H_0^1(\Omega) \text{ as } n \to \infty, \\ v_n &\to v \quad \text{strongly in } L^2(\Omega) \text{ as } n \to \infty, \\ v_n &\to v \quad \text{almost everywhere in } \Omega \text{ and } \|v\|_2 = 1. \end{aligned}$$

We claim that

$$\limsup_{n\to\infty} \int_\Omega \frac{F(x, t_n v_n)}{t_n^2} dx \leq \frac{1}{2} \int_{\{v>0\}} a_\infty v^2 dx. \tag{1.49}$$

Indeed, we have

$$\begin{aligned} \int_\Omega F(x, t_n v_n) dx = \int_{\{v\leq 0\}} F(x, t_n v_n^+) dx + \int_{\{v>0\}} F(x, t_n v_n^+) dx \\ + \int_{\{v_n\leq 0\}} F(x, t_n v_n) dx. \end{aligned} \tag{1.50}$$

Using $(f3)$ we can evaluate the first integral in the right-hand side of (1.50) as

$$\int_{\{v\leq 0\}} F(x, t_n v_n^+) dx \leq c \int_{\{v\leq 0\}} \left[(t_n v_n^+)^2 + 1\right] dx,$$

which leads to

$$\int_{\{v\leq 0\}} \frac{F(x, t_n v_n^+)}{t_n^2} dx \leq o(1) \quad \text{as } n \to \infty. \tag{1.51}$$

In the same manner we have

$$\int_{\{v_n\leq 0\}} F(x, t_n v_n) dx \leq c \int_\Omega t_n |v_n| dx$$

and so

$$\int_{\{v_n\leq 0\}} \frac{F(x, t_n v_n)}{t_n^2} dx \leq o(1) \quad \text{as } n \to \infty. \tag{1.52}$$

To evaluate the first integral of the right-hand side in (1.50) we first remark that

$$\lim_{t\to\infty} \frac{F(x,t)}{t^2} \leq \frac{1}{2} a_\infty(x) \quad \text{in } \Omega,$$

which implies

$$\lim_{n\to\infty} \frac{F(x, t_n v_n^+)}{t_n^2} \leq \frac{1}{2} a_\infty(x) v \quad \text{on the set } \{v > 0\}. \tag{1.53}$$

By (1.47) we also have

$$\frac{F(x,t_nv_n^+)}{t_n^2} \le c\left[(v_n^+)^2 + \frac{1}{t_n^2}\right] \quad \text{in } \Omega.$$

Because $v_n \to v$ in $L^2(\Omega)$, by Fatou's lemma we deduce

$$\limsup_{n\to\infty} \int_{\{v>0\}} \frac{F(x,t_nv_n^+)}{t_n^2}dx \le \frac{1}{2}\int_{\{v>0\}} a_\infty v^2 dx. \tag{1.54}$$

Combining (1.50) through (1.53) we obtain the claim. Furthermore, from $E(u_n) \le C$ we deduce that

$$\frac{1}{2}\int_\Omega |\nabla u_n|^2 dx \le C + \int_\Omega F(x,u_n)dx.$$

Dividing by $\|u_n\|_2$ and passing to the limit in the previous inequality, by (1.49) it follows that

$$\frac{1}{2}\int_\Omega |\nabla v|^2 dx \le \frac{1}{2}\int_{\{v>0\}} v^2 dx. \tag{1.55}$$

On the other hand, by the definition of λ_∞ we have

$$\int_\Omega |\nabla v^+|^2 - \int_{\{v>0\}} v^2 dx \ge \lambda_\infty \|v^+\|_2^2. \tag{1.56}$$

From (1.55) and (1.56) we deduce $v = 0$, which is a contradiction because $\|v\|_2 = 1$. This proves (i).

(ii) The proof follows by Fatou's lemma and relation (1.47), which establishes that F has a quadratic growth.

(iii) Because $a_0 < 0$, there exists $w \in H_0^1(\Omega)$ such that

$$\int_\Omega |\nabla w|^2 dx - \int_{\{w\neq 0\}} a_0 w dx < 0.$$

Moreover, replacing w with w^+ and then truncating w^+, we may assume that $w \in L^\infty(\Omega)$ and $w > 0$ in Ω. Notice that

$$\lim_{t\searrow 0} \frac{F(x,t)}{t^2} \ge \frac{1}{2}a_0(x) \quad \text{in } \Omega.$$

Hence

$$\lim_{\varepsilon\to 0} \frac{F(x,\varepsilon w)}{\varepsilon^2} \ge \frac{1}{2}a_0(x)w^2(x) \quad \text{on the set } \{w \neq 0\}.$$

Let $c_0 > 0$ be such that $f(x,s) \ge -c_0 s$ for all $0 \le s \le \|w\|_\infty$. This yields

$$\frac{F(x,\varepsilon w)}{\varepsilon^2} \ge -c_0 w^2 \ge -c_0 \quad \text{in } \Omega.$$

By Fatou's lemma we deduce

$$\liminf_{\varepsilon\to 0}\int_{\{w\neq 0\}}\frac{F(x,\varepsilon w)}{\varepsilon^2}dx \geq \frac{1}{2}\int_{\{w\neq 0\}} a_0 w^2 dx.$$

Therefore

$$\liminf_{\varepsilon\to 0}\int_{\Omega}\frac{F(x,\varepsilon w)}{\varepsilon^2}dx \geq \frac{1}{2}\int_{\{w\neq 0\}} a_0 w^2 dx.$$

The last inequality combined with the fact that $\lambda_0 < 0$ shows that

$$\frac{1}{2}\int_{\Omega}|\nabla w| - \int_{\Omega}\frac{F(x,\varepsilon w)}{\varepsilon^2}dx < 0,$$

for $\varepsilon > 0$ that is small enough. Hence $v := \varepsilon w$ satisfies the claim of (iii). □

We now come back to the proof of Theorem 1.5.2.

By Lemma 1.5.3 we see that $\inf\{E(u) : u \in H_0^1(\Omega)\}$ is achieved by some $u \in H_0^1(\Omega)$. We can assume that $u \geq 0$ in Ω; otherwise, we replace u with u^+ and use the fact that $F(x,u) \leq F(x,u^+)$ because $F(x,t) = f(x,0)t \leq 0$ for all $t \leq 0$. This implies that $u \in H_0^1(\Omega)$ is a weak solution of (1.46). Standard regularity arguments imply that u is actually a $C^2(\overline{\Omega})$ solution of (1.46).

Step 3: Uniqueness. Let u_1 and u_2 be two solutions of (1.46). Remark that

$$\frac{\Delta u_2}{u_2} - \frac{\Delta u_1}{u_1} = \frac{f(u_1)}{u_1} - \frac{f(u_2)}{u_2} \quad \text{in } \Omega.$$

Multiplying by $u_1^2 - u_2^2$ in the previous relation and then integrating over Ω we find

$$\int_{\Omega}\left(\frac{\Delta u_2}{u_2} - \frac{\Delta u_1}{u_1}\right)(u_1^2 - u_2^2)dx = \int_{\Omega}\left(\frac{f(u_1)}{u_1} - \frac{f(u_2)}{u_2}\right)(u_1^2 - u_2^2)dx. \tag{1.57}$$

From Remark 1.5.1 we obtain that $u_1^2/u_2,\ u_2^2/u_1 \in L^\infty(\Omega)$, and by Green's first formula, the left-hand side reads

$$\int_{\Omega}\left(\frac{\Delta u_2}{u_2} - \frac{\Delta u_1}{u_1}\right)(u_1^2 - u_2^2)dx = \int_{\Omega}\left(\left|\nabla u_1 - \frac{u_1}{u_2}\nabla u_2\right|^2 + \left|\nabla u_1 - \frac{u_1}{u_2}\nabla u_2\right|^2\right)dx.$$

Thus, the left-hand side in (1.57) is nonnegative. This implies

$$\int_{\Omega}\left(\frac{f(u_1)}{u_1} - \frac{f(u_2)}{u_2}\right)(u_1^2 - u_2^2)dx = 0,$$

which yields $u_1 = u_2$. Hence, problem (1.46) has a unique solution. This concludes the proof. □

Remark 1.5.4 *In [37], the existence of a solution $u \in H_0^1(\Omega) \cap L^\infty(\Omega)$ of problem (1.46) is obtained under weaker assumptions on f,—namely,*

(i) *for all* $t \geq 0$, $x \longmapsto f(x,t)$ *belongs to* $L^\infty(\Omega)$;
(ii) *for almost everywhere* $x \in \Omega$ *the mapping* $[0,\infty) \ni t \longmapsto f(x,t)$ *is continuous;*
(iii) *for each* $\delta > 0$ *there exists a constant* $c_\delta \geq 0$ *such that* $f(x,t) \geq -c_\delta t$ *for all* $0 \leq t \leq \delta$ *and almost everywhere* $x \in \Omega$.

If f *satisfies*

$$a_0(x) := \liminf_{t \searrow 0} \frac{f(x,t)}{t} < 0 < a_\infty(x) := \limsup_{t \to \infty} \frac{f(t)}{t} \qquad \text{in } \Omega,$$

then (1.46) has at least one solution in $u \in H_0^1(\Omega) \cap L^\infty(\Omega)$.

Remark 1.5.5 *If, instead of condition* $(f1)$, *we assume*

$(f1)'$ *the mapping* $(0,\infty) \ni t \longmapsto \frac{f(t)}{t}$ *is nonincreasing,*

then the uniqueness does not necessarily hold. In this case we only obtain

$$\frac{\nabla u_1}{u_1} = \frac{\nabla u_1}{u_1} \quad \text{and} \quad \frac{f(x,u_1)}{u_1} = \frac{f(x,u_2)}{u_2} \quad \text{in } \Omega.$$

In particular, this implies that the quotient u_1/u_2 *is constant.*

1.6 Comments and historical notes

The premises of the sub- and supersolution method can be traced back to E. Picard. He applied, in the early 1880s, the method of *successive approximations* to argue the existence of solutions for nonlinear elliptic equations that are suitable perturbations of uniquely solvable linear problems. This is the starting point of the use of sub- and supersolutions in connection with monotone methods. Picard's techniques were applied later by Poincaré [166] in connection with problems arising in astrophysics. In particular, Poincaré established that the nonlinear equation $\Delta u = e^u$ has at least one solution. Motivated by this equation, which arises in the theory of Fuchsian functions as well as by many nonlinear problems describing various phenomena in mathematical physics, Poincaré formulated and developed the *continuity method*, which consists of solving nonlinear equations by embedding them in a one-parameter family of problems. In a first step, this tool consists of solving a simple problem and then extending the solvability by a step-by-step change in the parameter. Because of the major contributions of Lyapunov (1906), Schmidt (1908), and Leray and Schauder (1934), this method became a powerful tool in the bifurcation theory.

A major breakthrough in the understanding of monotone methods in connection with nonlinear differential equations is the result of the work by Scorza Dragoni [179], who considered the solvability of Sturm–Liouville problems of the type

$$\begin{cases} y''(x) + f(x, y(x), y'(x)) = 0 & \text{in } (a,b), \\ x(a) = A, \ x(b) = B, \end{cases} \tag{1.58}$$

by using comparison principles. The concepts of sub- and supersolution were introduced by Nagumo [146] in 1937 who proved, using the shooting method, the existence of at least one solution for problem (1.58).

A direct consequence of the mean value formula for harmonic functions $u : \Omega \to \mathbb{R}$ (where Ω is a domain in $\mathbb{R}^N$) is that u cannot achieve its maximum value at an interior point of Ω, unless it is constant. This result was extended by Picard and Lichtenstein to general linear uniformly elliptic operators. A pioneering contribution is a result of the work by Hopf [108], who proved that if $u \in C^2(\Omega) \cap C(\overline{\Omega})$ is a solution of the linear problem

$$\begin{cases} -\Delta u = f(x) \geq 0 & \text{in } \Omega, \\ \quad u = g(x) & \text{on } \partial\Omega, \end{cases} \tag{1.59}$$

and if u attains a nonnegative minimum in an interior point of Ω, then u is constant. In particular, if $g \geq 0$ on $\partial\Omega$, then $u \geq 0$ in Ω. This is the first form of the maximum principle for superharmonic functions. An important consequence of this result is that problem (1.59) has at most one solution. Another property related to the maximum principle is a result of the work by Perron [159], who established the existence of solutions to the linear Dirichlet problem (1.59) if $f = 0$.

An immediate consequence of the maximum principle is that any superharmonic function on some domain Ω attains its minimum value on $\partial\Omega$. An additional qualitative result states that if $u \in C^2(\Omega) \cap C^1(\partial\Omega)$ is superharmonic in Ω and $x_0 \in \partial\Omega$ such that $u(x_0) = \inf_{x\in\Omega} u(x)$, then $\partial u(x_0)/\partial n < 0$, provided $\partial\Omega$ has the interior sphere condition in x_0. According to Stampacchia's maximum principle, the same conclusion holds if u is a solution of the problem

$$-\Delta u \geq \lambda u \quad \text{in } \Omega,$$

where $\lambda < \lambda_1$ (λ_1 stands for the least eigenvalue of the Laplace operator $(-\Delta)$ in $H_0^1(\Omega)$). Clément and Peletier [54] (see also Hess [105]) studied the same problem in the case when λ is larger than the first eigenvalue of the Laplace operator. Their result is stated in the following antimaximum principle.

Theorem 1.6.1 *Let $\Omega \subset \mathbb{R}^N$ be a bounded domain with a smooth boundary and assume that f is a positive and continuous function on $\overline{\Omega}$. Then there exists $\delta = \delta(f) > 0$ such that if $\lambda_1 < \lambda < \lambda_1 + \delta$ and if $u \in C^2(\Omega) \cap C^1(\partial\Omega)$ is such that*

$$-\Delta u - \lambda u = f \quad \text{in } \Omega\,,$$

then $u < 0$ in Ω and $\frac{\partial u}{\partial n} > 0$ on $\partial\Omega$.

We refer to Brezis and Browder [31] for further historical comments on the role of the maximum principle in the qualitative analysis of nonlinear partial differential equations.

The Keller–Osserman condition around the origin (1.19) in the statement of Vázquez's maximum principle may be reformulated, equivalently, as

$$\int_0^1 \frac{dt}{\sqrt{tf(t)}} = \infty. \tag{1.60}$$

In particular, this hypothesis asserts that the nonlinearity f must not be too large near the origin. Indeed, condition (1.60) is fulfilled provided $f(t)/t = O(1)$ as $t \searrow 0$. The linear case corresponds to Stampacchia's maximum principle, but Vázquez's result also covers cases when f is not necessarily bounded around the origin, for instance if $f(t)/\left(t\log^2 t\right) = O(1)$ as $t \searrow 0$. We have already seen in this chapter that dead core solutions appear, provided that condition (1.60) is not fulfilled, for instance if $f(t) = t^p$, with $p \in (0,1)$.

The existence result established in Theorem 1.5.2 corresponding to sublinear Dirichlet boundary value problems on bounded domains was extended to the whole space by Brezis and Kamin [32]. They considered the problem

$$-\Delta u = V(x)u^p \qquad \text{in } \mathbb{R}^N,\ N \geq 3, \tag{1.61}$$

where $0 < p < 1$ and $V \in L^\infty_{\text{loc}}(\mathbb{R}^N)$ is a nonnegative and nontrivial potential. Under these assumptions, Brezis and Kamin [32] proved that the *nonlinear* problem (1.61) has a bounded positive solution if and only if the *linear* problem

$$-\Delta v = V(x) \quad \text{in } \mathbb{R}^N$$

has a bounded solution. Moreover, in such a case, problem (1.61) has a minimal positive solution.

An interesting conjecture raised by Sattinger [178] concerns the existence of a solution for the problem

$$\begin{cases} -\Delta u = f(x,u) & \text{in } \Omega, \\ u = 0 & \text{on } \partial\Omega \end{cases}$$

in the presence of sub- and supersolutions that do not satisfy the natural relation $\underline{u} \leq \overline{u}$ in Ω. In such a case, Gossez and Omari [97] proved the existence of a solution in the case when $f(x,t)/t$ remains, within small perturbations, between the first two eigenvalues of the linearized operator. Habets and Omari [100] extended this work by providing a general nonresonance condition with respect to the second curve of the Fučik spectrum.

Most of the results developed in this chapter can be extended to quasilinear elliptic equations described by the p–Laplace operator $\Delta_p u := \operatorname{div}\left(|\nabla u|^{p-2}\nabla u\right)$, where $1 < p < \infty$. If $p = 2$, then Δ_p becomes the linear Laplace operator, whereas Δ_p is nonlinear and degenerates at the zeros of ∇u, provided $p \neq 2$. That is why in such a case the solutions need not to be smooth, nor even C^1, and the corresponding equations should be understood in a weak sense.

PART II

BLOW-UP SOLUTIONS

2

BLOW-UP SOLUTIONS FOR SEMILINEAR ELLIPTIC EQUATIONS

> Truth lies neither in the narrow frame of doctrines nor outside them. Attempt to find it must carry us beyond all boundaries and dogmatic limitations.
>
> Stelian Mihalaş, 1975

In this chapter we are concerned with a class of singular solutions for logistic-type equations. We focus on positive solutions with blow-up boundary behavior and we formulate several sufficient conditions for the existence of such solutions. One of the main results we describe here establishes a necessary and sufficient condition for the existence of blow-up boundary solutions in the singular case of a potential that vanishes on the boundary.

The exact blow-up rate of solutions near the boundary is essential to state uniqueness results. The approach we develop to study the asymptotic behavior of solutions is original and is based on Karamata's regular variation theory. A key role in all arguments in this chapter is played by the Keller–Osserman condition, which describes the growth of the nonlinearity at infinity. Our asymptotic analysis is based on the observation that the nonlinearity f fulfills the Keller–Osserman condition, provided f' has a positive Karamata regular variation. Several recent works illustrate the importance of the Karamata theory, which is a tool for the interplay between classical mathematical analysis and modern nonlinear analysis.

2.1 Introduction

Consider the equation

$$\Delta u = f(u) \quad \text{in } \Omega, \tag{2.1}$$

where $\Omega \subseteq \mathbb{R}^N$ $(N \geq 1)$ is a smooth domain and $f \in C^1[0,\infty)$ is a nonnegative function that satisfies

$(f1)$ $f(0) = 0$, $f' \geq 0$ and $f > 0$ on $(0, \infty)$.

Definition 2.1.1 *A nonnegative solution $u \in C^2(\Omega)$ of equation (2.1) is called a* blow-up (large, explosive) solution *if $u(x) \to \infty$ as $\operatorname{dist}(x, \partial\Omega) \to 0$ in case Ω bounded and if $u(x) \to \infty$ as $|x| \to \infty$ in case Ω is an exterior domain or $\Omega = \mathbb{R}^N$. If $\Omega = \mathbb{R}^N$, such a solution is called an* entire blow-up solution.

In view of Corollary 1.4.4, if f satisfies $(f1)$, then any blow-up solution of (2.1) is positive.

The first contributions in this direction are the result of the work by Keller [118] and Osserman [154]. They supplied the following result for the existence of blow-up solutions to (2.1) in bounded domains.

Theorem 2.1.2 *Assume that Ω is bounded, f satisfies $(f1)$, and*

$(f2)$ $\displaystyle\int_1^\infty \frac{dt}{\sqrt{F(t)}} < \infty$, *where* $\displaystyle F(t) := \int_0^t f(s)\,ds.$

Then, equation (2.1) has a positive blow-up solution.

Proof We first need the following auxiliary result.

Lemma 2.1.3 *Assume that f satisfies $(f1)$ and $(f2)$. Then*

(i) $\displaystyle\lim_{t\to\infty} \frac{f(t)}{t} = \infty$;

(ii) $\displaystyle\lim_{t\to\infty} \int_t^\infty \frac{ds}{\sqrt{F(s)-F(t)}} = 0.$

Proof (i) Let $g(t) = F(t)^{-1/2}$, $t \geq 1$. Then

$$tg(t) = \int_1^t g(s)ds + \int_1^t sg'(s)ds + g(1), \quad \text{for all } t \geq 1.$$

Because g is positive and $g' \leq 0$, both integrals in the right-hand side of the previous equality define monotone functions. Thus, there exists $\lim_{t\to\infty} tg(t)$. This means that $\lim_{t\to\infty} F(t)/t^2$ exists in $\mathbb{R}\cup\{\infty\}$. By $(f2)$ we finally deduce that $\lim_{t\to\infty} F(t)/t^2 = \infty$. Because $F(t) \leq tf(t)$ for all $t \geq 0$, we find $\lim_{t\to\infty} f(t)/t = \infty$.

(ii) Set $G(t) := \int_t^\infty ds/\sqrt{F(s)-F(t)}$, $t \geq 1$. Notice that G is well defined because, by virtue of $(f2)$, we have

$$G(t) = \int_1^\infty \frac{ds}{\sqrt{F(s+t-1)-F(t)}} < \infty.$$

Using $(f1)$ and the mean value theorem we deduce that

$$F(s) - F(t) \geq F(s-t) \quad \text{for all } s \geq t \geq 1,$$
$$F(s) - F(t) \geq (s-t)f(t) \quad \text{for all } s \geq t \geq 1.$$

Hence, for all $t \geq 1$ we have

$$\begin{aligned} G(t) &= \int_t^{2t} \frac{ds}{\sqrt{F(s)-F(t)}} + \int_{2t}^\infty \frac{ds}{\sqrt{F(s)-F(t)}} \\ &\leq \frac{1}{\sqrt{f(t)}} \int_t^{2t} \frac{ds}{\sqrt{s-t}} + \int_{2t}^\infty \frac{ds}{\sqrt{F(s-t)}} \\ &= 2\sqrt{\frac{t}{f(t)}} + \int_t^\infty \frac{d\tau}{\sqrt{F(\tau)}}. \end{aligned}$$

Now the assumption $(f2)$ together with (i) allows us to conclude. □

By virtue of Theorem 1.4.1, for all $n \geq 1$ there exists a unique solution $u_n \in C^2(\overline{\Omega})$ of the problem

$$\begin{cases} \Delta u = f(u) & \text{in } \Omega, \\ u \geq 0 & \text{in } \Omega, \\ u = n & \text{on } \partial\Omega. \end{cases}$$

By the weak maximum principle we also have $u_n \leq u_{n+1}$ in Ω, for all $n \geq 1$. We claim that $(u_n)_{n\geq 1}$ is uniformly bounded in any compact subset of Ω. For this purpose, we show that for any $x_0 \in \Omega$ there exists an open set $\omega \subset \Omega$ such that $x_0 \in \omega$, and $(u_n)_{n\geq 1}$ is uniformly bounded on ω.

Fix $x_0 \in \Omega$ and let $B \subset\subset \Omega$ be an open ball centered at x_0 with a radius $R > 0$. Consider the problem

$$\begin{cases} \Delta v = f(v) & \text{in } B, \\ v \geq 0 & \text{in } B, \\ v = n & \text{on } \partial B. \end{cases} \tag{2.2}$$

According to Theorem 1.4.1, problem (2.2) has a unique positive solution $v_n \in C^2(\overline{B})$. Moreover, because u_n is subharmonic, we obtain $u_n \leq v_n$ on ∂B and, by the weak maximum principle, it follows that

$$u_n \leq v_n \quad \text{in } B \text{ for all } n \geq 1. \tag{2.3}$$

Because of the uniqueness of the solution to (2.2), v_n is radially symmetric. Hence, $v_n(x) = v_n(r)$, $0 \leq r = |x - x_0| \leq R$ and

$$\begin{cases} v_n''(r) + \dfrac{N-1}{r} v_n'(r) = f(v_n(r)) & \text{for all } 0 \leq r < R, \\ v_n(r) > 0 & \text{for all } 0 \leq r < R, \\ v_n(R) = n. \end{cases} \tag{2.4}$$

Integrating in (2.4), by continuity arguments it follows that

$$v_n'(r) = r^{-N+1} \int_0^r t^{N-1} f(v_n(t))dt \quad \text{for all } 0 \leq r \leq R.$$

Therefore v_n is nondecreasing and

$$v_n'(r) \leq \frac{f(v_n(r))}{r^{N-1}} \int_0^r t^{N-1}dt = \frac{r}{N} f(v_n(r)) \quad \text{for all } 0 \leq r \leq R. \tag{2.5}$$

Using (2.5) in (2.4) we obtain

$$\frac{f(v_n(r))}{N} \leq v_n''(r) \quad \text{for all } 0 \leq r \leq R. \tag{2.6}$$

We now multiply by $v_n'(r)$ in (2.6) and then we integrate over $[\rho, r]$. We obtain

$$\frac{2}{N}\left(F(v_n(r)) - F(v_n(\rho))\right) \le v_n'^2(r) \quad \text{for all } \rho \le r \le R.$$

This yields

$$\frac{2}{N} \le \frac{v_n'^2(r)}{F(v_n(r)) - F(v_n(\rho))} \quad \text{for all } \rho \le r \le R.$$

By integrating over $[\rho, R]$ in the previous inequality we deduce

$$\begin{aligned} \frac{2(R-\rho)}{\sqrt{2N}} &\le \int_{v_n(\rho)}^{n} \frac{dt}{\sqrt{F(t) - F(v_n(\rho))}} \\ &\le \int_{v_n(\rho)}^{\infty} \frac{dt}{\sqrt{F(t) - F(v_n(\rho))}} \quad \text{for all } 0 \le \rho < R. \end{aligned}$$

Now Lemma 2.1.3 (ii) implies that the sequence $(v_n(\rho))_{n\ge1}$ is uniformly bounded. Hence, $(v_n)_{n\ge1}$ is uniformly bounded in $\omega := B(x_0, \rho)$. By virtue of (2.3) it follows that the sequence $(u_n)_{n\ge1}$ is uniformly bounded in ω. Therefore, there exists $u(x) := \lim_{n\to\infty} u_n(x)$, for all $x \in \Omega$. By standard elliptic arguments, $u \in C^2(\Omega)$ is a classical solution of (2.1). Because $u_n \le u_{n+1} \le u$ in Ω and $u_n = n$ on $\partial\Omega$, it follows that u blows up at the boundary. This finishes the proof. □

2.2 Blow-up solution for elliptic equations with vanishing potential

We are concerned in this section with the existence of positive blow-up solutions of

$$\Delta u = p(x) f(u) \quad \text{in } \Omega, \tag{2.7}$$

where $\Omega \subseteq \mathbb{R}^N$ $(N \ge 1)$ is either a smooth bounded domain or $\Omega = \mathbb{R}^N$ and $f \in C^1[0,\infty)$ satisfies $(f1)$ and $(f2)$. We also assume that p is a nonnegative function such that $p \in C^{0,\gamma}(\overline{\Omega})$ if Ω is bounded, and $p \in C^{0,\gamma}_{\rm loc}(\Omega)$, otherwise. The main feature is that the potential may vanish at some points in $\overline{\Omega}$.

2.2.1 *Existence results in bounded domains*

Lemma 2.2.1 *Assume that conditions $(f1)$ and $(f2)$ are fulfilled. Then,*

(i) $\displaystyle\int_0^1 \frac{dt}{f(t)} = \infty;$

(ii) $\displaystyle\int_1^\infty \frac{dt}{f(t)} < \infty.$

Proof (i) By virtue of $(f1)$ there exist c, $\delta > 0$ such that $f(t) \le ct$ for all $0 < t < \delta$. This yields $\int_0^1 dt/f(t) = \infty$.

(ii) Fix $R > 0$ and let B be the open ball centered at the origin with a radius R. By Theorem 2.1.2, there exists a positive blow-up solution v of (2.1) with $\Omega = B$. Using Theorem 1.4.1, there exists a unique $u_n \in C^2(\Omega)$ such that

$$\begin{cases} \Delta u_n = f(u_n) & \text{in } B, \\ u_n > 0 & \text{in } B, \\ u_n = n & \text{on } \partial B. \end{cases} \tag{2.8}$$

Because f is nondecreasing, by the weak maximum principle it follows that

$$u_n \le u_{n+1} \le v \quad \text{in } B.$$

This implies that $(u_n)_{n\ge1}$ is uniformly bounded in every compact subdomain of B. Hence, for all $x \in B$ there exists $u(x) := \lim_{n\to\infty} u_n(x)$. By standard regularity arguments, u is a blow-up solution of (2.1). Notice that by the uniqueness of (2.8), u_n is radially symmetric in B, which implies that u is also radially symmetric (in this sense we could also apply a well-known result of Gidas, Ni, and Nirenberg [93]). Therefore, $u(x) = u(r)$, $0 \le r = |x| \le R$, and u satisfies in the r variable the equation

$$u''(r) + \frac{N-1}{r} u'(r) = f(u(r)) \quad \text{for all } 0 \le r < R.$$

Integrating in the previous equality we obtain

$$u'(r) = r^{1-N} \int_0^r s^{N-1} f(u(s))\, ds \quad \text{for all } 0 \le r < R.$$

Similar to the proof in Theorem 2.1.2, we note that u is nondecreasing and

$$\frac{f(u(r))}{N} \le u''(r) \quad \text{for all } 0 \le r < R,$$

which yields

$$\frac{u'(r)}{f(u(r))} \le rN \quad \text{for all } 0 \le r < R.$$

Thus,

$$\int_{u(0)}^{u(r)} \frac{dt}{f(t)} \le \frac{r^2}{2N} \quad \text{for all } 0 \le r < R.$$

Letting $r \nearrow R$ we find

$$\int_{u(0)}^{\infty} \frac{dt}{f(t)} \le \frac{R^2}{2N}.$$

This ends the proof. □

Assume that p satisfies the assumption

($p1$) for every $x_0 \in \Omega$ with $p(x_0) = 0$ there exists a domain $\Omega_0 \subset\subset \Omega$ such that $x_0 \in \Omega_0$ and $p > 0$ on $\partial\Omega_0$.

In this case we have the following result.

Theorem 2.2.2 *Assume that Ω is bounded and conditions ($f1$), ($f2$), and ($p1$) are satisfied. Then equation (2.7) has a positive blow-up solution.*

Proof By Theorem 1.4.1, for any $n \geq 1$, the boundary value problem

$$\begin{cases} \Delta u = p(x)f(u) & \text{in } \Omega, \\ u > 0 & \text{in } \Omega, \\ u = n & \text{on } \partial\Omega \end{cases} \tag{2.9}$$

has a unique solution $u_n \in C^2(\overline{\Omega})$. Furthermore, by the weak maximum principle, the sequence $(u_n)_{n\geq 1}$ is nondecreasing.

We claim that $(u_n)_{n\geq 1}$ is uniformly bounded in every compact subset of Ω. To this aim it is enough to show that for all $x_0 \in \Omega$ there exists an open set $\mathcal{O} \subset\subset \Omega$ containing x_0 such that $(u_n)_{n\geq 1}$ is uniformly bounded in $\mathcal{O}$. Let $x_0 \in \Omega$ be fixed.

In our analysis, two cases may occur.

Case 1: $p(x_0) > 0$. By the continuity of p, there exists a ball $B = B(x_0, r) \subset\subset \Omega$ such that $m := \min\{p(x) : x \in \overline{B}\} > 0$. Let v be a positive blow-up solution of $\Delta v = mf(v)$ in B. By the maximum principle it follows that $u_n \leq v$ in B. Furthermore, v is bounded in $\overline{B}(x_0, r/2)$ so that it suffices to take $\mathcal{O} := B(x_0, r/2)$.

Case 2: $p(x_0) = 0$. Assumption ($p1$) and the boundedness of Ω imply the existence of a domain $\mathcal{O} \subset\subset \Omega$ that contains x_0 such that $p > 0$ on $\partial\mathcal{O}$. Case 1 shows that for any $x \in \partial\mathcal{O}$, there exist a ball $B(x, r_x)$ strictly contained in Ω and a constant $M_x > 0$ such that $u_n \leq M_x$ in $B(x, r_x/2)$, for any $n \geq 1$. Because $\partial\mathcal{O}$ is compact, it follows that it can be covered by a finite number of such balls, say $B(x_i, r_{x_i}/2)$, $i = 1, \ldots, k_0$. Setting $M := \max\{M_{x_1}, \ldots, M_{x_{k_0}}\}$ we have $u_n \leq M$ on $\partial\mathcal{O}$, for any $n \geq 1$. Applying the maximum principle we obtain $u_n \leq M$ in $\mathcal{O}$ and the claim follows.

Therefore, for all $x \in \Omega$ there exists $u(x) := \lim_{n\to\infty} u_n(x)$. Standard elliptic regularity arguments show that u is a solution of problem (2.7). Because $u_n \leq u_{n+1} \leq u$ in Ω, it follows that u blows up at the boundary. This concludes the proof. □

2.2.2 *Existence results in the whole space*

In this subsection we prove the existence of a positive entire blow-up solution of equation (2.7). We require the following condition on p:

($p1$)′ There exists a sequence of smooth bounded domains $(\Omega_n)_{n\geq 1}$ such that $\overline{\Omega}_n \subset \Omega_{n+1}$, $\mathbb{R}^N = \cup_{n=1}^{\infty}\Omega_n$, and ($p1$) holds in Ω_n, for any $n \geq 1$;

(p2) $\int_0^\infty r\phi(r)\,dr < \infty$, where $\phi(r) := \max\{p(x) : |x| = r\}$.

Remark that $(p1)'$ is automatically satisfied if $p(x) > 0$ for $|x|$ sufficiently large. An example of function p that satisfies both $(p2)$ and $(p1)'$, with p vanishing in every neighborhood of infinity, is provided here:

$$\begin{cases} p(r) = 0 \quad \text{for} \quad r = |x| \in [n - 1/3, n + 1/3], \; n \geq 1; \\ p(r) > 0 \quad \text{in} \quad \mathbb{R}_+ \setminus \bigcup_{n=1}^{\infty} [n - 1/3, n + 1/3]; \\ p \in C^1[0,\infty) \quad \text{and} \quad \max_{n \leq r \leq n+1} p(r) = \dfrac{2}{n^2(2n+1)}. \end{cases}$$

Obviously $(p1)'$ is fulfilled for $\Omega_n := B(0, n + 1/2)$. On the other hand, condition $(p2)$ is also satisfied because

$$\begin{aligned} \int_1^\infty rp(r)\,dr &= \sum_{n=1}^{\infty} \int_n^{n+1} rp(r)\,dr \\ &\leq \sum_{n=1}^{\infty} \int_n^{n+1} \frac{2r}{n^2(2n+1)}\,dr \\ &= \sum_{n=1}^{\infty} \frac{1}{n^2} < \infty. \end{aligned}$$

In this case we have the following existence result.

Theorem 2.2.3 *Let $\Omega = \mathbb{R}^N$ and assume that conditions $(f1)$, $(f2)$ and $(p1)'$, $(p2)$ are satisfied. Then equation (2.7) has a positive entire blow-up solution.*

Proof By Theorem 2.2.2, the boundary value problem

$$\begin{cases} \Delta u = p(x)f(u) & \text{in } \Omega_n, \\ u > 0 & \text{in } \Omega_n, \\ u = \infty & \text{on } \partial\Omega_n \end{cases} \tag{2.10}$$

has a solution $u_n \in C^2(\Omega)$. Because $\Omega_n \subset\subset \Omega_{n+1}$, we derive that $u_n \geq u_{n+1}$ in Ω_n, for all $n \geq 1$. Thus, for all $x \in \mathbb{R}^N$ we can define $U(x) := \lim_{n\to\infty} u_n(x)$. By standard regularity arguments we find that $U \in C^{2,\gamma}_{\text{loc}}(\mathbb{R}^N)$ and $\Delta U = p(x)f(U)$ in $\mathbb{R}^N$.

Let us now prove that U blows up at infinity. To this aim, it suffices to find a positive function $w \in C(\mathbb{R}^N)$ such that $U \geq w$ in $\mathbb{R}^N$ and $w(x) \to \infty$ as $|x| \to \infty$. This will also imply that U is positive.

Set

$$\Phi(r) := \int_0^r t^{1-N} \int_0^t s^{N-1}\phi(s)\,ds\,dr \qquad \text{for all } r \geq 0.$$

Obviously, $\Phi \in C^2[0,\infty)$. Furthermore, we have the following lemma.

Lemma 2.2.4 *$\lim_{r\to\infty} \Phi(r)$ is finite if and only if $\int_0^\infty r\phi(r)dr$ is finite.*

Proof For any $r > 0$ we have

$$\begin{aligned}\Phi(r) &= \frac{1}{N-2}\left(\int_0^r t\phi(t)\,dt - \frac{1}{r^{N-2}}\int_0^r t^{N-1}\phi(t)\,dt\right)\\ &\le \frac{1}{N-2}\int_0^r t\phi(t)\,dt.\end{aligned} \tag{2.11}$$

On the other hand,

$$\begin{aligned}\int_0^r t\phi(t)dt - \frac{1}{r^{N-2}}\int_0^r t^{N-1}\phi(t)dt &= \frac{1}{r^{N-2}}\int_0^r \left(r^{N-2} - t^{N-2}\right) t\phi(t)dt\\ &\ge \frac{r^{N-2} - (r/2)^{N-2}}{r^{N-2}}\int_0^{r/2} t\phi(t)dt.\end{aligned}$$

The last inequality combined with (2.11) yields

$$\frac{1}{N-2}\int_0^r t\phi(t)dt \ge \Phi(r) \ge \frac{1-(1/2)^{N-2}}{N-2}\int_0^{r/2} t\phi(t)dt.$$

The conclusion follows now by letting $r \to \infty$ in the previous estimates. This finishes the proof. □

By virtue of hypothesis $(p2)$ and Lemma 2.2.4 it follows that $L := \lim_{r\to\infty}\Phi(r)$ is finite. Then, $z(r) := L - \Phi(r)$, $r \ge 0$ is positive and verifies

$$\begin{cases} -\Delta z = \phi(r) & r := |x| \ge 0,\\ z(x) \to 0 & \text{as } |x| \to \infty. \end{cases} \tag{2.12}$$

According to Lemma 2.2.1, the mapping

$$(0,\infty) \ni t \longmapsto \int_t^\infty \frac{ds}{f(s)} \in (0,\infty)$$

is decreasing and bijective. We implicitely define w as

$$z(x) = \int_{w(x)}^{\infty} \frac{dt}{f(t)} \qquad \text{for all } x \in \mathbb{R}^N. \tag{2.13}$$

Note that by (2.12) we have $w \in C^2(\mathbb{R}^N)$ and $w \to \infty$ as $|x| \to \infty$. Furthermore, we have

$$-\Delta z(x) = \frac{1}{f(w)}\Delta w(x) - \frac{f'(w)}{f^2(w)}|\nabla w|^2 \quad \text{in } \mathbb{R}^N.$$

Hence,

$$\Delta w \geq p(x)f(w) \quad \text{in } \mathbb{R}^N.$$

We claim that $w \leq u_n$ in Ω_n for all $n \geq 1$. Obviously this inequality is true on $\partial\Omega_n$. Using the same arguments as in the proof of inequality (1.35) in Theorem 1.4.1 (with Ω replaced with Ω_n), we obtain that for any $\varepsilon > 0$ and $n \geq 1$ we have

$$w(x) \leq u_n(x) + \varepsilon(1 + |x|^2)^{-1/2} \qquad \text{in } \Omega_n.$$

Passing to the limit with $\varepsilon \to 0$ we derive $w \leq u_n$ in Ω_n. Consequently, $U \geq w$ in $\mathbb{R}^N$ and, by (2.13), $w(x) \to \infty$ as $|x| \to \infty$. This completes the proof. □

2.3 Blow-up solutions for logistic equations

In this section we discuss the existence of blow-up solutions to the following semilinear elliptic equation:

$$\Delta u + \lambda u = p(x)f(u) \quad \text{in } \Omega. \tag{2.14}$$

We assume that $\Omega \subseteq \mathbb{R}^N$ ($N \geq 3$) is a smooth bounded domain, $\lambda \in \mathbb{R}$, and $f \in C^1[0, \infty)$ is a nonnegative function that satisfies $(f2)$ and

$(f3)$ the mapping $(0, \infty) \ni t \longmapsto \dfrac{f(t)}{t}$ is increasing.

We also assume that $p \in C^{0,\gamma}(\overline{\Omega})$ $(0 < \gamma < 1)$ is nonnegative and $p \not\equiv 0$ in Ω. Notice that we allow p to vanish in Ω (or even on $\partial\Omega$).

Remark 2.3.1 *If f satisfies $(f2)$ and $(f3)$, by Lemma 2.1.3 and l'Hospital's rule we deduce that* $\lim_{t\to\infty} f(t)/t = \lim_{t\to\infty} f'(t) = \infty$.

Remark 2.3.2 *Typical examples of nonlinearities satisfying $(f2)$ and $(f3)$ are*

(i) $f(t) = e^t - 1$.
(ii) $f(t) = t^p$, $p > 1$.
(iii) $f(t) = t[\ln(t+1)]^p$, $p > 2$.

2.3.1 *The case of positive potentials*

We discuss here the existence of blow-up solutions to (2.14) under the additional hypothesis $p > 0$ in $\overline{\Omega}$. More generally, we consider the problem

$$\Delta u + a(x)u + b(x) = p(x)f(u) \quad \text{in } \Omega \tag{2.15}$$

where $a, b \in C^{0,\gamma}(\overline{\Omega})$, $b \geq 0$. We first need the following result.

Proposition 2.3.3 *Assume that $p > 0$ in $\overline{\Omega}$ and f satisfies $(f2)$, $(f3)$. Let $\phi \in C^{0,\gamma}(\partial\Omega)$ be a nonnegative function such that $\phi \not\equiv 0$. Then the boundary value problem*

$$\begin{cases} \Delta u + a(x)u + b(x) = p(x)f(u) & \text{in } \Omega, \\ u > 0 & \text{in } \Omega, \\ u = \phi & \text{on } \partial\Omega \end{cases} \tag{2.16}$$

has a unique solution $u \in C^2(\overline{\Omega})$.

Proof The uniqueness follows directly from Theorem 1.3.17. We will focus on the existence part. Applying Theorem 1.4.1, the boundary value problem

$$\begin{cases} \Delta u = \|a\|_\infty u + \|p\|_\infty f(u) & \text{in } \Omega, \\ u > 0 & \text{in } \Omega, \\ u = \phi & \text{on } \partial\Omega \end{cases} \tag{2.17}$$

has a unique solution $\underline{u} \in C^2(\overline{\Omega})$. Clearly $\underline{u}$ is a subsolution of (2.17). To provide a supersolution of (2.17), let us remark that for all $n \geq 1$, the mapping

$$\Phi_n(x,t) := a(x)t + b(x) - p(x)f(t) + \frac{1}{n}, \quad (x,t) \in \overline{\Omega} \times (0,\infty),$$

satisfies the hypotheses of Theorems 1.2.5 and 1.3.17. Hence, there exists a unique solution $u_n \in C^2(\overline{\Omega})$ of the problem

$$\begin{cases} \Delta u + a(x)u + b(x) = p(x)f(u) - \dfrac{1}{n} & \text{in } \Omega, \\ u > 0 & \text{in } \Omega, \\ u = \phi & \text{on } \partial\Omega. \end{cases}$$

By Theorem 1.3.17 we deduce $0 < \underline{u} \leq u_{n+1} \leq u_n$ in Ω. If $u(x) := \lim_{n\to\infty} u_n(x)$, $x \in \overline{\Omega}$, by standard elliptic regularity arguments we derive that $u \in C^2(\overline{\Omega})$ is a solution of (2.16). This finishes the proof. □

Under the assumptions of Proposition 2.3.3, we obtain the following theorem.

Theorem 2.3.4 *There exists a positive blow-up solution of equation (2.15).*

Proof Let $\Phi(t) := p_0 f(t) - \|a\|_\infty t - \bar{b}$, $t \geq 0$, where $p_0 = \inf_{\overline{\Omega}} p > 0$ and $\bar{b} := \sup_\Omega b + 1 > 0$. Let also τ be the unique positive solution of the equation $\Phi(t) = 0$. By Remark 2.3.1 we deduce that $\lim_{t\to\infty} \Phi(t)/f(t) = p_0 > 0$. By (A_2), we further derive that the mapping $(0,\infty) \ni t \longmapsto \Phi(t+\tau)$ satisfies the assumptions of Theorem 2.1.2. Hence, there exists a positive blow-up solution of $\Delta v = \Phi(v+\tau)$ in Ω. Thus, $U(x) := v(x) + \tau$, $x \in \Omega$, satisfies

$$\Delta U + \|a\|_\infty U + \bar{b} = p_0 f(U) \quad \text{in } \Omega,$$

and blows up at the boundary of Ω.

Let u_n be the unique solution of (2.16) with $\phi \equiv n$. By Theorem 1.3.17, $u_n \leq u_{n+1} \leq U$ in Ω. It follows that $u(x) := \lim_{n\to\infty} u_n(x)$, $x \in \Omega$, exists and defines a positive blow-up solution of (2.15). Moreover, every positive blow-up solution v of (2.15) satisfies $u_n \leq v$ in Ω, which yields $u \leq v$ in Ω. This means that u is a minimal blow-up solution of (2.15). This concludes the proof. □

2.3.2 *The case of vanishing potentials*

We assume in this section that p vanishes in Ω. Set

$$\Omega_0 := \text{int}\,\{x \in \Omega : p(x) = 0\} \tag{2.18}$$

and suppose that $\overline{\Omega}_0 \subset \Omega$ and $p > 0$ in $\Omega \setminus \overline{\Omega}_0$.

Let $\mathcal{L}$ be the unique self-adjoint operator associated with the quadratic form $\psi(u) = \int_\Omega |\nabla u|^2\,dx$ on

$$H^1_D(\Omega_0) := \{u \in H^1_0(\Omega) : u(x) = 0 \quad \text{for almost everywhere } x \in \Omega \setminus \Omega_0\}.$$

If $\partial\Omega_0$ satisfies the exterior cone condition then, according to Alama and Tarantello [4], $H^1_D(\Omega_0)$ coincides with $H^1_0(\Omega_0)$ and $\mathcal{L}$ is the classical Laplace operator with the Dirichlet condition on $\partial\Omega_0$.

Let $\lambda_{\infty,1}$ be the first Dirichlet eigenvalue of $\mathcal{L}$ in Ω_0. Set $\ell := \lim_{t\searrow 0} f(t)/t$ and denote by $\lambda_{0,1}$ the first eigenvalue of the operator $-\Delta + \ell p(x)$ in $H^1_0(\Omega)$.

Consider the following nonlinear Dirichlet problem:

$$\begin{cases} \Delta u + \lambda u = p(x)f(u) & \text{in } \Omega, \\ u > 0 & \text{in } \Omega, \\ u = 0 & \text{on } \partial\Omega. \end{cases} \tag{2.19}$$

Alama and Tarantello [4] established the following necessary and sufficient condition for the existence of solutions to problem (2.19).

Theorem 2.3.5 *Assume f satisfies assumption $(f3)$. Then problem (2.19) has a solution if and only if $\lambda_{0,1} < \lambda < \lambda_{\infty,1}$. Moreover, in this case the solution is unique.*

A related interesting problem is to establish a necessary and sufficient condition for the existence of blow-up boundary solutions of equation (2.14). The following result asserts that such a solution exists for small values of the real parameter λ, up to a certain range. However, if the variable potential is positive, then (2.14) has a blow-up boundary solution for any value of λ. We observe that our assumptions on the potential p allow that it may vanish on $\partial\Omega$. This is the most interesting case, because it corresponds to the "competition" between a vanishing potential and a blow-up nonlinearity.

Theorem 2.3.6 *Assume that f satisfies conditions $(f2)$ and $(f3)$. Then problem (2.14) has a positive blow-up solution if and only if $\lambda \in (-\infty, \lambda_{\infty,1})$.*

Proof NECESSARY CONDITION. Let u be a blow-up solution of equation (2.14). As we have already argued, u is positive in Ω. Suppose $\lambda_{\infty,1}$ is finite and let us assume by contradiction that $\lambda \geq \lambda_{\infty,1}$. Fix $\lambda_{0,1} < \mu < \lambda_{\infty,1}$ and denote by v a positive solution of problem (2.19) with $\lambda = \mu$. If $M := \max\{\max_{\overline{\Omega}} v/\min_\Omega u;\, 1\}$, then we have

$$\begin{cases} \Delta(Mu) + \lambda_{\infty,1}(Mu) \leq p(x)f(Mu) & \text{in } \Omega, \\ Mu \geq v & \text{in } \Omega, \\ Mu = \infty & \text{on } \partial\Omega. \end{cases}$$

Hence (v, Mu) is an ordered pair of sub- and supersolution of problem (2.19) with $\lambda = \lambda_{\infty,1}$. Thus, problem (2.19) with $\lambda = \lambda_{\infty,1}$ has at least one positive solution (between v and Mu), which is a contradiction. So, necessarily, $\lambda \in (-\infty, \lambda_{\infty,1})$.

SUFFICIENT CONDITION. We first need a similar result to that in Proposition 2.3.3.

Proposition 2.3.7 *Assume that $p > 0$ on $\partial\Omega$ and let $\phi \in C^{0,\gamma}(\partial\Omega)$ be a non-negative function such that $\phi \not\equiv 0$. Then the boundary value problem*

$$\begin{cases} \Delta u + \lambda u = p(x)f(u) & \text{in } \Omega, \\ u > 0 & \text{in } \Omega, \\ u = \phi & \text{on } \partial\Omega \end{cases} \tag{2.20}$$

has a solution if and only if $\lambda \in (-\infty, \lambda_{\infty,1})$. Moreover, in this case, the solution is unique.

Proof The first part follows exactly in the same way as in the proof of the necessary condition in Theorem 2.3.6.

For the sufficient condition, fix $\lambda < \lambda_{\infty,1}$. By Theorem 1.4.1 there exists a unique classical solution $\underline{u}$ of the problem

$$\begin{cases} \Delta u = |\lambda| u + \|p\|_\infty f(u) & \text{in } \Omega, \\ u > 0 & \text{in } \Omega, \\ u = \phi & \text{on } \partial\Omega. \end{cases} \tag{2.21}$$

Obviously $\underline{u}$ is a subsolution of (2.20). Let $\lambda_{\infty,1} > \lambda_* > \max\{\lambda, \lambda_{0,1}\}$ and u_* be the unique solution of (2.19) with $\lambda = \lambda_*$.

Let Ω_i $(i = 1, 2)$ be two subdomains of Ω such that

$$\Omega_0 \subset\subset \Omega_1 \subset\subset \Omega_2 \subset\subset \Omega$$

and $\Omega \setminus \overline{\Omega}_1$ is smooth.

Because $p > 0$ in $\Omega \setminus \overline{\Omega}_1$, by Theorem 2.3.4 there exists $U \in C^2(\Omega \setminus \overline{\Omega}_1)$, a positive solution of (2.14) in $\Omega \setminus \overline{\Omega}_1$.

We define $v \in C^2(\Omega)$ as a positive function in Ω such that $v = U$ in $\Omega \setminus \Omega_2$ and $u = u_*$ in Ω_1. Using Remark 2.3.1 and the fact that $\inf_{\Omega_2 \setminus \Omega_1} p > 0$, it is easy to check that $\overline{u} := Cv$ satisfies

$$\begin{cases} \Delta \overline{u} + \lambda \overline{u} \le p(x)f(\overline{u}) & \text{in } \Omega, \\ \overline{u} \ge \max_{\partial\Omega} \phi & \text{in } \Omega, \\ \overline{u} = \infty & \text{on } \partial\Omega, \end{cases} \tag{2.22}$$

provided $C > 0$ is large enough. It is clear that $\overline{u}$ is a positive supersolution of (2.20) and $\underline{u} \le \max_{\partial\Omega} \phi \le \overline{u}$ in Ω. Therefore, by the sub- and supersolution method, problem (2.20) has at least a solution $u \in C^2(\overline{\Omega})$ such that $\underline{u} \le u \le \overline{u}$ in Ω.

Now let $-\infty < \lambda < \lambda_{\infty,1}$ and u_1, u_2 be two solutions of (2.20). Because p may vanish in Ω we cannot apply Theorem 1.3.17 directly. However, we can adapt the arguments in the proof of Theorem 1.5.2 to obtain

$$\int_\Omega p(x)\left(\frac{f(u_2)}{u_2} - \frac{f(u_1)}{u_1}\right)(u_1^2 - u_2^2)dx = 0.$$

This yields $u_1 = u_2$ on the set $\{x \in \Omega : p(x) > 0\}$—that is, $p(x)f(u_1) = p(x)f(u_2)$ in Ω. Furthermore, by the strong maximum principle we obtain $u_1 = u_2$ in Ω. In the same manner we obtain that problem (2.19) has at most one solution. The proof of Proposition 2.3.7 is now complete. □

Proof of Theorem 2.3.6 completed. Fix $\lambda \in (-\infty, \lambda_{\infty,1})$. In our further analysis, two cases may occur:

CASE 1: $p > 0$ on $\partial\Omega$. Denote by u_n the unique solution of (2.20) with $\phi \equiv n$. For $\phi \equiv 1$, let $\underline{u}$ and $\overline{u}$ be the solutions of problems (2.21) and (2.22), respectively. The sub- and supersolution method combined with the uniqueness of the solution to (2.20) shows that $\underline{u} \le u_n \le u_{n+1} \le \overline{u}$ in Ω. Hence $U(x) := \lim_{n\to\infty} u_n(x)$, $x \in \Omega$, exists and is a positive blow-up solution of (2.14).

CASE 2: p vanishes on $\partial\Omega$. According to Proposition 2.3.3, for all $n \ge 1$ there exists a unique $v_n \in C^2(\overline{\Omega})$ such that

$$\begin{cases} \Delta v_n + \lambda v_n = \left(p(x) + \dfrac{1}{n}\right) f(v_n) & \text{in } \Omega,\\ v_n > 0 & \text{in } \Omega,\\ v_n = n & \text{on } \partial\Omega. \end{cases}$$

By Theorem 1.3.17, the sequence $(v_n)_{n\ge1}$ is nondecreasing. Moreover, $(v_n)_{n\ge1}$ is uniformly bounded on every compact subdomain of Ω. Indeed, let $K \subset\subset \Omega$ be an arbitrary compact set and $d := \operatorname{dist}(K, \partial\Omega) > 0$. Choose $\delta \in (0, d)$ small enough such that $\overline{\Omega}_0 \subset C_\delta$, where $C_\delta = \{x \in \Omega : \operatorname{dist}(x, \partial\Omega) > \delta\}$. Because $p > 0$ on ∂C_δ, by the results in Case 1 there exists a positive blow-up solution V of (2.14) for $\Omega = C_\delta$. Using Theorem 1.3.17 in C_δ, we deduce $v_n \le V$ in C_δ, for all $n \ge 1$. Hence, $(v_n)_{n\ge1}$ is uniformly bounded in K. By the monotonicity of $(v_n)_{n\ge1}$ we conclude that up to a subsequence $v_n \to v$ in $C(K)$. Finally, standard elliptic regularity arguments show that v is a positive blow-up solution of (2.14). This concludes the proof. □

2.4 An equivalent criterion to the Keller–Osserman condition

Our aim in this section is to supply an equivalent criterion to the Keller–Osserman condition $(f2)$. We assume that f satisfies the following assumption:

$(f4)$ f fulfills $(f3)$ and there exists $\lim_{t\to\infty}\left(\frac{F}{f}\right)'(t) := \alpha$.

Consider the set $\mathcal{G}$ of all functions $g : (0, \infty) \to \mathbb{R}$ satisfying the hypotheses

(i) there exists $\delta = \delta(g) > 0$ such that $g \in C^2(0, \delta)$ and $g'' > 0$ on $(0, \delta)$;
(ii) $\lim_{t \searrow 0} g(t) = \infty$;
(iii) there exists $\lim_{t \searrow 0} \frac{g'(t)}{g''(t)}$.

Remark first that $\mathcal{G}$ is nonempty. Indeed, let $\theta \in C^2(0, \infty)$ be a convex function such that $\lim_{t \searrow 0} \theta(t) = \infty$. Because θ' is nondecreasing, it follows that $\lim_{t \searrow 0} \theta'(t) = -\infty$. Thus,

$$\left| \frac{\theta'(t)}{(\theta'(t))^2 + \theta''(t)} \right| \leq \frac{1}{|\theta'(t)|} \to 0 \quad \text{as } t \searrow 0,$$

which proves that $e^\theta \in \mathcal{G}$.

Remark 2.4.1 $\lim_{t \searrow 0} \dfrac{g(t)}{g''(t)} = \lim_{t \searrow 0} \dfrac{g'(t)}{g''(t)} = 0$ *for any function* $g \in \mathcal{G}$.

Indeed, if $g \in \mathcal{G}$ is chosen arbitrarily, then $\lim_{t \searrow 0} g'(t) = -\infty$. Hence, the mapping $\psi(t) = \ln |g'(t)|$ is decreasing in the neighborhood of the origin and by (ii) there exists $\lim_{t \searrow 0} \psi'(t)$. Because $\lim_{t \searrow 0} \psi(t) = \infty$, it follows that $\lim_{t \searrow 0} \psi'(t) = -\infty$. Hence,

$$\lim_{t \searrow 0} \frac{g''(t)}{g'(t)} = \lim_{t \searrow 0} \psi'(t) = -\infty,$$

and then by l'Hospital's rule we find

$$\lim_{t \searrow 0} \frac{g(t)}{g'(t)} = \lim_{t \searrow 0} \frac{g'(t)}{g''(t)} = 0.$$

Lemma 2.4.2 *Assume that f fulfills $(f4)$. Then, the following hold:*
(i) $\alpha \geq 0$;
(ii) $\alpha \leq 1/2$ *provided that $(f2)$ is fulfilled.*

Proof (i) Suppose that $\alpha < 0$. Then, there exists $t_1 > 0$ such that

$$\left(\frac{F}{f} \right)' (t) \leq \frac{\alpha}{2} < 0 \qquad \text{for any } t \geq t_1.$$

Integrating this inequality over (t_1, ∞) we obtain a contradiction. It follows that $\alpha \geq 0$.

(ii) Using the definition of α we find

$$\lim_{t \to \infty} \frac{F(t) f'(t)}{f^2(t)} = 1 - \alpha.$$

By Remark 2.3.1 and l'Hospital's rule we obtain

$$\lim_{t \to \infty} \frac{F(t)}{f^2(t)} = \lim_{t \to \infty} \frac{1}{2f'(t)} = 0$$

and

$$0 \leq \lim_{t\to\infty} \frac{\sqrt{F(t)}/f(t)}{\int_t^\infty ds/\sqrt{F(s)}} = -\frac{1}{2} + \lim_{t\to\infty} \frac{F(t)f'(t)}{f^2(t)} = \frac{1}{2} - \alpha. \tag{2.23}$$

This concludes our proof. □

Lemma 2.4.3 *Assume that f fulfills $(f4)$. Then the Keller–Osserman condition $(f2)$ holds if and only if*

(A_g) $\displaystyle\lim_{t\searrow 0} \frac{tf(g(t))}{g''(t)} = \infty$ *for some function $g \in \mathcal{G}$.*

Proof Necessary condition. Because $(f2)$ holds, we can define the positive function g as follows:

$$\int_{g(t)}^\infty \frac{ds}{\sqrt{F(s)}} = t^\vartheta \quad \text{for all } t > 0, \tag{2.24}$$

where $\vartheta > 3/2$ is arbitrary. Obviously, $g \in C^2(0,\infty)$ and $\lim_{t\searrow 0} g(t) = \infty$. We claim that $g \in \mathcal{G}$ and condition (A_g) is fulfilled. For this purpose, we divide our argument into three steps.

Step 1: $\displaystyle\lim_{t\searrow 0} \frac{g'(t)}{t^{2\vartheta-1}f(g(t))} = \vartheta\left(\alpha - \frac{1}{2}\right).$

We derive twice relation (2.24) and we obtain

$$g'(t) = -\vartheta t^{\vartheta-1}\sqrt{F(g(t))}\,, \tag{2.25}$$

$$\begin{aligned} g''(t) &= \frac{\vartheta-1}{t}g'(t) + \frac{\vartheta^2}{2}t^{2\vartheta-2}f(g(t)) \\ &= \frac{\vartheta^2}{2}t^{2\vartheta-2}f(g(t))\left(\frac{2(\vartheta-1)}{\vartheta^2}\frac{g'(t)}{t^{2\vartheta-1}f(g(t))} + 1\right). \end{aligned} \tag{2.26}$$

By (2.23) and (2.25) we find

$$\begin{aligned} \lim_{t\searrow 0} \frac{g'(t)}{t^{2\vartheta-1}f(g(t))} &= \lim_{t\searrow 0} \frac{-\vartheta t^{\vartheta-1}\sqrt{F(g(t))}}{t^{2\vartheta-1}f(g(t))} \\ &= \lim_{t\searrow 0} -\vartheta \frac{\sqrt{F(g(t))}/f(g(t))}{\int_{g(t)}^\infty ds/\sqrt{F(s)}} \\ &= \lim_{u\to\infty} -\vartheta \frac{\sqrt{F(u)}/f(u)}{\int_u^\infty ds/\sqrt{F(s)}} = \vartheta\left(\alpha - \frac{1}{2}\right). \end{aligned}$$

Step 2: $g'' > 0$ in $(0,\delta)$ for δ small enough.

Because $\alpha \geq 0$, by the result obtained in Step 1 we find

$$\lim_{t\searrow 0} \frac{2(\vartheta-1)}{\vartheta^2}\frac{g'(t)}{t^{2\vartheta-1}f(g(t))} = \frac{2(\vartheta-1)}{\vartheta}\left(\alpha - \frac{1}{2}\right) \geq \frac{1}{\vartheta} - 1 > -1. \tag{2.27}$$

In view of (2.26), the assertion of this step follows.

Step 3: $\lim_{t\searrow 0} \frac{g'(t)}{g''(t)} = 0$ and $\lim_{t\searrow 0} \frac{tf(g(t))}{g''(t)} = \infty$.

Taking into account (2.26) and (2.27) we find

$$\lim_{t\searrow 0} \frac{g'(t)}{g''(t)} = \lim_{t\searrow 0} \frac{2t}{\vartheta^2} \frac{g'(t)}{t^{2\vartheta-1}f(g(t))} \frac{1}{\frac{2(\vartheta-1)}{\vartheta^2} \frac{g'(t)}{t^{2\vartheta-1}f(g(t))} + 1} = 0,$$

and, for any $t \in (0, \delta)$ where $\delta > 0$ is given by Step 2, we have

$$\frac{tf(g(t))}{g''(t)} = \frac{tf(g(t))}{\frac{\vartheta-1}{t} g'(t) + \frac{\vartheta^2}{2} t^{2\vartheta-2} f(g(t))} \geq \frac{tf(g(t))}{\frac{\vartheta^2}{2} t^{2\vartheta-2} f(g(t))} = \frac{2}{\vartheta^2 t^{2\vartheta-3}}.$$

Sending t to 0, the claim of Step 3 follows.

Sufficient condition. Let $g \in \mathcal{G}$ be chosen so that (A_g) is fulfilled. By l'Hospital's rule we find

$$\lim_{t\searrow 0} \frac{(g'(t))^2}{F(g(t))} = 2 \lim_{t\searrow 0} \frac{g''(t)}{f(g(t))} = 0.$$

We choose $\delta > 0$ small enough such that $g'(t) < 0$ and $g''(t) > 0$ for all $t \in (0, \delta)$. It follows that

$$\begin{aligned} \int_{g(\delta)}^{\infty} \frac{dt}{\sqrt{F(t)}} &= \lim_{t\searrow 0} \int_{g(\delta)}^{g(t)} \frac{ds}{\sqrt{F(s)}} \\ &= \lim_{t\searrow 0} \int_{t}^{\delta} \frac{-g'(s)\,ds}{\sqrt{F(g(s))}} \\ &\leq \delta \sup_{t\in(0,\delta)} \frac{-g'(t)}{\sqrt{F(g(t))}} < \infty. \end{aligned}$$

Hence, the growth condition (A_2) holds. This ends the proof. □

2.5 Singular solutions of the logistic equation on domains with holes

Denote by $\mathcal{D}$ and $\mathcal{R}$ the boundary operators

$$\mathcal{D}u := u \qquad \text{and} \qquad \mathcal{R}u := \frac{\partial u}{\partial n} + \beta(x)u,$$

where n is the unit outward normal to $\partial\Omega$, and $\beta \in C^{1,\gamma}(\partial\Omega)$ is a nonnegative function. Hence, $\mathcal{D}$ is the *Dirichlet* boundary operator and $\mathcal{R}$ is either the *Neumann* boundary operator, if $\beta \equiv 0$, or the *Robin* boundary operator, if $\beta \not\equiv 0$. Throughout, $\mathcal{B}$ can define any of these boundary operators.

We are concerned in this section with the following boundary blow-up problem:

$$\begin{cases} \Delta u + \lambda u = p(x)f(u) & \text{in } \Omega, \\ \mathcal{B}u = 0 & \text{on } \partial\Omega, \\ u = \infty & \text{on } \partial\Omega_0, \end{cases} \tag{2.28}$$

where Ω_0 is the interior of the zero set of p as defined in (2.18). We assume that $p > 0$ on $\partial\Omega$, $\Omega_0 \subset\subset \Omega$ is nonempty, connected, and has a smooth boundary.

In (2.28) condition $u = \infty$ on $\partial\Omega_0$ means that $u(x) \to \infty$ as $x \in \Omega \setminus \overline{\Omega}_0$ and $d(x) := \operatorname{dist}(x, \Omega_0) \to 0$.

Because of the mixed boundary condition imposed on $\partial\Omega$, a suitable comparison principle is needed. To this aim, we first provide the following result.

Lemma 2.5.1 *Assume that $p > 0$ in $\Omega \setminus \overline{\Omega}_0$, f satisfies (f3), and let u_1, $u_2 \in C^2(\overline{\Omega} \setminus \overline{\Omega}_0)$ be such that*

(i) $\Delta u_1 + \lambda u_1 - p(x)f(u_1) \le 0 \le \Delta u_2 + \lambda u_2 - p(x)f(u_2)$ *in* $\Omega \setminus \overline{\Omega}_0$;
(ii) $u_1, u_2 > 0$ *in* $\overline{\Omega} \setminus \overline{\Omega}_0$ *and* $\mathcal{B}u_1 \ge 0 \ge \mathcal{B}u_2$ *on* $\partial\Omega$;
(iii) $\limsup_{\operatorname{dist}(x,\partial\Omega_0)\to 0}(u_2 - u_1)(x) \le 0$.

Then $u_1 \ge u_2$ in $\overline{\Omega} \setminus \Omega_0$.

Proof If $\mathcal{B} = \mathcal{D}$, the conclusion follows by Theorem 1.3.17. We next assume that $\mathcal{B} = \mathcal{R}$. Let ϕ_1, ϕ_2 be two nonnegative C^2 functions in $\overline{\Omega} \setminus \Omega_0$ vanishing near $\partial\Omega_0$. Multiplying the first inequality in (i) by ϕ_1, the second one by ϕ_2, and applying integration by parts together with (ii), we deduce that

$$
\begin{aligned}
-\int_{\widetilde{\Omega}} (\nabla u_2 \cdot \nabla \phi_2 - \nabla u_1 \cdot \nabla \phi_1)\, dx - \int_{\partial\Omega} \beta(x)(u_2\phi_2 - u_1\phi_1)\, d\sigma(x) \\
\ge \int_{\widetilde{\Omega}} p(x)(f(u_2)\phi_2 - f(u_1)\phi_1)\, dx + \lambda \int_{\widetilde{\Omega}} (u_1\phi_1 - u_2\phi_2)\, dx,
\end{aligned}
\tag{2.29}
$$

where $\widetilde{\Omega} := \Omega \setminus \overline{\Omega}_0$. Let $\varepsilon_1 > \varepsilon_2 > 0$ and denote

$$
\begin{aligned}
&\Omega_+(\varepsilon_1, \varepsilon_2) := \{x \in \widetilde{\Omega} : u_2(x) + \varepsilon_2 > u_1(x) + \varepsilon_1\}. \\
&v_i := (u_i + \varepsilon_i)^{-1} \left((u_2 + \varepsilon_2)^2 - (u_1 + \varepsilon_1)^2\right)^+, \quad i = 1, 2.
\end{aligned}
$$

Because v_i can be closely approximated in the $H^1 \cap L^\infty$ topology on $\overline{\Omega} \setminus \Omega_0$ by nonnegative C^2 functions vanishing near $\partial\Omega_0$, it follows that (2.29) holds for v_i taking the place of ϕ_i. Because v_i vanishes outside the set $\Omega_+(\varepsilon_1, \varepsilon_2)$, relation (2.29) becomes

$$
\begin{aligned}
-\int_{\Omega_+(\varepsilon_1,\varepsilon_2)} (\nabla u_2 \cdot \nabla v_2 - \nabla u_1 \cdot \nabla v_1)\, dx - \int_{\partial\Omega} \beta(x)(u_2 v_2 - u_1 v_1)\, d\sigma(x) \\
\ge \int_{\Omega_+(\varepsilon_1,\varepsilon_2)} p(x)(f(u_2)v_2 - f(u_1)v_1)\, dx + a \int_{\Omega_+(\varepsilon_1,\varepsilon_2)} (u_1 v_1 - u_2 v_2)\, dx.
\end{aligned}
\tag{2.30}
$$

A straightforward computation shows that the first integral in the left-hand side of (2.30) equals

$$
-\int_{\Omega_+(\varepsilon_1,\varepsilon_2)} \left(\left| \nabla u_2 - \frac{u_2 + \varepsilon_2}{u_1 + \varepsilon_1} \nabla u_1 \right|^2 + \left| \nabla u_1 - \frac{u_1 + \varepsilon_1}{u_2 + \varepsilon_2} \nabla u_2 \right|^2 \right) dx \le 0.
$$

Passing to the limit as $0 < \varepsilon_2 < \varepsilon_1 \to 0$, the second term on the left-hand side of (2.30) converges to 0. Also, the first term in the right-hand side of (2.30) converges to

$$\int_{\Omega_+(0,0)} p(x) \left(\frac{f(u_2)}{u_2} - \frac{f(u_1)}{u_1} \right) (u_2^2 - u_1^2)\, dx,$$

whereas the other term converges to 0. Hence, we avoid a contradiction only in the case that $\Omega_+(0,0)$ has measure 0, which means that $u_1 \geq u_2$ in $\widetilde{\Omega}$. This concludes the proof. □

We start the study of (2.28) with the following auxiliary result.

Proposition 2.5.2 *Assume that (f2), (f3) hold and $p > 0$ on $\partial\Omega$. Then, for any positive function $\phi \in C^{2,\gamma}(\partial\Omega_0)$ and $\lambda \in \mathbb{R}$, the problem*

$$\begin{cases} \Delta u + \lambda u = p(x) f(u) & \text{in } \Omega, \\ \mathcal{B}u = 0 & \text{on } \partial\Omega, \\ u = \phi & \text{on } \partial\Omega_0 \end{cases} \tag{2.31}$$

has a unique positive solution.

Proof By Lemma 2.5.1 we find that problem (2.31) has at most one positive solution. To prove the existence of a positive solution to (2.31) we use the sub- and supersolution method.

Let $\omega \subset\subset \Omega_0$ be such that the first Dirichlet eigenvalue of $(-\Delta)$ in the smooth domain $\Omega_0 \setminus \overline{\omega}$ is greater than λ. Let $q \in C^{0,\gamma}(\overline{\Omega})$ be such that

$$\begin{aligned} q(x) &= p(x) && \text{in } \overline{\Omega} \setminus \Omega_0, \\ q(x) &= 0 && \text{in } \overline{\Omega}_0 \setminus \omega, \\ q(x) &> 0 && \text{in } \omega. \end{aligned}$$

By virtue of Proposition 2.3.7, the problem

$$\begin{cases} \Delta v + \lambda v = q(x) f(v) & \text{in } \Omega, \\ v = 1 & \text{on } \partial\Omega \end{cases}$$

has a unique positive solution v. Let us choose Ω_1 and Ω_2 as two subdomains of Ω such that $\Omega_0 \subset\subset \Omega_1 \subset\subset \Omega_2 \subset\subset \Omega$.

Define $w \in C^2(\overline{\Omega} \setminus \Omega_0)$ so that

$$w = 1 \text{ in } \overline{\Omega} \setminus \Omega_2, \quad w = v \text{ in } \overline{\Omega}_1 \setminus \Omega_0,$$

and let $m := \min_{\overline{\Omega} \setminus \Omega_0} w > 0$. We claim that Mw is a supersolution of problem (2.31), provided $M \geq 1$ is large enough.

We first remark that for all $M > 1$ we have

$$\begin{aligned} -\Delta(Mw) &= \lambda M v - M q(x) f(v) \\ &\geq \lambda(Mw) - p(x) f(Mw) \quad \text{in } \overline{\Omega}_1 \setminus \overline{\Omega}_0. \end{aligned} \tag{2.32}$$

Let $a := \sup_{\Omega\setminus\Omega_1}(\lambda w + \Delta w)$. By Remark 2.3.1, there exists $c \geq 1$ such that

$$f(mM) \geq \frac{Ma}{\min_{\overline{\Omega}\setminus\Omega_1} p} \quad \text{for all } M \geq c.$$

Therefore, for all $x \in \Omega \setminus \overline{\Omega}_1$ and $M \geq c$ we have

$$p(x)f(Mw) \geq (\min_{\overline{\Omega}\setminus\Omega_1} p)\, f(mM) \geq M(\lambda w + \Delta w),$$

which can be rewritten as

$$-\Delta(Mw) \geq \lambda(Mw) - p(x)f(Mw) \quad \text{for } x \in \Omega \setminus \overline{\Omega}_1 \text{ and } M \geq c. \tag{2.33}$$

By (2.32) and (2.33) it follows that

$$-\Delta(Mw) \geq \lambda(Mw) - p(x)f(Mw) \quad \text{in } \Omega \setminus \overline{\Omega}_0, \text{ for any } M \geq c.$$

On the other hand,

$$\mathcal{B}(Mw) \geq M \min\{1, \min_{x\in\partial\Omega} \beta(x)\} \geq 0 \quad \text{on } \partial\Omega, \text{ for every } M > 0.$$

Now the claim follows by taking $M \geq \max\{\max_{\partial\Omega_0} \phi/m; c\}$.

In view of Theorem 1.4.1, the boundary value problem

$$\begin{cases} \Delta u = |\lambda| u + \|p\|_\infty f(u) & \text{in } \Omega, \\ u > 0 & \text{on } \partial\Omega, \\ u = \phi & \text{on } \partial\Omega_0 \end{cases} \tag{2.34}$$

has a unique nonnegative solution $\underline{u}$ such that $\underline{u} > 0$ in $\Omega \setminus \overline{\Omega}_0$. Because $\underline{u} = 0$ on $\partial\Omega$, we find that $\mathcal{R}\underline{u} = \partial\underline{u}/\partial n \leq 0$ on $\partial\Omega$. It is easy to see that $\underline{u}$ is a subsolution of (2.31) and $\underline{u} \leq Mw$ in $\overline{\Omega}\setminus\Omega_0$ for M large enough. The conclusion of Proposition 2.5.2 follows now by the sub- and supersolution method. This ends the proof. □

Corollary 2.5.3 *For $m \geq 1$ sufficiently large, set*

$$\Omega_m := \{x \in \Omega : \ d(x) < 1/m\}. \tag{2.35}$$

If Ω_0 is replaced by Ω_m, then the statement of Lemma 2.5.2 holds.

Proof The proof is very easy in this case. The construction of a subsolution is made as before. As a supersolution we can choose any number $M \geq 1$ large enough. □

We are now ready to prove the main result of this section.

Theorem 2.5.4 *Let (f2) and (f4) hold. Then, for any $\lambda \in \mathbb{R}$, problem (2.28) has a minimal (respectively a maximal) positive solution $\underline{U}_\lambda$ (respectively $\overline{U}_\lambda$).*

Proof We first prove the existence of the minimal positive solution for problem (2.28). For any $n \geq 1$, let u_n be the unique positive solution of

$$\begin{cases} \Delta u + \lambda u = p(x)f(u) & \text{in } \Omega, \\ \mathcal{B}u = 0 & \text{on } \partial\Omega, \\ u = n & \text{on } \partial\Omega_0. \end{cases}$$

By Lemma 2.5.1, $u_n(x)$ increases with n for all $x \in \overline{\Omega}\backslash\overline{\Omega}_0$. Moreover, the following result holds true.

Lemma 2.5.5 *The sequence $(u_n(x))_{n\geq 1}$ is bounded from above by some function $V(x)$, which is uniformly bounded on all compact subsets of $\overline{\Omega} \setminus \overline{\Omega}_0$.*

Proof Let q be a C^2 function on $\overline{\Omega} \setminus \Omega_0$ such that

$$0 < q(x) \leq p(x) \qquad \text{for all } x \in \overline{\Omega} \setminus \overline{\Omega}_0.$$

For x bounded away from $\partial\Omega_0$, it is not a problem to find such a function q. If x is close to the boundary of Ω_0 we chose $\delta > 0$ small enough such that $x \mapsto d(x)$ is a C^2 function in the set $\Omega_\delta = \{x \in \Omega : 0 < d(x) < \delta\}$. Then, for $x \in \Omega_\delta$ we can define

$$q(x) = \int_0^{d(x)} \int_0^t \min_{d(z)\geq s} b(z)\, ds\, dt.$$

Let $g \in \mathcal{G}$ be a function such that (A_g) holds. The existence of g is guaranteed by Lemma 2.4.3. Because $q(x) \to 0$ as $d(x) \searrow 0$, by virtue of hypothesis $(f1)$ and Remark 2.4.1, there exists $\delta > 0$ such that for all $x \in \Omega$ with $0 < d(x) < \delta$ and $M > 1$ we have

$$\frac{q(x)f(Mg(q(x)))}{Mg''(q(x))} > \sup_{\overline{\Omega}\backslash\Omega_0} |\nabla q(x)|^2 + \frac{g'(q(x))}{g''(q(x))} \inf_{\overline{\Omega}\backslash\Omega_0} (\Delta q(x)) + \lambda \frac{g(q(x))}{g''(q(x))}. \tag{2.36}$$

Here, $\delta > 0$ is taken sufficiently small so that $g'(q(x)) < 0$ and $g''(q(x)) > 0$ for all x with $0 < d(x) < \delta$.

For $n_0 \geq 1$ fixed, define V^* as follows:

(i) $V^*(x) = u_{n_0}(x) + 1$ for $x \in \overline{\Omega}$ and near $\partial\Omega$;
(ii) $V^*(x) = g(q(x))$ for x satisfying $0 < d(x) < \delta$;
(iii) $V^* \in C^2(\overline{\Omega} \setminus \overline{\Omega}_0)$ is positive in $\overline{\Omega} \setminus \overline{\Omega}_0$.

We show that for $M > 1$ large enough the upper bound of the sequence $(u_n(x))_{n\geq 1}$ can be taken as $V := MV^*$. Because

$$\mathcal{B}V = M\,\mathcal{B}V^* \geq M \min\{1, \beta(x)\} \geq 0 \quad \text{on } \partial\Omega$$

and

$$\lim_{d(x)\searrow 0} (u_n(x) - V(x)) = -\infty < 0,$$

to conclude that $u_n \leq V$ in $\overline{\Omega} \setminus \overline{\Omega}_0$ it is sufficient to show that

$$-\Delta V - \lambda V + p(x)f(V) \geq 0 \quad \text{in } \Omega \setminus \overline{\Omega}_0. \tag{2.37}$$

For $x \in \Omega$ satisfying $0 < d(x) < \delta$ and $M > 1$ we have

$$\begin{aligned}-\Delta V - \lambda V + p(x)f(V) &= M\Big(- \Delta g(q(x)) - \lambda g(q(x)) + p(x)f(g(q(x)))\Big) \\ &\geq Mg''(q(x)) \left(- \frac{g'(q(x))}{g''(q(x))} \Delta q(x) - |\nabla q(x)|^2 \right. \\ &\quad \left. - \lambda \frac{g(q(x))}{g''(q(x))} + q(x) \frac{f(Mg(q(x)))}{Mg''(q(x))} \right).\end{aligned}$$

By virtue of (2.36) we thus derive

$$-\Delta V - \lambda V + p(x)f(V) \geq 0 \quad \text{for all } x \in \overline{\Omega} \setminus \overline{\Omega}_0 \text{ with } 0 < d(x) < \delta.$$

For $x \in \overline{\Omega}$ satisfying $d(x) \geq \delta$ we also have

$$-\Delta V - \lambda V + p(x)f(V) = M \left(-\Delta V^* - \lambda V^* + p(x) \frac{f(MV^*)}{M} \right) \geq 0,$$

for $M > 0$ sufficiently large. It follows that (2.37) is fulfilled provided M is large enough. This finishes the proof of the lemma. □

We now come back to the proof of Theorem 2.5.4. By Lemma 2.5.5, $\underline{U}_\lambda(x) := \lim_{n\to\infty} u_n(x)$ exists, for any $x \in \overline{\Omega} \setminus \overline{\Omega}_0$. Moreover, $\underline{U}_\lambda$ is a positive solution of problem (2.28). Using Lemma 2.5.1 once more, we find that any positive solution u of (2.28) satisfies $u \geq u_n$ on $\overline{\Omega} \setminus \overline{\Omega}_0$, for all $n \geq 1$. Hence $\underline{U}_\lambda$ is the minimal positive solution of problem (2.28).

To achieve the existence of the maximal positive solution to problem (2.28), we use essentially the same argument that we use earlier.

Lemma 2.5.6 *If Ω_0 is replaced by Ω_m defined in (2.35), then problem (2.28) has a minimal positive solution provided that $(f2)$ and $(f3)$ are fulfilled.*

Proof The argument used here, which is easier because $p > 0$ on $\overline{\Omega} \setminus \Omega_m$, is similar to that in the proof of Lemma 2.5.5. The only difference that appears here (except the replacement of Ω_0 with Ω_m) is related to the construction of $V^*(x)$ for x near $\partial\Omega_m$. For this purpose, instead of Lemma 2.4.3 we use Theorem 2.3.6, which says that for any $\lambda \in \mathbb{R}$ there exists a positive blow-up solution u_λ of (2.14) in the domain $\Omega \setminus \overline{\Omega}_m$. We next define $V^*(x) = u_\lambda(x)$ for $x \in \Omega \setminus \overline{\Omega}_m$ and near $\partial\Omega_m$. For $M > 1$ and $x \in \Omega \setminus \overline{\Omega}_m$ near $\partial\Omega_m$ we have

$$\begin{aligned}-\Delta V - \lambda V + p(x)f(V) &= -M\Delta V^* - \lambda MV^* + p(x)f(MV^*) \\ &= p(x)\Big(f(MV^*) - Mf(V^*)\Big) \geq 0.\end{aligned}$$

This completes the proof of our lemma. □

Let v_m be the minimal positive solution of the problem considered in the statement of Lemma 2.5.6. It is easy to see that $v_m \geq v_{m+1} \geq u$ in $\overline{\Omega}\backslash\overline{\Omega}_m$, where u is any positive solution of (2.28). Hence $\overline{U}_\lambda(x) := \lim_{m\to\infty} v_m(x) \geq u(x)$, for all $x \in \Omega\backslash\overline{\Omega}_0$. A regularity and compactness argument show that $\overline{U}_\lambda$ is a positive blow-up solution of (2.28). Consequently, $\overline{U}_\lambda$ is the maximal positive solution. This concludes the proof of Theorem 2.5.4. □

2.6 Uniqueness of blow-up solution

We establish in this section a general uniqueness result for problem (2.28) under suitable conditions on p and f. Assume that there exists a positive increasing function $k \in C^1(0, \delta_0)$ for some $\delta_0 > 0$ such that p satisfies the following hypotheses:

$(p3)$ $\displaystyle\lim_{d(x)\searrow 0} \frac{p(x)}{k(d(x))} = c$ for some constant $c > 0$.

$(p4)$ $\displaystyle K(t) = \frac{\int_0^t \sqrt{k(s)}\, ds}{\sqrt{k(t)}} \in C^1[0, \delta_0).$

Also assume that there exist $\zeta > 0$ and $t_0 \geq 1$ such that

$(f5)$ $f(\xi t) \leq \xi^{1+\zeta} f(t)$, for all $\xi \in (0,1)$ and all $t \geq t_0/\xi$.

$(f6)$ The mapping $(0,1] \ni \xi \longmapsto A(\xi) = \lim_{t\to\infty} \dfrac{f(\xi t)}{\xi f(t)}$ is continuous and positive.

Remark that $\lim_{t\to\infty} f(t)/t^{1+\zeta}$ exists and is positive because the mapping $t \longmapsto f(t)/t^{1+\zeta}$ is nondecreasing in a neighborhood of infinity. Thus, the hypotheses $(f1)$ and $(f5)$ imply $(f2)$. By $(f6)$ we derive that A can be continuously extended in the whole $(0,\infty)$ by setting $A(1/\xi) = 1/A(\xi)$ for all $\xi \in (0,1)$. Moreover, we have the following lemma.

Lemma 2.6.1 *The function $A : (0,\infty) \to (0,\infty)$ is bijective, provided that $(f5)$ and $(f6)$ hold.*

Proof Because A is continuous and $A(1/\xi) = 1/A(\xi)$ for all $\xi > 0$, the surjectivity of A follows at once if we show that $\lim_{\xi\searrow 0} A(\xi) = 0$. To this aim, let $\xi \in (0,1)$ be fixed. Using $(f5)$ we find

$$\frac{f(\xi t)}{\xi f(t)} \leq \xi^\zeta \quad \text{for all } t \geq \frac{t_0}{\xi},$$

which yields $A(\xi) \leq \xi^\zeta$. Because $\xi \in (0,1)$ is arbitrary, it follows that $\lim_{\xi\searrow 0} A(\xi) = 0$.

We next show that the function $\xi \longmapsto A(\xi)$ is increasing on $(0,\infty)$, which concludes the proof of our lemma. Let $0<\xi_1<\xi_2<\infty$ be chosen arbitrarily. Using assumption $(f5)$ once more, we obtain

$$f(\xi_1 t) = f\left(\frac{\xi_1}{\xi_2}\xi_2 t\right) \le \left(\frac{\xi_1}{\xi_2}\right)^{1+\zeta} f(\xi_2 t) \quad \text{for all } t \ge t_0\,\frac{\xi_2}{\xi_1}.$$

It follows that

$$\frac{f(\xi_1 t)}{\xi_1 f(t)} \le \left(\frac{\xi_1}{\xi_2}\right)^{\zeta} \frac{f(\xi_2 t)}{\xi_2 f(t)} \quad \text{for all } t \ge t_0\,\frac{\xi_2}{\xi_1}.$$

Passing to the limit as $t\to\infty$ we find

$$A(\xi_1) \le \left(\frac{\xi_1}{\xi_2}\right)^{\zeta} A(\xi_2) < A(\xi_2),$$

which finishes the proof. □

Lemma 2.6.2 *Assume that $(f1)$, $(f2)$, $(f3)$ with $\alpha \neq 0$, (p_3), and (p_4) are fulfilled. Then, the following hold:*

(i) $K'(0)(1-2\alpha)+2\alpha \in (0,1]$.

(ii) $h \in \mathcal{G}$, *where h is the function defined by (2.40).*

Proof (i) Because $\alpha \neq 0$, by Lemma 2.4.2 we find $0<\alpha\le 1/2$. Therefore, the claim of (i) follows if we prove that $K'(0)\in[0,1]$. To this aim, we first remark that $K(0)=0$. Indeed, supposing the contrary, we obtain

$$\lim_{t\searrow 0}\left[\ln\left(\int_0^t \sqrt{k(s)}\,ds\right)\right]'(t) = \frac{1}{K(0)} \in (0,\infty),$$

which contradicts the fact that $\lim_{t\searrow 0} \ln\left(\int_0^t \sqrt{k(s)}\,ds\right) = -\infty$. Hence, $K(0)=0$, which yields $K'(0)\ge 0$. Because $K\in C^1[0,\delta_0)$, we have

$$K'(0) = \lim_{t\searrow 0}\left(\frac{\int_0^t \sqrt{k(s)}\,ds}{\sqrt{k(t)}}\right)',$$

so that

$$\lim_{t\searrow 0}\frac{k'(t)\int_0^t \sqrt{k(s)}\,ds}{k^{3/2}(t)} = 2\left(1-\lim_{t\searrow 0}\left(\frac{\int_0^t \sqrt{k(s)}\,ds}{\sqrt{k(t)}}\right)'\right) \qquad (2.38)$$
$$= 2(1-K'(0)).$$

Furthermore, $K'(0)\le 1$. Indeed, assuming the contrary, relation (2.38) yields $k'(t)<0$ in a positive small neighborhood of origin. But this is impossible, because $\lim_{t\searrow 0} k(t)=0$ and $k>0$ on $(0,\delta_0)$.

(ii) Using the definition of h, we deduce that $h \in C^2(0,\delta_0)$ and $\lim_{t\searrow 0} h(t) = \infty$. Then, by twice deriving relation (2.40), we find

$$h'(t) = -\sqrt{k(t)}\,\sqrt{2F(h(t))} \quad \text{for all } t \in (0,\delta_0)$$

and

$$\begin{aligned} h''(t) &= k(t)f(h(t)) - \frac{1}{\sqrt{2}}\frac{\sqrt{F(h(t))}}{\sqrt{k(t)}}\,k'(t) \\ &= k(t)f(h(t))\left(1 - \frac{k'(t)\int_0^t \sqrt{k(s)}\,ds}{k^{3/2}(t)}\,\frac{\sqrt{F(h(t))}/f(h(t))}{\int_{h(t)}^\infty ds/\sqrt{F(s)}}\right). \end{aligned}$$

Using (2.23) and (2.38), we obtain

$$\begin{aligned} \lim_{t\searrow 0}\frac{h'(t)}{h''(t)} &= \frac{-2}{K'(0)(1-2\alpha)+2\alpha}\lim_{t\searrow 0}\frac{\sqrt{F(h(t))}/f(h(t))}{\int_{h(t)}^\infty ds/\sqrt{F(s)}}\lim_{t\searrow 0}\frac{\int_0^t \sqrt{k(s)}\,ds}{\sqrt{k(t)}} \\ &= \frac{2\alpha-1}{K'(0)(1-2\alpha)+2\alpha}K(0) = 0 \end{aligned}$$

and

$$\lim_{t\searrow 0}\frac{h''(t)}{k(t)f(h(t))} = K'(0)(1-2\alpha)+2\alpha > 0, \tag{2.39}$$

which shows that h'' is positive in $(0,\delta_1)$ for some $\delta_1 > 0$. This concludes our proof. □

Theorem 2.6.3 *Assume that conditions $(f1)$ through $(f6)$, $\alpha \neq 0$, $(p3)$, and $(p4)$ hold. Then for any $\lambda \in \mathbb{R}$, problem (2.28) has a unique positive blow-up solution U_λ. Moreover,*

$$\lim_{d(x)\searrow 0}\frac{U_\lambda(x)}{h(d(x))} = \xi_0,$$

where h is defined by

$$\int_{h(t)}^\infty \frac{ds}{\sqrt{2F(s)}} = \int_0^t \sqrt{k(s)}\,ds \quad \textit{for all } 0 < t < \delta_0, \tag{2.40}$$

and ξ_0 is the unique positive solution of

$$A(\xi) = \frac{K'(0)(1-2\alpha)+2\alpha}{c}.$$

Proof By Lemma 2.6.2 we have $h \in \mathcal{G}$. For all $\xi > 0$ set

$$\Pi(\xi) := \lim_{d(x)\searrow 0} p(x)\frac{f(h(d(x))\xi)}{h''(d(x))\,\xi}.$$

Using $(p3)$ and (2.39) we find

$$\begin{aligned}\Pi(\xi) &= \lim_{d(x)\searrow 0} \frac{p(x)}{k(d(x))} \frac{k(d(x))f(h(d(x)))}{h''(d(x))} \frac{f(h(d(x))\xi)}{\xi f(h(d(x)))}\\ &= c \lim_{t\searrow 0} \frac{k(t)f(h(t))}{h''(t)} \lim_{t\to\infty} \frac{f(\xi t)}{\xi f(t)}\\ &= \frac{c}{K'(0)(1-2\alpha)+2\alpha} A(\xi).\end{aligned}$$

Thus, by Lemma 2.6.1 we deduce that $\Pi : (0,\infty) \to (0,\infty)$ is bijective. Let ξ_0 be the unique positive solution of the equation $\Pi(\xi) = 1$—that is,

$$A(\xi_0) = \frac{K'(0)(1-2\alpha)+2\alpha}{c}.$$

Let $\xi_1 := \Pi^{-1}(1-4\varepsilon)$, respectively $\xi_2 := \Pi^{-1}(1+4\varepsilon)$, for arbitrary $0 < \varepsilon < 1/4$. Using Remark 2.4.1, the hypothesis $(p3)$, and the regularity of $\partial\Omega_0$, we can choose $\delta > 0$ small enough such that

(i) $\operatorname{dist}(x, \partial\Omega_0)$ is a C^2 function on the set $\{x \in \Omega : \operatorname{dist}(x, \partial\Omega_0) \le 2\delta\}$;

(ii) $\left|\frac{h'(t)}{h''(t)} \Delta d(x) + a\,\frac{h(t)}{h''(t)}\right| < \varepsilon$ and $h''(t) > 0$ for all $0 < t < 2\delta$ and x satisfying $0 < d(x) < 2\delta$;

(iii) $(\Pi(\xi_2) - \varepsilon)\,\frac{h''(d(x))\,\xi_2}{f(h(d(x))\xi_2)} \le p(x) \le (\Pi(\xi_1) + \varepsilon)\,\frac{h''(d(x))\,\xi_1}{f(h(d(x))\xi_1)}$, for every x with $0 < d(x) < 2\delta$;

(iv) $p(y) < (1+\varepsilon)p(x)$, for every x, y with $0 < d(y) < d(x) < 2\delta$.

Let $\sigma \in (0,\delta)$ be arbitrary and define $\underline{v}_\sigma(x) := h(d(x)+\sigma)\xi_1$, for any x with $d(x)+\sigma < 2\delta$, respectively $\overline{v}_\sigma(x) := h(d(x)-\sigma)\xi_2$ for any x with $\sigma < d(x) < 2\delta$.

Using (ii), (iv), the first inequality in (iii), and the fact that $|\nabla d(x)| \equiv 1$, for $\sigma < d(x) < 2\delta$ we obtain

$$\begin{aligned}&-\Delta\overline{v}_\sigma(x) - \lambda\overline{v}_\sigma(x) + p(x)f(\overline{v}_\sigma(x))\\ &= \xi_2\, h''(d(x)-\sigma)\times\\ &\quad\times\left(-\frac{h'(d(x)-\sigma)}{h''(d(x)-\sigma)}\Delta d(x) - \lambda\,\frac{h(d(x)-\sigma)}{h''(d(x)-\sigma)} - 1 + \frac{p(x)f(h(d(x)-\sigma)\xi_2)}{h''(d(x)-\sigma)\,\xi_2}\right)\\ &\ge \xi_2\, h''(d(x)-\sigma)\times\\ &\quad\times\left(-\frac{h'(d(x)-\sigma)}{h''(d(x)-\sigma)}\Delta d(x) - \lambda\,\frac{h(d(x)-\sigma)}{h''(d(x)-\sigma)} - 1 + \frac{\Pi(\xi_2)-\varepsilon}{1+\varepsilon}\right)\\ &\ge 0.\end{aligned}$$

Similarly, using (ii), (iv), and the second inequality in (iii), when $d(x)+\sigma < 2\delta$ we find

$$
\begin{aligned}
&-\Delta \underline{v}_\sigma(x) - \lambda \underline{v}_\sigma(x) + p(x) f(\underline{v}_\sigma(x)) \\
&= \xi_1 \, h''(d(x)+\sigma) \times \\
&\quad \times \left(-\frac{h'(d(x)+\sigma)}{h''(d(x)+\sigma)} \Delta d(x) - \lambda \frac{h(d(x)+\sigma)}{h''(d(x)+\sigma)} - 1 + \frac{p(x) f(h(d(x)+\sigma)\xi_1)}{h''(d(x)+\sigma)\,\xi_1} \right) \\
&\le \xi_1 \, h''(d(x)+\sigma) \times \\
&\quad \times \left(-\frac{h'(d(x)+\sigma)}{h''(d(x)+\sigma)} \Delta d(x) - \lambda \frac{h(d(x)+\sigma)}{h''(d(x)+\sigma)} - 1 + (1+\varepsilon)(\Pi(\xi_1)+\varepsilon) \right) \\
&\le 0.
\end{aligned}
$$

Define $\Omega_\delta := \{x \in \Omega : \; d(x) < \delta\}$ and let $\omega \subset\subset \Omega_0$ be such that the first Dirichlet eigenvalue of $(-\Delta)$ in the smooth domain $\Omega_0 \setminus \overline{\omega}$ is strictly greater than λ. Also, let $q \in C^{0,\gamma}(\overline{\Omega}_\delta)$ be such that

$$
\begin{aligned}
0 < q(x) \le p(x) \qquad & \text{in } \overline{\Omega}_\delta \setminus \overline{\Omega}_0, \\
q(x) = 0 \qquad & \text{in } \overline{\Omega}_0 \setminus \omega, \\
q(x) > 0 \qquad & \text{in } \omega.
\end{aligned}
$$

Then, by Theorem 2.3.6 there exists a positive blow-up solution of

$$
-\Delta w = \lambda w - q(x) f(w) \quad \text{in } \Omega_\delta.
$$

Suppose that u is an arbitrary solution of (2.28) and let $v := u + w$. Then v satisfies

$$
-\Delta v \ge \lambda v - p(x) f(v) \quad \text{in } \Omega_\delta \setminus \overline{\Omega}_0.
$$

Because $v = \infty > \underline{v}_\sigma$ on $\partial\Omega_0$ and $v = \infty > \underline{v}_\sigma$ on $\partial\Omega_\delta$, we find

$$
u + w \ge \underline{v}_\sigma \text{ in } \Omega_\delta \setminus \overline{\Omega}_0. \tag{2.41}
$$

Similarly,

$$
\overline{v}_\sigma + w \ge u \text{ on } \Omega_\delta \setminus \overline{\Omega}_\sigma. \tag{2.42}
$$

Letting $\sigma \to 0$ in (2.41) and (2.42), we deduce

$$
h(d(x))\,\xi_2 + 2w \ge u + w \ge h(d(x))\,\xi_1 \quad \text{in } \Omega_\delta \setminus \overline{\Omega}_0.
$$

Because w is uniformly bounded on $\partial\Omega_0$, it follows that

$$
\xi_1 \le \liminf_{d(x)\searrow 0} \frac{u(x)}{h(d(x))} \le \limsup_{d(x)\searrow 0} \frac{u(x)}{h(d(x))} \le \xi_2. \tag{2.43}
$$

Letting $\varepsilon \to 0$ in (2.43) and looking at the definition of ξ_1 respectively ξ_2, we find

$$
\lim_{d(x)\searrow 0} \frac{u(x)}{h(d(x))} = \xi_0. \tag{2.44}
$$

This behavior of the solution will be speculated to achieve the uniqueness of solution to (2.28). Indeed, let u_1, u_2 be two positive solutions of (2.28). For any $\varepsilon > 0$, denote $\widetilde{u}_i := (1+\varepsilon)\, u_i$, $i = 1, 2$. By virtue of (2.44) we obtain

$$\lim_{d(x)\searrow 0} \frac{u_1(x) - \widetilde{u}_2(x)}{h(d(x))} = \lim_{d(x)\searrow 0} \frac{u_2(x) - \widetilde{u}_1(x)}{h(d(x))} = -\varepsilon\, \xi_0 < 0,$$

which implies

$$\lim_{d(x)\searrow 0} (u_1(x) - \widetilde{u}_2(x)) = \lim_{d(x)\searrow 0} (u_2(x) - \widetilde{u}_1(x)) = -\infty.$$

On the other hand, because the mapping $t \longmapsto f(t)/t$ is increasing for $t > 0$, we have

$$\begin{aligned} -\Delta \widetilde{u}_i &= -(1+\varepsilon)\, \Delta u_i = (1+\varepsilon)\, (\lambda\, u_i - p(x) f(u_i)) \\ &\geq \lambda\, \widetilde{u}_i - p(x) f(\widetilde{u}_i) \quad \text{in } \Omega \setminus \overline{\Omega}_0, \\ \mathcal{B}\widetilde{u}_i &= \mathcal{B} u_i = 0 \quad \text{on } \partial\Omega. \end{aligned}$$

By Lemma 2.5.1,

$$u_1(x) \leq \tilde{u}_2(x), \quad u_2(x) \leq \tilde{u}_1(x) \quad \text{in } \Omega \setminus \overline{\Omega}_0.$$

Letting $\varepsilon \to 0$ we obtain $u_1 = u_2$. Hence, problem (2.28) has a unique solution. This finishes the proof of Theorem 2.6.3. □

Remark 2.6.4 *Some classes of nonlinearities f satisfying the assumptions of Theorem 2.6.3 are*

(i) $f(t) = t^p$, $p > 1$.
(ii) $f(t) = t^p \ln(t+1)$, $p > 1$.
(iii) $f(t) = t^p \arctan t$, $p > 1$.
(iv) $f(t) = t \ln^p(t+1)$, $p > 2$.

Remark 2.6.5 *Assume that f satisfies $(f3)$ and $(f5)$. Then equation (2.14) with $\lambda = 0$ and $p \equiv 1$ has a unique blow-up solution $\widetilde{u}$. Moreover, $\widetilde{u}$ has the following asymptotic behavior:*

$$\lim_{\operatorname{dist}(x,\partial\Omega)\to 0} \frac{\tilde{u}(x)}{\Gamma(\operatorname{dist}(x,\partial\Omega))} = 1,$$

where Γ is the function defined as

$$\int_{\Gamma(t)}^{\infty} \frac{ds}{\sqrt{2F(s)}} = t \quad \textit{for all } t > 0.$$

Let $\Omega_1 \subset\subset \Omega$ be a connected subdomain with a smooth boundary and such that $\overline{\Omega}_0 \subset \Omega_1$. A direct consequence of Theorem 2.6.3 is seen in the following corollary.

Corollary 2.6.6 *Let $(f6)$ be added to the assumptions of Remark 2.6.5. Then, for any $\lambda \in \mathbb{R}$, problem (2.28) with $p \equiv 1$ on $\partial\Omega_1$ and Ω_0 replaced with Ω_1 has a unique positive solution U_λ. Moreover, U_a behaves on $\partial\Omega_1$ exactly in the same manner as $\tilde{u}$ on $\partial\Omega$—that is,*

$$\lim_{\operatorname{dist}(x,\partial\Omega_1)\to 0} \frac{U_\lambda(x)}{\Gamma(\operatorname{dist}(x,\partial\Omega_1))} = 1.$$

Proof We use the argument of Lemma 2.5.6 to deduce the existence of a positive solution for the problem considered here. Concerning the uniqueness, let us remark that conditions $(p3)$ and $(p4)$ are fulfilled by taking $c = 1$ and $k \equiv 1$ on $(0,\infty)$. It follows that h defined by (2.40) coincides with Γ. Notice that $\Gamma'(t) = -\sqrt{2F(\Gamma(t))}$ and $\Gamma''(t) = f(\Gamma(t))$ for any $t > 0$. Thus, we obtain $\Gamma \in \mathcal{G}$ (without calling Lemma 2.6.2) and $\Pi(\xi) = A(\xi)$, for all $\xi > 0$. So, by Lemma 2.6.1, $\Pi : (0,\infty) \to (0,\infty)$ is bijective. From now on, we proceed as in the proof of Theorem 2.6.3, which only requires us to replace h with Γ and Ω_0 with Ω_1. This concludes the proof. □

2.7 A Karamata theory approach for uniqueness of blow-up solution

In this section we provide a different approach for the uniqueness of a positive blow-up solution of the logistic equation. We are concerned with the following boundary blow-up problem:

$$\begin{cases} \Delta u + \lambda u = p(x) f(u) & \text{in } \Omega, \\ u > 0 & \text{in } \Omega, \\ u = \infty & \text{on } \partial\Omega, \end{cases} \tag{2.45}$$

where $\lambda \in \mathbb{R}$, $p \in C^{0,\gamma}(\overline{\Omega})$ $(0 < \gamma < 1)$, such that

$$\Omega_0 := \operatorname{int}\{x \in \Omega : \ p(x) = 0\}$$

is smooth (possibly empty), $\overline{\Omega}_0 \subset \Omega$, and $p > 0$ in $\Omega \setminus \overline{\Omega}_0$.

The uniqueness of the solution to (2.45) will be achieved using the Karamata regular variation theory, originally introduced by Karamata [113]. We start this section with the following definition.

Definition 2.7.1 *A positive measurable function $f : [m,\infty) \to (0,\infty)$, $m > 0$, is called* regularly varying (at infinity) with index $\rho \in \mathbb{R}$ *if*

$$\lim_{t\to\infty} \frac{f(\xi t)}{f(t)} = \xi^\rho \quad \textit{for all } \xi > 0.$$

When the index of regular variation ρ is zero, we say that the function f is slowly varying.

We denote by $\mathbb{R}_\rho$ the class of regular varying functions with index $\rho \in \mathbb{R}$. The canonical ρ-varying function is $f(t) = t^\rho$, $t > 0$. The functions

$$\ln(1+t),\ \ln\ln(e+t),\ \exp\{(\ln t)^\alpha\}\ (0<\alpha<1),$$

vary slowly, as well as any measurable function on $[m, \infty)$, $m > 0$, with a positive limit at infinity. Furthermore, any function $f \in \mathbb{R}_q$ can be written in terms of a slowly varying function. Indeed, set $f(t) = t^\rho g(t)$. From the previous definition we conclude that g varies slowly.

The basic properties of regular and slowly varying functions are summarized next. For further details, we refer the reader to Seneta [180].

Proposition 2.7.2 *We have the following:*

(i) *For any slowly varying function $g : [m, \infty) \to (0, \infty)$, $m > 0$, and any $q > 0$ we have $t^q g(t) \to \infty$ and $t^{-q} g(t) \to 0$ as $t \to \infty$.*

(ii) *Any positive function $g \in C^1([m, \infty))$ satisfying $tg'(t)/g(t) \to 0$ as $t \to \infty$ is slowly varying. Moreover, if the previous limit is $\rho \in \mathbb{R}$, then $g \in \mathbb{R}_\rho$.*

(iii) *Assume $f : [m, \infty) \to (0, \infty)$, $m > 0$, is measurable and locally integrable. Then f varies regularly if and only if there exists $q \in \mathbb{R}$ and*

$$\ell := \lim_{t\to\infty} \frac{t^{q+1} f(t)}{\int_m^t s^q f(s)ds} > 0.$$

In this case, $f \in \mathbb{R}_\rho$ with $\rho = \ell - q - 1$;

(iv) *If $f \in \mathbb{R}_\rho$ is Lebesgue integrable on each finite subinterval of $[m, \infty)$, then for all $q > -\rho - 1$ we have*

$$\lim_{t\to\infty} \frac{t^{q+1} f(t)}{\int_m^t s^q f(s)ds} = q + \rho + 1.$$

We also have the following equivalence for C^1 functions.

Lemma 2.7.3 *Let $f \in C^1[0, \infty)$ be a nonnegative function that fulfills $(f3)$. Then the following conditions are equivalent:*

(i) *$f' \in \mathbb{R}_\rho$ for some $\rho \in \mathbb{R}$.*

(ii) *There exists $\vartheta := \lim_{t\to\infty} \frac{tf'(t)}{f(t)} < \infty$.*

(iii) *There exists $\alpha := \lim_{t\to\infty} \left(\frac{F}{f}\right)'(t) > 0$.*

Moreover, $\rho \geq 0$ and $\alpha = 1/(\rho + 2) = 1/(\vartheta + 1)$.

Remark that if $f' \in \mathbb{R}_\rho$ with $\rho > 0$, then $\lim_{t\to\infty} f(t)/t^p = \infty$, for all $1 < p < 1+\rho$. This implies that f fulfills $(f2)$. The converse implication is not necessarily true; we can take, for instance, $f(t) = u \ln^4(t+1)$, $t > 0$. Also notice that $(f2)$ may fail when $\rho = 0$. This is illustrated by $f(t) = t$ or $f(t) = t\ln(t+1)$, $t > 0$.

Proof (i)⇒(ii). Let $f \in \mathbb{R}_\rho$. Then $f'(t) = t^\rho g(t)$ where g varies slowly. If $\rho < 0$ then, by Proposition 2.7.2 (i) and l'Hospital's rule we find $\lim_{t\to\infty} f(t)/t = \lim_{t\to\infty} f'(t) = 0$, which contradicts $(f3)$. Hence $\rho \geq 0$ and by Proposition 2.7.2 (iv) (with $q = 0$) we obtain $\lim_{t\to\infty} tf'(t)/f(t) = \rho + 1$.

(ii)⇒(i). Using the assumption $(f3)$ we have $tf'(t) \geq f(t)$ for all $t > 0$, which yields $\vartheta \geq 1$. By Proposition 2.7.2 (iii) we now obtain $f \in \mathbb{R}_{\vartheta-1}$.

(ii)⇒(iii). By l'Hospital rule we deduce $\lim_{r\to\infty} F(t)/(tf(t)) = 1/(1+\vartheta)$ and

$$\begin{aligned}\lim_{t\to\infty}\left(\frac{F}{f}\right)' &= 1 - \lim_{t\to\infty}\frac{F(t)f'(t)}{f^2(t)}\\ &= 1 - \lim_{t\to\infty}\frac{F(t)}{tf(t)}\cdot\frac{tf'(t)}{f(t)}\\ &= \frac{1}{1+\vartheta}.\end{aligned}$$

Hence $\alpha = 1/(1+\vartheta)$.

(iii)⇒(ii). From $\lim_{t\to\infty}(F/f)'(t) = \alpha > 0$ we have $\alpha < 1$ and

$$\lim_{t\to\infty}\frac{F(t)f'(t)}{f^2(t)} = 1 - \alpha. \tag{2.46}$$

Let us choose $t_0 > 0$ such that $(F/f)'(t) \geq \alpha/2$, for all $t \geq t_0$. Then,

$$\frac{F(t)}{f(t)} \geq \frac{(t-t_0)\alpha}{2} + \frac{F(t_0)}{f(t_0)} \quad \text{for all } t \geq t_0.$$

Passing to the limit with $t \to \infty$, we find $\lim_{t\to\infty} F(t)/f(t) = \infty$. By l'Hospital's rule we deduce

$$\lim_{t\to\infty}\frac{tf(t)}{F(t)} = \frac{1}{\alpha}. \tag{2.47}$$

Multiplying (2.46) and (2.47) we obtain $\lim_{t\to\infty} tf'(t)/f(t) = (1-\alpha)/\alpha$. According to Proposition 2.7.2 (iii), it follows that $f' \in \mathbb{R}_\rho$, where $\rho = (1-2\alpha)/\alpha$. This completes the proof. □

Inspired by the definition of α, we denote by $\mathcal{K}$ the *Karamata class*, consisting of all positive, increasing C^1 functions k defined on $(0,\nu)$, for some $\nu > 0$, which satisfy

$$\ell_i := \lim_{t\searrow 0}\left(\frac{\int_0^t k(s)ds}{k(t)}\right)^{(i)}, \; i = 0, 1.$$

We observe that $\ell_0 = 0$ and $\ell_1 \in [0,1]$, for every $k \in \mathcal{K}$. Our next result gives examples of functions $k \in \mathcal{K}$ with $\lim_{t\searrow 0} k(t) = 0$, for every $\ell_1 \in [0,1]$.

Lemma 2.7.4 *Let $f \in C^1[m,\infty)$, $m > 0$, be such that $f' \in \mathbb{R}_\rho$, for some $\rho > -1$. Then we have the following:*

(i) *If $k(t) = \exp\{-f(1/t)\}$, $0 < t \leq 1/m$, then $k \in \mathcal{K}$ with $\ell_1 = 0$.*

(ii) *If* $k(t) = 1/f(1/t)$, $t \leq 1/m$, *then* $k \in \mathcal{K}$ *with* $\ell_1 = 1/(\rho+2) \in (0,1)$.
(iii) *If* $k(t) = 1/\ln f(1/t)$, $0 < t \leq 1/m$, *then* $k \in \mathcal{K}$ *with* $\ell_1 = 1$.

Proof Because $\rho > -1$, from Proposition 2.7.2 (i) and (iv) we have

$$\lim_{t\to\infty} tf'(t) = \infty$$

and

$$\lim_{t\to\infty} \frac{tf'(t)}{f(t)} = \rho + 1 > 0. \tag{2.48}$$

Therefore, in any of the cases (i), (ii), or (iii) we have $\lim_{t\searrow 0} k(t) = 0$ and k is a C^1 increasing function on $(0,\nu)$, for $\nu > 0$ sufficiently small.

(i) Using (2.48) we have

$$\lim_{t\searrow 0} \frac{tk'(t)}{k(t)\ln k(t)} = \lim_{t\searrow 0} -\frac{1/tf'(1/t)}{tf(1/t)} = -(\rho+1).$$

Thus, by l'Hospital's rule we obtain

$$\begin{aligned}
\lim_{t\searrow 0} \frac{tk(t)}{\ln k(t)\int_0^t k(s)ds} &= \lim_{t\searrow 0} \frac{tk(t)/\ln k(t)}{\int_0^t k(s)ds} \\
&= \lim_{t\searrow 0} \frac{(tk'(t)+k(t))\ln k(t) - tk'(t)}{k(t)\ln^2 k(t)} \\
&= \lim_{t\searrow 0} \left\{ \frac{tk'(t)}{k(t)\ln k(t)} + \frac{1}{\ln k(t)} - \frac{tk'(t)}{k(t)\ln^2 k(t)} \right\} \\
&= -(\rho+1).
\end{aligned}$$

Hence,

$$\lim_{t\searrow 0} \frac{\left(\int_0^t k(s)\,ds\right)k'(t)}{k^2(t)} = \lim_{t\searrow 0} \frac{\ln k(t)\int_0^t k(s)ds}{tk(t)} \cdot \frac{tk'(t)}{k(t)} = 1,$$

which implies

$$\ell_1 = 1 - \lim_{t\searrow 0} \frac{\left(\int_0^t k(s)ds\right)k'(t)}{k^2(t)} = 0.$$

(ii) It is easy to see that

$$\lim_{t\searrow 0} \frac{tk'(t)}{k(t)} = \lim_{t\searrow 0} \frac{1/tf'(1/t)}{f(1/t)} = \rho + 1.$$

By l'Hospital's rule we obtain $\lim_{t\searrow 0} \int_0^t k(s)ds/(tk(t)) = 1/(\rho+2)$. Thus,

$$\ell_1 = 1 - \lim_{t\searrow 0} \frac{\int_0^t k(s)ds}{tk(t)} \cdot \frac{tk'(t)}{k(t)} = \frac{1}{\rho+2}.$$

(iii) In a similar manner we have $\lim_{t\searrow 0} tk'(t)/k^2(t) = \rho + 1$. By l'Hospital's rule, $\lim_{t\searrow 0} \int_0^t k(s)ds/(tk(t)) = 1$. Thus,

$$\ell_1 = 1 - \lim_{t\searrow 0} \frac{\int_0^t k(s)ds}{t} \cdot \frac{tk'(t)}{k^2(t)} = 1.$$

This concludes the proof. □

As a consequence of Proposition 2.7.2 and Lemma 2.7.4 we obtain the following corollary.

Corollary 2.7.5 *Let $f \in C^1[m,\infty)$, $m > 0$. Then $f' \in \mathbb{R}_\rho$ with $\rho > -1$ if and only if there exist $q > 0$, $C > 0$, and $B > D$ such that*

$$f(t) = Ct^q \exp\left\{\int_B^u \frac{y(t)}{t}\,dt\right\} \qquad \textit{for all } t \geq B, \tag{2.49}$$

where $y \in C[B,\infty)$ satisfies $\lim_{t\to\infty} y(t) = 0$. In this case, $f' \in \mathbb{R}_\rho$ with $\rho = q-1$.

The core result of this section is the following.

Theorem 2.7.6 *Let $f \in C^1[0,\infty)$ be such that $(f3)$ holds and $f' \in \mathbb{R}_\rho$ with $\rho > 0$. Assume that $p \equiv 0$ on $\partial\Omega$ satisfies*

$$p(x) = c\,k^2(d(x)) + o(k^2(d(x))) \qquad \textit{as } d(x) \searrow 0,$$

for some constant $c > 0$ and $k \in \mathcal{K}$. Then, for any $\lambda \in (-\infty, \lambda_{\infty,1})$, problem (2.45) has a unique blow-up solution u_λ. Moreover,

$$\lim_{d(x)\searrow 0} \frac{u_\lambda(x)}{h(d(x))} = \xi_0 \qquad \textit{where } \xi_0 = \left(\frac{2+\ell_1\rho}{c(2+\rho)}\right)^{1/\rho}, \tag{2.50}$$

and h is defined by

$$\int_{h(t)}^\infty \frac{ds}{\sqrt{2F(s)}} = \int_0^t k(s)ds \qquad \textit{for all } t \in (0,\nu). \tag{2.51}$$

By Corollary 2.7.5, the assumption $f' \in \mathbb{R}_\rho$ with $\rho > 0$ holds if and only if f satisfies (2.49) with $q = \rho + 1$ and for some $B, C > 0$. If B is large enough (for instance, if $y > -\rho$ on $[B,\infty)$), then $f(t)/t$ is increasing on $[B,\infty)$. Thus, to obtain the whole range of functions f for which Theorem 2.7.6 applies, we have only to "paste" a suitable smooth function on $[0,B]$ in accordance with condition $(f3)$. A simple way to do this is to define $f(t) = t^{\rho+1}\exp\{\int_0^t z(s)/s ds\}$, for all $t \geq 0$, where $z \in C[0,\infty)$ is nonnegative and such that $\lim_{t\searrow 0} z(t)/t \in [0,\infty)$ and $\lim_{t\searrow\infty} z(t) = 0$. Clearly, $f(t) = t^q$, $f(t) = t^q\ln(t+1)$, and $f(t) = t^q \arctan t$ $(t > 1)$ fall into this category.

Lemma 2.7.4 provides a practical method to find functions k which can be considered in the statement of Theorem 2.7.6. Here are some examples: $k(t) =$

$\exp\{-1/t^\alpha\}$, $k(t) = \exp\{-\ln(1+1/t)/t^\alpha\}$, $k(t) = \exp\{-[\arctan(1/t)]/t^\alpha\}$, $k(t) = -1/\ln t$, $k(t) = t^\alpha/\ln(1+1/t)$, $k(t) = t^\alpha$, for some $\alpha > 0$.

As we have already seen in the previous section, the uniqueness lies upon the crucial observation (2.50), which shows that all explosive solutions have the same boundary behavior.

Proof Fix $\lambda \in (-\infty, \lambda_{\infty,1})$. By Theorem 2.3.6, the problem (2.45) has at least one blow-up solution.

The uniqueness will follow at once if we prove that (2.50) holds for any solution u_λ of (2.45). Indeed, if u_1 and u_2 are two arbitrary blow-up solutions of (2.45), then (2.50) yields $\lim_{d(x)\searrow 0} u_1(x)/u_2(x) = 1$. Hence, for any $0 < \varepsilon < 1$, there exists $\delta = \delta(\varepsilon) > 0$ such that

$$(1-\varepsilon)u_2(x) \le u_1(x) \le (1+\varepsilon)u_2(x), \tag{2.52}$$

for all $x \in \Omega$ with $0 < d(x) \le \delta$. Choosing a smaller $\delta > 0$ if necessary, we can assume that $\overline{\Omega}_0 \subset C_\delta$, where $C_\delta := \{x \in \Omega :\ d(x) > \delta\}$.

It is clear that u_1 is a positive solution of the boundary value problem

$$\begin{cases} \Delta\phi + \lambda\phi = p(x)f(\phi) & \text{in } C_\delta, \\ \phi = u_1 & \text{on } \partial C_\delta. \end{cases} \tag{2.53}$$

By $(f3)$ and (2.52), we see that $\phi^- := (1-\varepsilon)u_2$ is a subsolution and $\phi^+ := (1+\varepsilon)u_2$ is a supersolution of problem (2.53). By the sub- and supersolution method, (2.53) has a positive solution ϕ_1 satisfying $\phi^- \le \phi_1 \le \phi^+$ in C_δ. Because $p > 0$ in $\overline{C}_\delta \setminus \overline{\Omega}_0$, by Lemma 2.3.7 we derive that (2.53) has a unique positive solution, (i.e., $u_1 \equiv \phi_1$ in C_δ). This yields $(1-\varepsilon)u_2(x) \le u_1(x) \le (1+\varepsilon)u_2(x)$ in C_δ, so that (2.52) holds in Ω. Passing to the limit with $\varepsilon \to 0$, we conclude that $u_1 \equiv u_2$.

To establish (2.50) we first state some useful properties about h.

Lemma 2.7.7 *We have*

(i) $h \in C^2(0,\nu)$ *and* $\lim_{t\searrow 0} h(t) = \infty$;

(ii) *for all* $\xi > 0$ *there holds* $\lim_{t\searrow 0} \dfrac{h''(t)}{k^2(t)f(h(t)\xi)} = \dfrac{1}{\xi^{\rho+1}}\dfrac{2+\rho\ell_1}{2+\rho}$, *so,* $h'' > 0$ *on* $(0, 2\delta)$, *for* $\delta > 0$ *small enough;*

(iii) $\lim_{t\searrow 0} h(t)/h''(t) = \lim_{t\searrow 0} h'(t)/h''(t) = 0$.

Proof (i) Follows directly from (2.51).

(ii) Because $f \in \mathbb{R}_{\rho+1}$, it suffices to prove the claim for $\xi = 1$.

Clearly $h'(t) = -k(t)\sqrt{2F(h(t))}$ and

$$h''(t) = k^2(t)f(h(t))\left(1 - 2\frac{k'(t)\left(\int_0^t k(s)ds\right)}{k^2(t)} \cdot \frac{\sqrt{F(h(t))}}{f(h(t))\int_{h(t)}^\infty \sqrt{F(s)}ds}\right), \tag{2.54}$$

for all $0 < t < \nu$. It is easy to be seen that $\lim_{t\to\infty} \sqrt{F(t)}/f(t) = 0$. Thus, from l'Hospital's rule and Lemma 2.7.3 we infer that

$$\lim_{t\to\infty} \frac{\sqrt{F(t)}}{f(t)\int_t^\infty F(s)^{-1/2}ds} = \frac{1}{2} - \alpha = \frac{\rho}{2(\rho+2)}. \tag{2.55}$$

Using (2.54) and (2.55) we derive (ii) and also

$$\begin{aligned}\lim_{t\searrow 0} \frac{h'(t)}{h''(t)} &= \frac{-2(2+\rho)}{2+\ell_1\rho} \cdot \lim_{t\searrow 0} \frac{\int_0^t k(s)ds}{k(t)} \cdot \lim_{t\to\infty} \frac{\sqrt{F(t)}}{f(t)\int_t^\infty \sqrt{F(s)}ds} \\ &= \frac{-\rho\ell_0}{2+\ell_1\rho} = 0.\end{aligned} \tag{2.56}$$

From (i) and (ii) we derive that $\lim_{t\searrow 0} h'(t) = -\infty$. So, l'Hospital's rule and (2.56) yield $\lim_{t\searrow 0} h(t)/h'(t) = 0$. This and (2.56) lead to $\lim_{t\searrow 0} h(t)/h''(t) = 0$, which proves (iii). This completes the proof of the lemma. □

Let us come back to the proof of Theorem 2.7.6.

We fix $\varepsilon \in (0, c/2)$. Because $p \equiv 0$ on $\partial\Omega$, we may take $\delta > 0$ such that

(i) $d(x)$ is a C^2 function on the set $\{x \in \mathbb{R}^N : d(x) < 2\delta\}$;
(ii) k^2 is increasing on $(0, 2\delta)$;
(iii) $(c-\varepsilon)k^2(d(x)) < p(x) < (c+\varepsilon)k^2(d(x))$, for all $x \in \Omega$ with $0 < d(x) < 2\delta$;
(iv) $h''(t) > 0$ for all $t \in (0, 2\delta)$.

Let $\sigma \in (0, \delta)$ be arbitrary and define $\xi^{\pm} := [(2+\ell_1\rho)/(c \mp 2\varepsilon)(2+\rho)]^{1/\rho}$ and $v_\sigma^-(x) := h(d(x)+\sigma)\xi^-$, for all x with $d(x)+\sigma < 2\delta$ respectively, $v_\sigma^+(x) := h(d(x)-\sigma)\xi^+$, for all x with $\sigma < d(x) < 2\delta$.

Using (i)-(iv) and the fact that $|\nabla d(x)| \equiv 1$, when $\sigma < d(x) < 2\delta$ we obtain

$$\begin{aligned}\Delta v_\sigma^+ + \lambda v_\sigma^+ - p(x)f(v_\sigma^+) \le \xi^+ h''(d(x)-\sigma) \Bigg(&\frac{h'(d(x)-\sigma)}{h''(d(x)-\sigma)}\Delta d(x) + \\ \lambda\frac{h(d(x)-\sigma)}{h''(d(x)-\sigma)} + 1 - (c-\varepsilon)&\frac{k^2(d(x)-\sigma)f(h(d(x)-\sigma)\xi^+)}{h''(d(x)-\sigma)\xi^+}\Bigg).\end{aligned}$$

Similarly, when $d(x)+\sigma < 2\delta$ we find

$$\begin{aligned}\Delta v_\sigma^- + \lambda v_\sigma^- - p(x)f(v_\sigma^-) \ge \xi^- h''(d(x)+\sigma) \Bigg(&\frac{h'(d(x)+\sigma)}{h''(d(x)+\sigma)}\Delta d(x) + \\ \lambda\frac{h(d(x)+\sigma)}{h''(d(x)+\sigma)} + 1 - (c+\varepsilon)&\frac{k^2(d(x)+\sigma)f(h(d(x)+\sigma)\xi^-)}{h''(d(x)+\sigma)\xi^-}\Bigg).\end{aligned}$$

Using Lemma 2.7.7 (ii) and (iii) and by taking $\delta > 0$ sufficiently small, we can assume that

$$\Delta v_\sigma^+(x) + \lambda v_\sigma^+(x) - p(x)f(v_\sigma^+(x)) \le 0 \quad \text{for all } x \text{ with } \sigma < d(x) < 2\delta,$$

$$\Delta v_\sigma^-(x) + \lambda v_\sigma^-(x) - p(x)f(v_\sigma^-(x)) \geq 0 \quad \text{for all } x \text{ with } d(x) + \sigma < 2\delta.$$

Let Ω_1 and Ω_2 be two smooth bounded domains such that $\Omega \subset\subset \Omega_1 \subset\subset \Omega_2$ and the first Dirichlet eigenvalue of $(-\Delta)$ in the domain $\Omega_1 \setminus \overline{\Omega}$ is greater than λ. Let also $a \in C^{0,\gamma}(\overline{\Omega}_2)$ be such that

$$\begin{aligned} 0 < a(x) \leq p(x) &\qquad \text{in } \Omega \setminus C_{2\delta}, \\ a = 0 &\qquad \text{in } \overline{\Omega}_1 \setminus \Omega, \\ a > 0 &\qquad \text{in } \Omega_2 \setminus \overline{\Omega}_1. \end{aligned}$$

By Theorem 2.3.6, there exists a solution w of the problem

$$\begin{cases} \Delta w + \lambda w = a(x)f(w) & \text{in } \Omega \setminus C_{2\delta}, \\ w > 0 & \text{in } \Omega \setminus C_{2\delta}, \\ w = \infty & \text{on } \partial(\Omega \setminus C_{2\delta}). \end{cases} \tag{2.57}$$

Let now u_λ be an arbitrary blow-up solution of (2.45) and set $v := u_\lambda + w$. Then v satisfies

$$\Delta v + \lambda v - p(x)f(v) \leq 0 \quad \text{in } \Omega \setminus \overline{C}_{2\delta}.$$

Because $v = \infty > v_\sigma^-$ on $\partial\Omega$ and $v = \infty > v_\sigma^-$ on $\partial C_{2\delta}$, it follows that

$$u_\lambda + w \geq v_\sigma^- \quad \text{in } \Omega \setminus \overline{C}_{2\delta}, \tag{2.58}$$

and similarly

$$v_\sigma^+ + w \geq u_a \quad \text{in } C_\sigma \setminus \overline{C}_{2\delta}. \tag{2.59}$$

Letting $\sigma \to 0$ in (2.58) and (2.59), we deduce

$$h(d(x))\xi^+ + 2w \geq u_\lambda + w \geq h(d(x))\xi^- \quad \text{for all } x \in \Omega \setminus \overline{C}_{2\delta}.$$

Because w is uniformly bounded on $\partial\Omega$, we have

$$\xi^- \leq \liminf_{d(x)\searrow 0} \frac{u_\lambda(x)}{h(d(x))} \leq \limsup_{d(x)\searrow 0} \frac{u_\lambda(x)}{h(d(x))} \leq \xi^+.$$

Letting $\varepsilon \to 0$ in the previous inequality we obtain (2.50). This concludes the proof. □

2.8 Comments and historical notes

Problems related to large solutions have a long history and are studied by many authors and in many contexts. Singular value problems of this type go back to the pioneering work by Bieberbach [26] in 1916 on the equation $\Delta u = e^u$ in a smooth bounded domain $\Omega \subset \mathbb{R}^2$. He showed that there exists a unique solution such that $u(x) - \log(d(x)^{-2})$ is bounded as $d(x) \to 0$. Problems of this type arise in Riemannian geometry. If a Riemannian metric of the form $|ds|^2 = \exp(2u(x))|dx|^2$

has constant Gaussian curvature $-c^2$, then $\Delta u = c^2 e^{2u}$. We also recall that the equation $\Delta u = e^u$ was studied by Max von Laue[1] (1918) who established, using statistical mechanics, that the density in equilibrium of ideal gases at any point is proportional to an exponential function, which can be described explicitly in terms of the electrostatic potential. Motivated by a problem in mathematical physics, Rademacher [173] continued the study of Bieberbach on smooth bounded domains in $\mathbb{R}^3$. These kinds of problems were later studied under the general form $\Delta u = f(u)$ in N–dimensional domains. We refer the reader to [13]–[16], [49], [50], [53], [65], [70], [128], [129], [133]–[136], and [191].

We point out that the study of blow-up boundary solutions is motivated by many natural phenomena. For instance, Keller [117] was concerned with an electrohydrodynamic model and his conclusions are the following: as the mass of uniformly charged gas contained in a container $\Omega \subset \mathbb{R}^3$ increases to infinity, then its density remains bounded at any interior point, but becomes infinite on the surface $\partial\Omega$.

In a celebrated paper, Loewner and Nirenberg [133] linked the uniqueness of the blow-up solution to the growth rate at the boundary. Motivated by certain geometric problems, they established the uniqueness of blow-up solutions for

$$\Delta u = u^{(N+2)/(N-2)}, \quad N \geq 3.$$

Bandle and Marcus [15] give results on asymptotic behavior and uniqueness of the blow-up solution for more general nonlinearities including $f(t) = t^p$, where $p > 1$.

The approach in Section 2.2 is the result of the work by Cîrstea and Rădulescu [49]. Similar results were obtained by Cheng and Ni [41] or Marcus [135] but under the stronger assumption $p > 0$ on $\partial\Omega$ instead of $(p1)$.

In Section 2.5 we were concerned with the existence of blow-up solutions for semilinear elliptic problems with a mixed boundary condition. Note that the Robin condition $\mathcal{R} = 0$ arises in heat flow problems in a body with constant temperature in the surrounding medium. More generally, if α and β are smooth functions on $\partial\Omega$ such that $\alpha, \beta \geq 0$, $\alpha + \beta > 0$, then the boundary condition $\mathcal{B}u = \alpha\partial u/\partial n + \beta u = 0$ represents the exchange of heat at the surface of the reactant by Newtonian cooling. Moreover, the boundary condition $\mathcal{B}u = 0$ is called an *isothermal (Dirichlet) condition* if $\alpha \equiv 0$, and it becomes an *adiabatic (Neumann) condition* if $\beta \equiv 0$. An intuitive meaning of the condition $\alpha + \beta > 0$ on $\partial\Omega$ is that, for the diffusion process described by problem (2.14), either the reflection phenomenon or the absorption phenomenon may occur at each point of the boundary. If $f(t) = t^p$ $(p > 1)$, then the semilinear elliptic problem

$$\begin{cases} \Delta u + \lambda u = a(x)u^p & \text{in } \Omega, \\ \mathcal{B}u = 0 & \text{on } \partial\Omega \end{cases} \tag{2.60}$$

[1]The Nobel Prize in physics, 1914, "for his discovery of the diffraction of X-rays by crystals".

is a basic population model (see, for example, [106]) and is also related to some prescribed curvature problems in Riemannian geometry (see, for example, [155] and [115]). The existence of positive solutions of (2.60) has been intensively studied (see, for instance, [4], [7], [58], [62], [79] and [155]).

If the variable potential $a(x)$ is positive in $\overline{\Omega}$, then

$$\Delta u + \lambda u = a(x)u^p \quad \text{in } \Omega,$$

is known as the *logistic problem.* This equation has been proposed as a model for population density of a steady-state single species $u(x)$ when Ω is fully surrounded by inhospitable areas. The understanding of the asymptotics for positive solutions of the degenerate logistic equation leads to the study of blow-up boundary solutions.

The two approaches for the uniqueness of the blow-up solution to the logistic equation in Sections 2.6 and 2.7 are the result of the work by Cîrstea and Rădulescu [50], [51].

We have seen in this chapter that if $f \in C^1[0,\infty)$ is an *increasing* function such that $f(0) = 0$, then the problem

$$\begin{cases} \Delta u = f(u) & \text{in } \Omega, \\ u > 0 & \text{in } \Omega, \\ u = \infty & \text{on } \partial\Omega \end{cases} \tag{2.61}$$

has a solution if and only if the Keller–Osserman condition

$$\int_1^\infty \frac{dt}{\sqrt{F(t)}} < \infty$$

is fulfilled, where $F(t) := \int_0^t f(s)\,ds$. A natural question is to establish a related result provided f fails to exhibit monotone behavior. This problem was studied by Dumont, Dupaigne, Goubet, and Rădulescu [71], who formulated an answer in terms of two Keller–Osserman–type conditions introduced by Aftalion and Reichel [1]. In a general setting, we say that a function $f : [0,\infty) \to [0,\infty)$ (not necessarily increasing!) satisfies the *strong Keller–Osserman condition* whenever there exists $\alpha > 0$ such that

$$\Phi(\alpha) := \int_\alpha^\infty \frac{dt}{\sqrt{F(t) - F(\alpha)}} < \infty .$$

Accordingly, we say that f satisfies the *weak Keller–Osserman condition* whenever

$$\liminf_{\alpha\to\infty} \Phi(\alpha) = 0.$$

For instance, the function $f(t) = t^2(1+\cos t)$ satisfies the weak Keller–Osserman condition, even if $\limsup_{\alpha\to\infty} \Phi(\alpha) = \infty$.

The result established by Dumont, Dupaigne, Goubet, and Rădulescu [71] is the following.

Theorem 2.8.1 *The following statements are equivalent:*

(i) f *satisfies the strong Keller–Osserman condition.*

(ii) f *satisfies the weak Keller–Osserman condition.*

(iii) *There exists a ball* $\Omega = B_R$ *such that problem (2.61) admits at least a solution.*

(iv) *Given any smooth domain* Ω*, problem (2.61) admits at least a solution.*

In particular, Theorem 2.8.1 implies existence of blow-up boundary solutions for oscillating functions such as $f(u) = u^2(1 + \cos u)$.

The results developed in this chapter also apply to problems on Riemannian manifolds if Δ is replaced by the Laplace–Beltrami operator

$$\Delta_B = \frac{1}{\sqrt{c}} \frac{\partial}{\partial x_i} \left(\sqrt{c}\, a_{ij}(x) \frac{\partial}{\partial x_i} \right), \qquad c := \det\,(a_{ij}),$$

with respect to the metric $ds^2 = c_{ij}\, dx_i dx_j$, where (c_{ij}) is the inverse of (a_{ij}). In this case the results developed in this chapter apply to concrete problems arising in Riemannian geometry. For instance (cf. Loewner and Nirenberg [133]), if Ω is replaced by the standard N dimensional sphere (S^N, g_0), Δ is the Laplace–Beltrami operator Δ_{g_0}, $f(u) = (N-2)/[4(N-1)]\, u^{(N+2)/(N-2)}$, and $\lambda = N(N-2)/4$, we find the prescribing scalar curvature equation on S^N. For further applications and connections with other fields of mathematics such as Brownian motion and potential theory, we refer the reader to Dynkin [73], Iscoe [110], and Le Gall [130].

3

ENTIRE SOLUTIONS BLOWING UP AT INFINITY FOR ELLIPTIC SYSTEMS

> The object of pure physics is the unfolding of the laws of the intelligible world; the object of pure mathematics that of unfolding the laws of human intelligence.
>
> James J. Sylvester (1814–1897)

In this chapter, we are concerned with entire radially symmetric solutions of logistic-type systems in anisotropic media. In terms of the growth of the variable potential functions, we establish conditions such that the solutions are either bounded or blow up at infinity. An important role in the results we establish in this chapter is played by the properties of the *central value set.*

3.1 Introduction

Consider the following semilinear elliptic system

$$\begin{cases} \Delta u = p(x)f(v) & \text{in } \mathbb{R}^N, \\ \Delta v = q(x)g(u) & \text{in } \mathbb{R}^N, \\ u > 0,\ v > 0 & \text{in } \mathbb{R}^N, \end{cases} \tag{3.1}$$

where $N \geq 3$ and $p,\, q \in C^{0,\gamma}_{\text{loc}}(\mathbb{R}^N)$ $(0 < \gamma < 1)$ are nonnegative and radially symmetric potentials. We also assume that $f,\, g \in C^{0,\gamma}_{\text{loc}}[0, \infty)$ $(0 < \gamma < 1)$ are positive and nondecreasing on $(0, \infty)$.

We are concerned in this chapter with the existence of entire radially symmetric solutions of (3.1) that blow up at infinity—that is, radially symmetric solutions (u, v) satisfying $u(x) \to \infty$ and $v(x) \to \infty$ as $|x| \to \infty$.

Notice that radially symmetric solutions of (3.1) satisfy the ordinary differential system

$$\begin{cases} u''(r) + \dfrac{N-1}{r}u'(r) = p(r)g(v(r)), \\ v''(r) + \dfrac{N-1}{r}v'(r) = q(r)f(u(r)), \\ u(r) > 0,\ v(r) > 0, \end{cases} \tag{3.2}$$

for all $r \geq 0$.

Our aim is to give a characterization of the *central value set* $\mathcal{G}$ of all $(a, b) \in \mathbb{R}_+ \times \mathbb{R}_+$, where $\mathbb{R}_+ = (0, \infty)$ such that there exists an entire radially symmetric solution (u, v) of (3.1) with the property $u(0) = a$ and $v(0) = b$. Such a pair will be called the *central value* of the solution (u, v) of system (3.1).

Thus, any radially symmetric solution (u, v) of (3.1) with the central value $(a, b) \in \mathbb{R}_+ \times \mathbb{R}_+$ satisfies the following system of integral equations:

$$\begin{cases} u(r) = a + \int_0^r t^{1-N} \int_0^t s^{N-1} p(s) g(v(s)) ds dt, & r \geq 0, \\ v(r) = b + \int_0^r t^{1-N} \int_0^t s^{N-1} q(s) f(u(s)) ds dt, & r \geq 0. \end{cases} \tag{3.3}$$

3.2 Characterization of the central value set

3.2.1 *Bounded or unbounded entire solutions*

We are first concerned with a framework corresponding to an unbounded central value set. In such a case, taking into account both the decay of the variable potentials and the growth rate of the nonlinear terms, we establish sufficient conditions such that the solutions of the nonlinear system (3.1) are either bounded or blow up at infinity.

Theorem 3.2.1 *Assume that*

$$\lim_{t \to \infty} \frac{g(cf(t))}{t} = 0 \quad \textit{for all } c > 0. \tag{3.4}$$

Then $\mathcal{G} = \mathbb{R}_+ \times \mathbb{R}_+$. Moreover, the following properties are valid:

(i) *If p and q satisfy*

$$\int_0^\infty tp(t)\, dt = \infty \quad \textit{and} \quad \int_0^\infty tq(t)\, dt = \infty, \tag{3.5}$$

then all radially symmetric solutions of (3.1) blow up at infinity.

(ii) *If p and q satisfy*

$$\int_0^\infty tp(t)\, dt < \infty \quad \textit{and} \quad \int_0^\infty tq(t)\, dt < \infty, \tag{3.6}$$

then all radially symmetric solutions of (3.1) are bounded.

For any $r > 0$ define

$$A(r) := \int_0^r t^{1-N} \int_0^t s^{N-1} p(s) ds dt,$$

$$B(r) := \int_0^r t^{1-N} \int_0^t s^{N-1} q(s) ds dt.$$

Remark 3.2.2 *Using Lemma 2.2.4 we deduce the following properties:*

(i) *Condition (3.5) holds if and only if* $\lim_{r\to\infty} A(r) = \lim_{r\to\infty} B(r) = \infty$.

(ii) *Condition (3.6) holds if and only if both* $\lim_{r\to\infty} A(r)$ *and* $\lim_{r\to\infty} B(r)$ *are finite.*

Proof Fix $(a, b) \in \mathbb{R}_+ \times \mathbb{R}_+$. A radially symmetric solution (u, v) of (3.1) with the central value (a, b) will be obtained by means of a sequence of approximations $\{(u_k, v_k)\}_{k\geq 1}$, as follows. Set $v_0 \equiv b$ and let $(u_k)_{k\geq 1}$ and $(v_k)_{k\geq 1}$ be two sequences of functions defined as

$$\begin{cases} u_k(r) = a + \displaystyle\int_0^r t^{1-N} \int_0^t s^{N-1} p(s) g(v_{k-1}(s)) ds dt, \\ v_k(r) = b + \displaystyle\int_0^r t^{1-N} \int_0^t s^{N-1} q(s) f(u_k(s)) ds dt, \end{cases} \tag{3.7}$$

for all $r \geq 0$. Because $v_1(r) \geq b$, we find $u_2(r) \geq u_1(r)$ for all $r \geq 0$. This implies $v_2(r) \geq v_1(r)$, which further produces $u_3(r) \geq u_2(r)$ for all $r \geq 0$. Proceeding in the same manner we conclude that

$$u_k(r) \leq u_{k+1}(r) \quad \text{and} \quad v_k(r) \leq v_{k+1}(r) \quad \text{for all } r \geq 0 \text{ and } k \geq 1.$$

We claim that the nondecreasing sequences $(u_k(r))_{k\geq 1}$ and $(v_k(r))_{k\geq 1}$ are bounded from above on bounded sets. Indeed, for all $r \geq 0$ we have

$$u_k(r) \leq u_{k+1}(r) \leq a + g(v_k(r))A(r) \tag{3.8}$$

and

$$v_k(r) \leq b + f\left(u_k(r)\right) B(r). \tag{3.9}$$

Let $R > 0$ be arbitrary. By (3.8) and (3.9) we find

$$u_k(R) \leq a + g\left(b + f\left(u_k(R)\right) B(R)\right) A(R) \quad \text{for all } k \geq 1,$$

or, equivalently,

$$1 \leq \frac{a}{u_k(R)} + \frac{g\left(b + f\left(u_k(R)\right) B(R)\right)}{u_k(R)} A(R) \quad \text{for all } k \geq 1. \tag{3.10}$$

By the monotonicity of $(u_k(R))_{k\geq 1}$, there exists $L(R) := \lim_{k\to\infty} u_k(R)$. We claim that $L(R)$ is finite. Indeed, assuming the contrary, letting $k \to \infty$ in (3.10) and using (3.4) we obtain a contradiction.

Because $u_k'(r), v_k'(r) \geq 0$ it follows that the map $(0, \infty) \ni R \longmapsto L(R)$ is nondecreasing in $(0, \infty)$, and for all $k \geq 1$ we have

$$u_k(r) \leq u_k(R) \leq L(R) \quad \text{for all } 0 \leq r \leq R, \tag{3.11}$$

$$v_k(r) \leq b + f\left(L(R)\right) B(R) \quad \text{for all } 0 \leq r \leq R. \tag{3.12}$$

Thus, there exists $\overline{L} := \lim_{R\to\infty} L(R) \in (0, \infty]$ and the sequences $(u_k(r))_{k\geq 1}$, $(v_k(r))_{k\geq 1}$ are bounded from above on bounded sets. Therefore, for all $r \geq 0$ we can define $u(r) := \lim_{k\to\infty} u_k(r)$ and $v(r) := \lim_{k\to\infty} v_k(r)$. Passing to the limit in (3.7) we obtain that (u, v) is a solution of (3.1) with the central value (a, b).

(i) Assume that p and q fulfill the condition (3.6). By Remark 3.2.2 this also implies that $\overline{A} := \lim_{r\to\infty} A(r) < \infty$ and $\overline{B} := \lim_{r\to\infty} B(r) < \infty$. Let (u, v) be a radially symmetric solution of (3.1) with the central value (a, b). This yields

$$\begin{aligned} u(r) &\leq a + g(v(r))A(r) && \text{for all } r \geq 0, \\ v(r) &\leq b + f(u(r))\,B(r) && \text{for all } r \geq 0. \end{aligned} \tag{3.13}$$

Hence,

$$u(r) \leq a + g(b + \overline{B}u(r))\overline{A} \quad \text{for all } r \geq 0.$$

Therefore,

$$1 \leq \frac{a}{u(r)} + \frac{g(b + \overline{B}u(r))}{u(r)}\overline{A} \quad \text{for all } r \geq 0. \tag{3.14}$$

Now, if u is unbounded we deduce that $u(r) \to \infty$ as $r \to \infty$. Passing to the limit in (3.14) with $r \to \infty$, we obtain a contradiction. Hence, u is bounded and, using (3.13), v is also bounded in $[0, \infty)$.

(ii) Assume now that condition (3.5) holds. In this case, $\lim_{r\to\infty} A(r) = \lim_{r\to\infty} B(r) = \infty$. Let (u, v) be a radially symmetric solution of (3.1) with central value (a, b). For all $r \geq 0$ we have

$$u(r) \geq a + g(b)\,A(r) \quad \text{and } v(r) \geq b + f(a)\,B(r).$$

Letting $r \to \infty$ we derive that both u and v blow up at infinity. This concludes the proof of Theorem 3.2.1. □

Some examples of nonlinearities f and g satisfying the assumptions of Theorem 3.2.1 are stated here:

(i) Let a_j, b_k, α_j, $\beta_k > 0$ and $\alpha\beta < 1$, where $\alpha = \max_{1\leq j\leq l} \alpha_j$ and $\beta = \max_{1\leq k\leq m} \beta_k$. Define

$$f(t) = \begin{cases} 0 & \text{if } t < 0, \\ \displaystyle\sum_{j=1}^{l} a_j t^{\alpha_j} & \text{if } t \geq 0, \end{cases} \quad \text{and } g(t) = \begin{cases} 0 & \text{if } t < 0, \\ \displaystyle\sum_{k=1}^{m} b_k t^{\beta_j} & \text{if } t \geq 0. \end{cases}$$

(ii) $f(t) = (1 + t^2)^{\alpha/2}$ and $g(t) = (1 + t^2)^{\beta/2}$ for $t \in \mathbb{R}$, with α, $\beta > 0$ and $\alpha\beta < 1$.

(iii) Let α, $\beta > 0$ such that $\alpha\beta < 1$ and define

$$f(t) = \begin{cases} 0 & \text{if } t < 0, \\ t^\alpha & \text{if } 0 \leq t < 1, \\ t^\beta & \text{if } t \geq 1, \end{cases} \quad \text{and } g(t) = \begin{cases} 0 & \text{if } t < 0, \\ t^\beta & \text{if } 0 \leq t < 1, \\ t^\alpha & \text{if } t \geq 1. \end{cases}$$

(iv) Let $g(t) = t$ for $t \in \mathbb{R}$, $f(t) = 0$ for $t \le 0$ and

$$f(t) = t\left(-\ln\left(\left(\frac{2}{\pi}\right)\arctan t\right)\right)^{\alpha} \quad \text{for } t > 0,$$

where $\alpha \in (0, 1/2)$.

Concerning the bounded radially symmetric solutions of (3.1), we establish the following auxiliary result.

Proposition 3.2.3 *Assume that condition (3.6) holds and let f and g be locally Lipschitz functions on $(0,\infty)$ that fulfill (3.4). Let (u,v) and $(\widetilde{u},\widetilde{v})$ be two bounded radially symmetric solutions of (3.1). Then there exists a positive constant C such that for all $r \ge 0$ there holds*

$$\max\{|u(r)-\widetilde{u}(r)|, |v(r)-\widetilde{v}(r)|\} \le C \max\{|u(0)-\widetilde{u}(0)|, |v(0)-\widetilde{v}(0)|\}.$$

Proof Set $K := \max\{|u(0)-\widetilde{u}(0)|, |v(0)-\widetilde{v}(0)|\}$. From the first equation in (3.3) we find

$$u'(r) - \widetilde{u}'(r) = r^{1-N}\int_0^r s^{N-1}p(s)(g(v(s)) - g(\widetilde{v}(s)))\,ds.$$

Hence,

$$|u(r)-\widetilde{u}(r)| \le K + \int_0^r t^{1-N}\int_0^t s^{N-1}p(s)|g(v(s)) - g(\widetilde{v}(s))|\,ds\,dt. \tag{3.15}$$

Because (u,v) and $(\widetilde{u},\widetilde{v})$ are bounded, for any $r \ge 0$ we have

$$|g(v(r)) - g(\widetilde{v}(r))| \le m|v(r)-\widetilde{v}(r)|,$$
$$|f(u(r)) - f(\widetilde{u}(r))| \le m|u(r)-\widetilde{u}(r)|,$$

where m denotes a Lipschitz constant for both functions f and g. Therefore, by (3.15) we find

$$|u(r)-\widetilde{u}(r)| \le K + m\int_0^r t^{1-N}\int_0^t s^{N-1}p(s)|v(s)-\widetilde{v}(s)|\,ds\,dt, \tag{3.16}$$

for all $r \ge 0$. In the same manner we obtain

$$|v(r)-\widetilde{v}(r)| \le K + m\int_0^r t^{1-N}\int_0^t s^{N-1}q(s)|u(s)-\widetilde{u}(s)|\,ds\,dt, \tag{3.17}$$

for all $r \ge 0$. Define next

$$X(r) := K + m\int_0^r t^{1-N}\int_0^t s^{N-1}p(s)|v(s)-\tilde{v}(s)|\,ds\,dt,$$

$$Y(r) := K + m\int_0^r t^{1-N}\int_0^t s^{N-1}q(s)|u(s)-\tilde{u}(s)|\,ds\,dt.$$

Clearly X and Y are nondecreasing functions and $X(0) = Y(0) = K$.

A simple computation together with (3.16) and (3.17) leads us to

$$\begin{aligned}(r^{N-1}X')'(r) &= mr^{N-1}p(r)|v(r)-\tilde{v}(r)| \le mr^{N-1}p(r)Y(r),\\ (r^{N-1}Y')'(r) &= mr^{N-1}q(r)|u(r)-\tilde{u}(r)| \le mr^{N-1}q(r)X(r),\end{aligned} \tag{3.18}$$

for all $r \ge 0$. Because Y is nondecreasing, we have

$$\begin{aligned}X(r) &\le K + mY(r)A(r)\\ &\le K + \frac{m}{N-2}Y(r)\int_0^r tp(t)\,dt\\ &\le K + mC_pY(r) \qquad \text{for all } r \ge 0,\end{aligned} \tag{3.19}$$

where

$$C_p = \frac{1}{N-2}\int_0^\infty tp(t)\,dt.$$

Using (3.19) in the second inequality of (3.18) we find

$$(r^{N-1}Y')'(r) \le mr^{N-1}q(r)(K + mC_pY(r)) \qquad \text{for all } r \ge 0.$$

Integrating twice from 0 to r in the previous inequality we obtain

$$Y(r) \le K(1+mC_q) + \frac{m^2}{N-2}C_p\int_0^r tq(t)Y(t)\,dt \qquad \text{for all } r \ge 0,$$

where

$$C_q = \frac{1}{N-2}\int_0^\infty tq(t)\,dt.$$

Next we apply the Gronwall inequality in the following form: assume that $0 < T \le +\infty$ and let φ, α, $\beta : [0,T) \to \mathbb{R}$ be continuous functions, with α nondecreasing and $\beta \ge 0$. If

$$\varphi(t) \le \alpha(t) + \int_0^t \beta(s)\varphi(s)ds \qquad \text{for all } t \in [0,T),$$

then

$$\varphi(t) \le \alpha(t)\exp\left(\int_0^t \beta(s)ds\right) \qquad \text{for all } t \in [0,T).$$

Thus, by Gronwall's inequality, we deduce that for all $r \ge 0$,

$$\begin{aligned}Y(r) &\le K(1+mC_q)\exp\left(m^2(N-2)^{-1}C_p\int_0^r tq(t)\,dt\right)\\ &\le K(1+mC_q)\exp\left(m^2C_pC_q\right),\end{aligned} \tag{3.20}$$

and similarly for X. The conclusion now follows from (3.20), (3.16), and (3.17). This finishes the proof. □

3.2.2 *Role of the Keller–Osserman condition*

We assume in this section that f and g satisfy the stronger regularity $f,\, g \in C^1[0,\infty)$. We drop assumption (3.4) and we require in turn that f and g fulfill

$(H1)$ $f(0) = g(0) = 0$ and there exists $\sigma := \liminf_{t\to\infty} \frac{f(t)}{g(t)} > 0$;

$(H2)$ g satisfies the Keller–Osserman condition

$$\int_1^{\infty} \frac{dt}{\sqrt{G(t)}} < \infty, \text{ where } G(t) := \int_0^t g(s)\, ds.$$

Remark that assumptions $(H1)$ and $(H2)$ imply that f satisfies condition $(H2)$, too. Set $\nu := \max\{p(0), q(0)\} \geq 0$ and $\eta := \min\{p, q\}$.

Throughout this section we assume that $\nu > 0$, η is not identically zero at infinity, and condition (3.6) on p and q holds. Before the main result is stated, we provide several auxiliary results concerning the set of central value.

Lemma 3.2.4 $\mathcal{G} \neq \emptyset$.

Proof Notice that $p+q$ satisfies the assumptions $(p1)'$ and $(p2)$ in Section 2.2. Thus, by Theorem 2.2.3 and the fact that p and q are radially symmetric, there exists a radially symmetric solution $\psi \in C^2_{\text{loc}}(\mathbb{R}^N)$ of the problem

$$\begin{cases} \Delta\psi = (p+q)(x)(f+g)(\psi) & \text{in } \mathbb{R}^N, \\ \psi > 0 & \text{in } \mathbb{R}^N, \\ \psi = \infty & \text{as } |x| \to \infty. \end{cases}$$

Hence, ψ satisfies the integral equation

$$\psi(r) = \psi(0) + \int_0^r t^{1-N} \int_0^t s^{N-1}(p+q)(s)(f+g)(\psi(s))\, ds\, dt,$$

for all $r \geq 0$. We claim that $(0, \psi(0)] \times (0, \psi(0)] \subseteq \mathcal{G}$. To this aim, let us fix $0 < a, b \leq \psi(0)$. Define the sequences $(u_k)_{k\geq 1}$ and $(v_k)_{k\geq 0}$ by $v_0(r) := b$ for all $r \geq 0$ and

$$\begin{cases} u_k(r) = a + \displaystyle\int_0^r t^{1-N} \int_0^t s^{N-1} p(s) g(v_{k-1}(s))\, ds\, dt, \\ v_k(r) = b + \displaystyle\int_0^r t^{1-N} \int_0^t s^{N-1} q(s) f(u_k(s))\, ds\, dt, \end{cases} \tag{3.21}$$

for all $r \geq 0$ and $k \geq 1$. We first remark that $v_0 \leq v_1$, which produces $u_1 \leq u_2$. Consequently, $v_1 \leq v_2$, which further yields $u_2 \leq u_3$. With the same arguments, we obtain that $(u_k)_{k\geq 1}$ and $(v_k)_{k\geq 0}$ are nondecreasing sequences. Because $\psi'(r) \geq 0$ and $b = v_0 \leq \psi(0) \leq \psi(r)$ for all $r \geq 0$, we find

$$\begin{aligned} u_1(r) &\le a + \int_0^r t^{1-N} \int_0^t s^{N-1} p(s) g(\psi(s))\, ds\, dt \\ &\le \psi(0) + \int_0^r t^{1-N} \int_0^t s^{N-1} (p+q)(s)(f+g)(\psi(s))\, ds\, dt \\ &= \psi(r). \end{aligned}$$

Thus, $u_1 \le \psi$. It follows that

$$\begin{aligned} v_1(r) &\le b + \int_0^r t^{1-N} \int_0^t s^{N-1} q(s) f(\psi(s))\, ds\, dt \\ &\le \psi(0) + \int_0^r t^{1-N} \int_0^t s^{N-1} (p+q)(s)(f+g)(\psi(s))\, ds\, dt \\ &= \psi(r). \end{aligned}$$

A further induction argument yields

$$u_k(r) \le \psi(r) \quad \text{and} \quad v_k(r) \le \psi(r) \quad \text{for all } r \ge 0 \text{ and } k \ge 1.$$

Thus, $\{(u_k, v_k)\}_{k\ge1}$ converges and $(u(r), v(r)) := \lim_{k\to\infty}(u_k(r), v_k(r))$, $r \ge 0$ is a radially symmetric solution of (3.1) with central value (a, b). This completes the proof. □

A direct consequence of Lemma 3.2.4 is the following corollary.

Corollary 3.2.5 *If* $(a,b) \in \mathcal{G}$, *then* $(0,a] \times (0,b] \subseteq \mathcal{G}$.

Proof The process used before can be repeated by taking

$$\begin{cases} u_k(r) = a_0 + \int_0^r t^{1-N} \int_0^t s^{N-1} p(s) g(v_{k-1}(s))\, ds\, dt & \text{for all } r \ge 0, \\ v_k(r) = b_0 + \int_0^r t^{1-N} \int_0^t s^{N-1} q(s) f(u_k(s))\, ds\, dt & \text{for all } r \ge 0, \end{cases}$$

where $0 < a_0 \le a$, $0 < b_0 \le b$ and $v_0 \equiv b_0$.

Now let (U, V) be a radially symmetric solution of (3.1) with central value (a, b). As in Lemma 3.2.4, we obtain

$$u_k(r) \le u_{k+1}(r) \le U(r) \quad \text{for all } r \ge 0 \text{ and } k \ge 1,$$

$$v_k(r) \le v_{k+1}(r) \le V(r) \quad \text{for all } r \ge 0 \text{ and } k \ge 1.$$

Thus, for all $r \ge 0$ we can define $(u(r), v(r)) := \lim_{k\to\infty}(u_k(r), v_k(r))$. Then $u \le U$, $v \le V$ on $[0,\infty)$ and (u, v) is a radially symmetric solution of (3.1) with the central value (a_0, b_0). This shows that $(a_0, b_0) \in \mathcal{G}$, so that our assertion is proved. □

Lemma 3.2.6 $\mathcal{G}$ *is bounded.*

Proof Set $0 < \lambda < \min\{\sigma, 1\}$ and let $\delta = \delta(\lambda)$ be large enough so that

$$f(t) \geq \lambda g(t) \quad \text{for all } t \geq \delta. \tag{3.22}$$

Because η is radially symmetric and not identically zero at infinity, we can find an open ball B_R centered at the origin and of radius $R > 0$ such that $\eta > 0$ on ∂B_R. Furthermore, according to Theorem 2.2.2, there exists $\zeta \in C^2(B_R)$ such that

$$\begin{cases} \Delta\zeta = \lambda\eta(x)g\left(\dfrac{\zeta}{2}\right) & \text{in } B_R, \\ \zeta > 0 & \text{in } B_R, \\ \zeta \to \infty & \text{as } |x| \to R. \end{cases}$$

Arguing by contradiction, let us assume that $\mathcal{G}$ is not bounded. Then there exists $(a, b) \in \mathcal{G}$ such that $a + b > \max\{2\delta, \zeta(0)\}$. Let (u, v) be a radially symmetric solution of (3.1) with the central value (a, b). Because $u(x) + v(x) \geq a + b > 2\delta$ for all $x \in \mathbb{R}^N$, by (3.22) we find

$$f(u(x)) \geq f\left(\frac{u(x) + v(x)}{2}\right) \geq \lambda g\left(\frac{u(x) + v(x)}{2}\right) \quad \text{if } u(x) \geq v(x)$$

and

$$g(v(x)) \geq g\left(\frac{u(x) + v(x)}{2}\right) \geq \lambda g\left(\frac{u(x) + v(x)}{2}\right) \quad \text{if } v(x) \geq u(x).$$

It follows that

$$\begin{aligned} \Delta(u + v) &= p(x)g(v) + q(x)f(u) \\ &\geq \eta(x)(g(v) + f(u)) \\ &\geq \lambda\eta(x)g\left(\frac{u + v}{2}\right) \quad \text{in } \mathbb{R}^N. \end{aligned}$$

On the other hand, $\zeta(x) \to \infty$ as $|x| \to R$ and $u, v \in C^2(\overline{B}_R)$. Thus, by the weak maximum principle, we conclude that $u + v \leq \zeta$ in B_R. But this is impossible because $u(0) + v(0) = a + b > \zeta(0)$. This finishes the proof. □

Lemma 3.2.7 $F(\mathcal{G}) \subset \mathcal{G}$.

Proof Let $(a, b) \in F(\mathcal{G})$ and $n_0 \geq 1$ be such that $n_0 \min\{a, b\} > 1$. We claim that

$$(a - 1/n, b - 1/n) \in \mathcal{G} \quad \text{for all } n \geq n_0.$$

Indeed, if this is not true, by Corollary 3.2.5 we find

$$D := [a - 1/n, \infty) \times [b - 1/n, \infty) \subseteq (\mathbb{R}_+ \times \mathbb{R}_+) \setminus \mathcal{G},$$

for some $n \geq n_0$. Hence, we can find a small ball B centered in (a, b) such that $B \subset\subset D$; in other words, $B \cap \mathcal{G} = \emptyset$. But this contradicts the choice of (a, b).

Consequently, for all $n \geq n_0$ there exists (u_n, v_n), a radially symmetric solution of (3.1) with the central value $(a-1/n, b-1/n)$. Thus, for any $n \geq n_0$ and $r \geq 0$ we have

$$u_n(r) = a - \frac{1}{n} + \int_0^r t^{1-N} \int_0^t s^{N-1} p(s) g(v_n(s))\, ds\, dt,$$

$$v_n(r) = b - \frac{1}{n} + \int_0^r t^{1-N} \int_0^t s^{N-1} q(s) f(u_n(s))\, ds\, dt.$$

We observe that $(u_n)_{n \geq n_0}$ and $(v_n)_{n \geq n_0}$ are nondecreasing sequences. Next, we prove that $(u_n)_{n \geq n_0}$ and $(v_n)_{n \geq n_0}$ converge in $\mathbb{R}^N$. To this aim, let $x_0 \in \mathbb{R}^N$ be arbitrary. Because η is not identically zero at infinity, we may find $R > 0$ such that $x_0 \in B_R$ and $\eta > 0$ on ∂B_R.

Because $\sigma = \liminf_{t \to \infty} f(t)/g(t) > 0$, we find $\tau \in (0, 1)$ such that

$$f(t) \geq \tau g(t), \qquad \text{for all } t \geq \frac{a+b}{2} - \frac{1}{n_0}.$$

Therefore, on the set where $u_n \geq v_n$ we have

$$f(u_n) \geq f\left(\frac{u_n + v_n}{2}\right) \geq \tau g\left(\frac{u_n + v_n}{2}\right).$$

Similarly, on the set where $u_n \leq v_n$ we have

$$g(v_n) \geq g\left(\frac{u_n + v_n}{2}\right) \geq \tau g\left(\frac{u_n + v_n}{2}\right).$$

Now it is easy to see that for all $x \in \mathbb{R}^N$ we have

$$\Delta(u_n + v_n) \geq \tau \eta(x) g\left(\frac{u_n + v_n}{2}\right).$$

On the other hand, by Theorem 2.2.2 there exists $\zeta \in C^2(B_R)$ such that

$$\begin{cases} \Delta \zeta = \tau \eta(x) g\left(\dfrac{\zeta}{2}\right) & \text{in } B_R, \\ \zeta > 0 & \text{in } B_R, \\ \zeta \to \infty & \text{as } |x| \to R. \end{cases}$$

The weak maximum principle yields $u_n + v_n \leq \zeta$ in B_R. So, it makes sense to define $(u(x_0), v(x_0)) := \lim_{n \to \infty}(u_n(x_0), v_n(x_0))$. Because x_0 is arbitrary, the functions u, v exist on $\mathbb{R}^N$. Hence, (u, v) is a radially symmetric solution of (3.1) with central value (a, b)—that is, $(a, b) \in \mathcal{G}$. □

For $(c, d) \in (\mathbb{R}_+ \times \mathbb{R}_+) \setminus \mathcal{G}$, denote by $R_{c,d}$ the supremum over $r > 0$ such that there exists a radially symmetric solution of (3.1) in $B(0, r)$ so that $(u(0), v(0)) = (c, d)$.

Lemma 3.2.8 *For all* $(c,d) \in (\mathbb{R}_+ \times \mathbb{R}_+) \setminus \mathcal{G}$ *we have* $0 < R_{c,d} < \infty$.

Proof Because $\nu > 0$ and $p, q \in C[0,\infty)$, there exists $\varepsilon > 0$ such that $(p+q)(r) > 0$ for all $0 \le r < \varepsilon$. Let $0 < R < \varepsilon$ be arbitrary. Hence, there exists a positive radially symmetric solution of

$$\Delta\psi_R = (p+q)(x)(f+g)(\psi_R) \quad \text{in } B_R,$$

which blows up as $|x| \to R$. Furthermore, for any $0 \le r < R$ we have

$$\psi_R(r) = \psi_R(0) + \int_0^r t^{1-N} \int_0^t s^{N-1}(p+q)(s)(f+g)(\psi_R(s))\,ds\,dt.$$

Obviously, $\psi_R' \ge 0$. Thus, we find

$$\psi_R'(r) = r^{1-N} \int_0^r s^{N-1}(p+q)(s)(f+g)(\psi_R(s))\,ds \le C(f+g)(\psi_R(r)),$$

where $C > 0$ is a positive constant such that $\int_0^\varepsilon (p+q)(s)\,ds \le C$.

Because $f+g$ satisfies the hypotheses $(H1)$ and $(H2)$, by Lemma 2.2.1 (ii) we derive

$$\int_1^\infty \frac{dt}{(f+g)(t)} < \infty.$$

Therefore,

$$-\frac{d}{dr}\int_{\psi_R(r)}^\infty \frac{ds}{(f+g)(s)} = \frac{\psi_R'(r)}{(f+g)(\psi_R(r))} \le C \quad \text{for all } 0 < r < R.$$

Integrating in the last inequality and taking into account the fact that $\psi_R(r) \to \infty$ as $r \nearrow R$, we obtain

$$\int_{\psi_R(0)}^\infty \frac{ds}{(f+g)(s)} \le CR.$$

Now, we let $R \searrow 0$ in the previous relation and we have

$$\lim_{R \searrow 0} \int_{\psi_R(0)}^\infty \frac{ds}{(f+g)(s)} = 0.$$

This implies that $\psi_R(0) \to \infty$ as $R \searrow 0$. Thus, there exists $0 < \rho < \varepsilon$ such that $0 < c, d \le \psi_\rho(0)$. Set $v_0 \equiv d$ and define

$$\begin{cases} u_k(r) = c + \displaystyle\int_0^r t^{1-N} \int_0^t s^{N-1} p(s) g(v_{k-1}(s))\,ds\,dt, \\ v_k(r) = d + \displaystyle\int_0^r t^{1-N} \int_0^t s^{N-1} q(s) f(u_k(s))\,ds\,dt, \end{cases} \tag{3.23}$$

for all $r \geq 0$. As in the proof of Lemma 3.2.4 we deduce $(u_k)_{k\geq 1}$, and $(v_k)_{k\geq 0}$ are nondecreasing, and for all $k \geq 1$ there holds

$$\max\{u_k(r), v_k(r)\} \leq \psi_\rho(r) \quad \text{for all } 0 \leq r < \rho.$$

Thus, for any $0 \leq r < \rho$ there exists $(u(r), v(r)) := \lim_{k\to\infty}(u_k(r), v_k(r))$ that, moreover, is a radially symmetric solution of (3.1) in B_ρ such that $(u(0), v(0)) = (c, d)$. This shows that $R_{c,d} \geq \rho > 0$. By the definition of $R_{c,d}$ we also derive

$$\lim_{r\nearrow R_{c,d}} u(r) = \infty \quad \text{and} \quad \lim_{r\nearrow R_{c,d}} v(r) = \infty. \tag{3.24}$$

On the other hand, because $(c, d) \notin \mathcal{G}$, we conclude that $R_{c,d}$ is finite. This completes the proof. □

The main result in this section is the following.

Theorem 3.2.9 *Assume that $\nu > 0$, η is not identically zero at infinity and (H_1), (H_2), and (3.6) hold. Then any entire radially symmetric solution (u, v) of (3.1) with $(u(0), v(0)) \in F(\mathcal{G})$ blows up at infinity.*

Proof Let $(a, b) \in F(\mathcal{G})$. By Lemma 3.2.7, $(a, b) \in \mathcal{G}$ so that there exists (U, V), a radially symmetric solution of (3.1) with $(U(0), V(0)) = (a, b)$. Obviously, for any $n \geq 1$, $(a + 1/n, b + 1/n) \in (\mathbb{R}_+ \times \mathbb{R}_+) \setminus \mathcal{G}$. By Lemma 3.2.8, $R_n := R_{a+1/n,b+1/n}$ is a positive number. Let (U_n, V_n) be the radially symmetric solution of (3.1) in B_{R_n} with the central value $(a + 1/n, b + 1/n)$. Thus, for all $0 \leq r < R_n$, (U_n, V_n) is a solution of the system of integral equations

$$\begin{cases} U_n(r) = a + \dfrac{1}{n} + \displaystyle\int_0^r t^{1-N} \int_0^t s^{N-1} p(s) g(V_n(s))\, ds\, dt, \\ V_n(r) = b + \dfrac{1}{n} + \displaystyle\int_0^r t^{1-N} \int_0^t s^{N-1} q(s) f(U_n(s))\, ds\, dt. \end{cases} \tag{3.25}$$

In view of (3.24), we have

$$\lim_{r\nearrow R_n} U_n(r) = \lim_{r\nearrow R_n} V_n(r) = \infty \quad \text{for all } n \geq 1.$$

We claim that the sequence $(R_n)_{n\geq 1}$ is nondecreasing. Indeed, if $(u_k)_{k\geq 1}$ and $(v_k)_{k\geq 1}$ are the sequences of functions defined by (3.23) with $c = a + 1/(n + 1)$ and $d = b + 1/(n + 1)$, then for all $k \geq 1$ and $0 \leq r < R_n$ we have

$$u_k(r) \leq u_{k+1}(r) \leq U_n(r), \quad v_k(r) \leq v_{k+1}(r) \leq V_n(r). \tag{3.26}$$

This implies that $(u_k(r))_{k\geq 1}$ and $(v_k(r))_{k\geq 1}$ converge for any $0 \leq r < R_n$. Moreover, $(U_{n+1}, V_{n+1}) := \lim_{k\to\infty}(u_k, v_k)$ is a radially symmetric solution of (3.1) in B_{R_n} with the central value $(a+1/(n+1), b+1/(n+1))$. By the definition of R_{n+1}, it follows that $R_{n+1} \geq R_n$ for any $n \geq 1$.

Set $R := \lim_{n\to\infty} R_n$ and let $0 \le r < R$ be arbitrary. Then, there exists $n_1 = n_1(r)$ such that $r < R_n$ for all $n \ge n_1$. From (3.26) we derive that $U_{n+1} \le U_n$ and $V_{n+1} \le V_n$ on $[0, R_n)$, for all $n \ge n_1$. Thus, for all $0 \le r < R$ we can define

$$U(r) := \lim_{n\to\infty} U_n(r) \quad \text{and} \quad V(r) := \lim_{n\to\infty} V_n(r). \tag{3.27}$$

Because $U_n'(r) \ge 0$, from (3.25) we find

$$V_n(r) \le b + \frac{1}{n} + f(U_n(r)) \int_0^\infty t^{1-N} \int_0^t s^{N-1} q(s)\, ds\, dt.$$

This yields

$$V_n(r) \le C_1 U_n(r) + C_2 f(U_n(r)) \quad \text{for all } 0 \le r < R, \tag{3.28}$$

where C_1 is an upper bound of $(V(0) + 1/n)/(U(0) + 1/n)$ and

$$C_2 = \int_0^\infty t^{1-N} \int_0^t s^{N-1} q(s)\, ds\, dt \le \frac{1}{N-2} \int_0^\infty t q(t)\, dt < \infty.$$

Define $h(t) := g(C_1 t + C_2 f(t))$, $t \ge 0$. It is easy to see that h satisfies the hypotheses of Lemma 2.2.1. Hence, we may define

$$\Gamma(t) = \int_t^\infty \frac{ds}{h(s)} \quad \text{for all } t > 0.$$

Notice that U_n verifies

$$\Delta U_n = p(x) g(V_n) \quad \text{in } B_{R_n},$$

which combined with (3.28) implies

$$\Delta U_n \le p(x) h(U_n) \quad \text{in } B_{R_n}.$$

A simple computation yields

$$\begin{aligned} \Delta\Gamma(U_n) &= \Gamma'(U_n)\Delta U_n + \Gamma''(U_n)|\nabla U_n|^2 \\ &= \frac{-1}{h(U_n)}\Delta U_n + \frac{h'(U_n)}{(h(U_n))^2}|\nabla U_n|^2 \\ &\ge \frac{-1}{h(U_n)} p(x) h(U_n) \\ &= -p(x) \quad \text{in } B_{R_n}. \end{aligned}$$

Therefore,

$$\left(r^{N-1} \frac{d}{dr} \Gamma(U_n) \right)' \ge -r^{N-1} p(r) \quad \text{for any } 0 \le r < R_n.$$

Fix $0 \leq r < R$. Then $r < R_n$ for all $n \geq n_1$, provided n_1 is large enough. Integrating the previous inequality from 0 to r we deduce

$$\frac{d}{dr}\Gamma(U_n) \geq -r^{1-N}\int_0^r s^{N-1}p(s)\,ds \qquad \text{for all } 0 \leq r < R_n.$$

Integrating the last inequality over $[r, R_n - \varepsilon]$ we obtain

$$\Gamma(U_n(R_n-\varepsilon)) - \Gamma(U_n(r)) \geq -\int_r^{R_n-\varepsilon} t^{1-N}\int_0^t s^{N-1}p(s)\,ds\,dt \qquad \text{for all } n \geq n_1.$$

Recall that $U_n(R_n - \varepsilon) \to \infty$ as $\varepsilon \to 0$ and $\Gamma(t) \to 0$ as $t \to \infty$. Therefore, passing to the limit with $\varepsilon \to 0$ in the last inequality we deduce that

$$\Gamma(U_n(r)) \leq \int_r^{R_n} t^{1-N}\int_0^t s^{N-1}p(s)\,ds\,dt \qquad \text{for all } n \geq n_1.$$

By letting $n \to \infty$ in the previous relation, we obtain

$$\Gamma(U(r)) \leq \int_r^{R} t^{1-N}\int_0^t s^{N-1}p(s)\,ds\,dt \qquad \text{for all } 0 \leq r < R$$

or, equivalently,

$$U(r) \geq \Gamma^{-1}\left(\int_r^{R} t^{1-N}\int_0^t s^{N-1}p(s)\,ds\,dt\right) \qquad \text{for all } 0 \leq r < R.$$

Passing to the limit as $r \nearrow R$, and using the fact that $\lim_{t\searrow 0}\Gamma^{-1}(t) = \infty$, we deduce

$$\lim_{r\nearrow R} U(r) \geq \lim_{r\nearrow R}\Gamma^{-1}\left(\int_r^{R} t^{1-N}\int_0^t s^{N-1}p(s)\,ds\,dt\right) = \infty.$$

But (U, V) is an entire solution of (3.1), so that we conclude $R = \infty$ and $\lim_{r\to\infty} U(r) = \infty$. Using (3.6) and the fact that $V'(r) \geq 0$ we find

$$\begin{aligned} U(r) &\leq a + g(V(r))\int_0^\infty t^{1-N}\int_0^t s^{N-1}p(s)\,ds\,dt \\ &\leq a + g(V(r))\frac{1}{N-2}\int_0^\infty tp(t)\,dt, \end{aligned}$$

for all $r \geq 0$. This last estimate leads to $\lim_{r\to\infty} V(r) = \infty$. Consequently, (U, V) is an entire blow-up solution of (3.1). This concludes the proof. □

3.3 Comments and historical notes

We have provided in this chapter a characterization of the central value set $\mathcal{G}$ of radially symmetric solutions to a nonlinear elliptic system of logistic type. Our study has pointed out the role played by the Keller–Osserman condition in this characterization. First, if the nonlinear terms f and g have *almost* sublinear growth, then the central value set consists of entire $\mathbb{R}_+ \times \mathbb{R}_+$, and the existence of radially symmetric solutions blowing up at infinity depends on the growth of potentials p and q. In turn, if f and g are comparable, in the sense that $\liminf_{t\to\infty} f(t)/g(t) \in (0,\infty)$, and satisfy the Keller–Osserman condition at infinity, then the central value set is bounded. Furthermore, any element on the positive boundary of $\mathcal{G}$—that is, on $\partial\mathcal{G} \cap (\mathbb{R}_+ \times \mathbb{R}_+)$,—is a central value of an entire blow-up solution.

The results in this chapter are the work of Cîrstea and Rădulescu [52]. We also refer to Lair and Shaker [126] for the particular case of pure powers in the nonlinearities.

In Part II of this volume (Chapters 2 and 3) we have been concerned with boundary blow-up solutions of some classes of nonlinear elliptic equations and systems. Our approach was essentially based on the maximum principle, in combination with other ingredients, such as elliptic estimates, regularity arguments, and differential equations techniques. In all the results we have established in these two chapters, we studied only positive solutions, especially because of the physical meaning of the corresponding unknowns. A different approach was developed by Aftalion and Reichel [1], who argued the existence of multiple boundary blow-up solutions of the problem

$$\Delta u = f(u) \qquad \text{in } \Omega,$$

where Ω is a bounded and convex domain. Taking into account the growth of f, they distinguished two distinct situations:

(i) f is a sign-changing nonlinearity. In such a case there is both a positive and a sign-changing blow-up boundary solution.

(ii) $\inf f > 0$. In this case, Aftalion and Reichel [1] considered the bifurcation problem

$$\Delta u = \lambda f(u) \qquad \text{in } \Omega \tag{3.29}$$

and showed that there is some critical value $\lambda^\star > 0$ such that problem (3.29) has blow-up boundary solutions if and only if $0 < \lambda < \lambda^\star$.

PART III

ELLIPTIC PROBLEMS WITH SINGULAR NONLINEARITIES

4

SUBLINEAR PERTURBATIONS OF SINGULAR ELLIPTIC PROBLEMS

> In mathematics the art of proposing a question must be held of higher value than solving it.
>
> Georg Cantor (1845–1918)

From now on, we are interested in the qualitative analysis of solutions to semilinear elliptic equations or systems involving singular nonlinear terms, such as $u^{-\alpha}$ (with $\alpha > 0$) or gradient terms like $|\nabla u|^a$ (with $0 < a \leq 2$). We are concerned with existence and uniqueness properties, but also with the asymptotic behavior of solutions. A special feature will be played by the influence of one or several real parameters, which usually are referred as *bifurcation parameters.* The term *bifurcation* is one that is also used in topological dynamics and catastrophe theory. In such cases, it is usually considered a family of functions or vector fields dependent on parameters. The associated bifurcation values of parameters are those in which the system is not actually stable, in the sense that small variations of the parameters change the topological nature of the collection of orbits. Our setting in this volume is quite closely related to the study of stability phenomena, by means of the first eigenvalue of the associated linearized operator.

Singular problems arise in the study of non-Newtonian fluids, boundary layer phenomena for viscous fluids, chemical heterogeneous catalysts, as well as in the theory of heat conduction in electrically conducting materials. The associated singular stationary or evolution equations describe various physical phenomena. For instance, superdiffusivity equations of this type have been proposed by de Gennes [60] as a model for long-range Van der Waals interactions in thin films spreading on solid surfaces. Singular equations also appear in the study of cellular automata and interacting particle systems with self-organized criticality (see [39]), as well as to describe the flow over an impermeable plate (see [38]).

4.1 Introduction

This chapter is devoted to the study of elliptic problems of the following type:

$$\begin{cases} -\Delta u = \Phi(x, u, \lambda) & \text{in } \Omega, \\ u > 0 & \text{in } \Omega, \\ u = 0 & \text{on } \partial\Omega \end{cases} \tag{4.1}$$

in a smooth domain $\Omega \subseteq \mathbb{R}^N (N \geq 1)$, where $\Phi : \overline{\Omega} \times (0, \infty) \times \mathbb{R} \to \mathbb{R}$ is a Hölder continuous function. The main feature of this chapter, as well as of the following ones, is that we allow Φ to be unbounded near the origin with respect to its second variable—that is, Φ is a *singular nonlinearity*. Therefore, the singular character of our problem lies not in the prescribed behavior of the solution at the boundary, as we required in Chapters 2 and 3, but in the nonlinearities that govern problems of type (4.1).

To make these arguments more transparent, let us present a simple example. Consider the problem

$$\begin{cases} \Delta u = u^p & \text{in } \Omega, \\ u > 0 & \text{in } \Omega, \\ u = +\infty & \text{on } \partial\Omega. \end{cases}$$

If $p > 1$, by Theorem 2.1.3, the previous problem has a solution u. By the change of variable $u = 1/v$ we derive that v satisfies

$$\begin{cases} -\Delta v = v^{2-p} - \dfrac{2}{v} |\nabla v|^2 & \text{in } \Omega, \\ v > 0 & \text{in } \Omega, \\ v = 0 & \text{on } \partial\Omega. \end{cases} \tag{4.2}$$

This equation contains both singular nonlinearities, like v^{-1} or v^{2-p} (if $p > 2$), and a convection (gradient) term, denoted by $|\nabla v|^2$. The influence of the gradient term in problems like (4.2) will be emphasized in Chapter 9.

By classical solution of problem (4.1) we mean a function $u \in C^2(\Omega) \cap C(\overline{\Omega})$ that satisfies pointwise (4.1). Because of the presence of the singular term $\Phi(x, u, \lambda)$, solutions to (4.1) are not $C^2(\overline{\Omega})$ functions. However, in some particular cases, which will be discussed herein, the regularity of classical solutions to (4.1) can be improved up to the class $u \in C^{1,\gamma}(\overline{\Omega})$ and $\Delta u \in L^1(\Omega)$.

The simplest model that falls within the theory we develop in this chapter is

$$\begin{cases} -\Delta u = g(u) & \text{in } \Omega, \\ u > 0 & \text{in } \Omega, \\ u = 0 & \text{on } \partial\Omega, \end{cases} \tag{4.3}$$

where $\Omega \subset \mathbb{R}^N (N \geq 1)$ is a smooth bounded domain and $g \in C^1(0, \infty)$ is a decreasing function such that $\lim_{t \searrow 0} g(t) = \infty$.

Existence of a classical solution to (4.3) follows directly from Theorem 1.2.5. In particular, if $\Omega = (0, 1) \subset \mathbb{R}$ then there exists a unique $y \in C^2(0, 1) \cap C[0, 1]$ such that

$$\begin{cases} -y''(t) = g(y(t)) & 0 < t < 1, \\ y(t) > 0 & 0 < t < 1, \\ y(0) = y(1) = 0. \end{cases} \tag{4.4}$$

A surprising result is that boundary estimates for the solution u of (4.3) in higher dimensions are expressed in terms of y. More exactly, setting $d(x) := \operatorname{dist}(x, \partial\Omega)$, we have the following result.

Theorem 4.1.1 *Problem (4.3) has a unique solution $u \in C^2(\Omega) \cap C(\overline{\Omega})$ and there exist c, m, $M > 0$ such that*

$$my(cd(x)) \leq u(x) \leq My(cd(x)) \quad \textit{for all } x \in \Omega, \tag{4.5}$$

where y is the unique solution of (4.4). Moreover, if $g \in L^1(0,1)$, then the solution of problem (4.3) verifies

$$c_1 d(x) \leq u(x) \leq c_2 d(x) \quad \textit{for all } x \in \Omega \tag{4.6}$$

for some positive constants c_1 and c_2.

Proof Because y is concave, there exists $y'(0+) := \lim_{t \searrow 0} y(t) \in (0, \infty]$. Thus, we can find $0 < a < 1$ such that $y' > 0$ in $(0, a)$. That is, y is strictly increasing on $(0, a)$. From Corollary A.1.3 in Appendix A, there exist two positive constants, C_1 and C_2, such that

$$C_1\varphi_1(x) \leq d(x) \leq C_2\varphi_1(x) \quad \text{in } \Omega.$$

Let us take $c > 0$ such that $cC_2\varphi_1 \leq a/2$ in Ω. We first prove that there exists $M > 1$ large enough such that $\overline{u} := My(c\varphi_1)$ is a supersolution of (4.3). Indeed, we have

$$-\Delta\overline{u} = Mc^2|\nabla\varphi_1|^2 g(y(c\varphi_1)) + \lambda_1 Mc\varphi_1 y'(c\varphi_1) \quad \text{in } \Omega.$$

By virtue of the strong maximum principle, there exist $\Omega_0 \subset\subset \Omega$ and $\delta > 0$ such that

$$|\nabla\varphi_1| > \delta \quad \text{in } \Omega \setminus \Omega_0. \tag{4.7}$$

Thus, we can choose $M > 1$ such that

$$Mc|\nabla\varphi_1| > 1 \quad \text{in } \Omega \setminus \Omega_0. \tag{4.8}$$

Therefore,

$$-\Delta\overline{u} \geq g(y(c\varphi_1)) \geq g(My(c\varphi_1)) = g(\overline{u}) \quad \text{in } \Omega \setminus \Omega_0. \tag{4.9}$$

Because $\varphi_1 y'(c\varphi_1)$ is bounded away from zero in Ω_0, we can take $M > 1$ such that

$$Mc\lambda_1\varphi_1 y'(c\varphi_1) \geq g(y(c\varphi_1)) \quad \text{in } \Omega_0. \tag{4.10}$$

Hence,

$$-\Delta\overline{u} \geq g(y(c\varphi_1)) \geq g(My(c\varphi_1)) = g(\overline{u}) \quad \text{in } \Omega_0. \tag{4.11}$$

From relations (4.9) and (4.11), it follows that $\overline{u} = My(c\varphi_1)$ is a supersolution of problem (4.3), provided $M > 1$ satisfies (4.8) and (4.10). In a similar way we deduce the existence of a constant $0 < m < 1$ such that $\underline{u} := my(c\varphi_1)$ is a subsolution of (4.3). Thus, $\underline{u} \le u \le \overline{u}$ in $\overline{\Omega}$ and (4.5) follows.

We claim that if $g \in L^1(0,1)$ then $y'(0+) < \infty$ and $y'(1-) < \infty$—that is, $y \in C^1[0,1]$. To this aim, we multiply by $y'(t)$ in (4.4) and we integrate over $[\varepsilon, t]$, where $0 < \varepsilon < t < 1$. We find

$$-y'^2(t) + y'^2(\varepsilon) = 2\int_\varepsilon^t g(y(s))y'(s)ds = 2\int_{y(\varepsilon)}^{y(t)} g(\tau)d\tau,$$

for all $0 < \varepsilon < t < 1$. Because $\int_0^1 g(\tau)d\tau < \infty$, we can let $\varepsilon \to 0$ in the previous equality and we obtain $y'(0+) < \infty$. In the same manner we derive $y'(1-) < \infty$. Hence, there exists $c_1, c_2 > 0$ such that

$$c_1 t \le y(t) \le c_2 t \quad \text{for all } 0 \le t \le 1.$$

Using the previous estimate in (4.5) we easily deduce (4.6). The proof is now complete. □

Estimate (4.5) allows us to say more about the regularity of the solution to (4.3). To illustrate this matter, we consider the case $g(s) = s^{-\alpha}$, $\alpha > 1$. Thus, problem (4.3) becomes

$$\begin{cases} -\Delta u = u^{-\alpha} & \text{in } \Omega, \\ u > 0 & \text{in } \Omega, \\ u = 0 & \text{on } \partial\Omega. \end{cases} \tag{4.12}$$

Theorem 4.1.2 *For all $\alpha > 1$, problem (4.12) has a unique solution $u \in C^2(\Omega) \cap C(\overline{\Omega})$. Moreover, we have*

(i) $u \notin C^1(\overline{\Omega})$;

(ii) $u \in H_0^1(\Omega)$ if and only if $\alpha < 3$.

Proof (i) Existence follows from Theorem 1.2.5 (i). Note that if $\alpha > 1$ then

$$y(t) = \left(\frac{(2+\alpha)^2}{2(\alpha-1)}\right)^{2/(1+\alpha)} t^{2/(1+\alpha)} \quad \text{for all } 0 \le t \le 1,$$

is the unique solution of (4.4). Therefore, by (4.5) we have

$$c_1 \varphi_1^{2/(1+\alpha)} \le u \le c_2 \varphi_1^{2/(1+\alpha)} \quad \text{in } \Omega. \tag{4.13}$$

Fix $x_0 \in \partial\Omega$ and let n be the outer unit normal vector on $\partial\Omega$ at x_0.

Using (4.13) we have

$$\begin{aligned} \frac{\partial u}{\partial n}(x_0) &= \lim_{t \nearrow 0} \frac{u(x_0 + tn) - u(x_0)}{t} \\ &\le c_1 \lim_{t \nearrow 0} \frac{\varphi_1(x_0 + tn) - \varphi_1(x_0)}{t} \varphi_1^{(1-\alpha)/(1+\alpha)}(x_0 + tn) \\ &= -\infty. \end{aligned}$$

Hence, $u \notin C^1(\overline{\Omega})$.

(ii) Assume first that $1 < \alpha < 3$. Using Theorem 1.2.5, the approximated problem

$$\begin{cases} -\Delta u = (u+1/k)^{-\alpha} & \text{in } \Omega, \\ u > 0 & \text{in } \Omega, \\ u = 0 & \text{on } \partial\Omega \end{cases}$$

has a unique solution $u_k \in C^2(\overline{\Omega})$ for all $k \geq 1$. It is easy to see that

$$u_k \leq u_{k+1} \leq u \quad \text{in } \Omega. \tag{4.14}$$

Moreover, $v_k := u_k + 1/k$ verifies

$$\begin{cases} -\Delta v_k = v_k^{-\alpha} & \text{in } \Omega, \\ v_k > 0 & \text{in } \Omega, \\ v_k = 1/k & \text{on } \partial\Omega. \end{cases}$$

Hence,

$$u_k + \frac{1}{k} \geq u_{k+1} + \frac{1}{k+1} \geq u \quad \text{in } \Omega. \tag{4.15}$$

Therefore, by (4.14) and (4.15) we obtain

$$\int_\Omega |\nabla u_k|^2 dx = -\int_\Omega u_k \Delta u_k dx = \int_\Omega u_k (u_k + 1/k)^{-\alpha} dx \leq \int_\Omega u^{1-\alpha} dx.$$

By (4.13) and Corollary A.1.3 in Appendix A we have

$$\int_\Omega |\nabla u_k|^2 dx \leq c \int_\Omega \varphi_1^{-2(\alpha-1)/(\alpha+1)} dx < \infty.$$

Hence, $(u_k)_{k\geq 1}$ is bounded in $H_0^1(\Omega)$. Then, passing to a subsequence we see that $(u_k)_{k\geq 1}$ is weakly convergent in $H_0^1(\Omega)$ to some $v \in H_0^1(\Omega)$. Because $(u_k)_{k\geq 1}$ converges pointwise to u in Ω, we conclude that $u = v \in H_0^1(\Omega)$.

Assume now $\alpha \geq 3$ and that problem (4.12) has a solution $u \in H_0^1(\Omega)$. Taking into account estimate (4.13) and the fact that $2(\alpha-1)/(\alpha+1) \geq 1$, by Corollary A.1.3 (ii), it follows that

$$\int_\Omega u^{1-\alpha} dx = \infty. \tag{4.16}$$

Let $(w_k)_{k\geq 1} \subset C_0^\infty(\Omega)$ be such that $w_k \to u$ in $H_0^1(\Omega)$ as $k \to \infty$. For each $k \geq 1$ set $w_k^+ = \max\{w_k, 0\}$. Then $(w_k^+)_{k\geq 1} \subset H_0^1(\Omega)$ and $\nabla w_k^+ = \nabla w_k \chi_{\{w_k > 0\}}$. Passing to a subsequence if necessary, we may assume that $(w_k^+)_{k\geq 1}$ converges to u almost everywhere in Ω. By (4.16) and Fatou's lemma we find

$$\lim_{k\to\infty} \int_\Omega w_k^+ u^{-\alpha} dx = \infty.$$

Then,

$$\int_\Omega \nabla u \cdot \nabla w_k^+ dx = -\int_\Omega w_k^+ \Delta u dx = \int_\Omega w_k^+ u^{-\alpha} dx.$$

It follows that

$$\int_\Omega |\nabla u|^2 dx = \lim_{k\to\infty} \int_\Omega \nabla u \cdot \nabla w_k^+ dx = \infty,$$

which is a contradiction. Hence, $u \notin H_0^1(\Omega)$ and the proof of Theorem 4.1.2 is now complete. □

4.2 An ODE with mixed nonlinearities

To see more clearly the nature of the problems we are dealing with, let us consider the following ODE:

$$\begin{cases} -y''(t) = y^{-\alpha}(t) + y^p(t) & -1 < t < 1, \\ y(t) > 0 & -1 < t < 1, \\ y(-1) = y(1) = 0, \end{cases} \tag{4.17}$$

where $0 < \alpha, p < 1$. The existence of a solution to (4.17) follows by virtue of Theorem 1.2.5. To achieve the uniqueness, we effectively construct a solution y of (4.17) such that $y'' \in L^1(-1,1)$. Then, by Theorem 1.3.17, we derive that y is the unique solution of (4.17). We assume that y' is odd, which yields $y'(0) = 0$ and $y(-t) = y(t)$, for all $-1 \le t \le 1$. Multiplying by $y'(t)$ in (4.17) and then integrating over $[0,t]$, $0 < t < 1$, we obtain

$$(y')^2(t) = 2\int_{y(t)}^{y(0)} f(s)ds \quad \text{for all } 0 \le t < 1, \tag{4.18}$$

where $f(t) = t^{-\alpha} + t^p$. Because $\int_0^1 f(t)dt < \infty$, it follows that $y'(1-) = \lim_{t\searrow 1} y'(t)$ is finite. Similarly, we obtain $y'(-1+) \in \mathbb{R}$ so that $y \in C^1[-1,1]$. This yields

$$c_1(1-|t|) \le y(t) \le c_2(1-|t|) \quad \text{for all } -1 \le t \le 1,$$

where c_1 and c_2 are positive constants. Combining the last inequality with (4.17) we deduce that $y'' \in L^1(-1,1)$ and the uniqueness follows by Theorem 1.3.17. Actually, we shall see in the next section that $y \in C^2(-1,1) \cap C^{1,1-\alpha}[-1,1]$.

Let $c = y(0) > 0$ and $F(t) = \int_t^c f(s)ds$, $0 \le t \le c$. From (4.18) we have

$$y'(t) = -\sqrt{2F(y(t))} \quad \text{for all } 0 \le t \le 1.$$

It follows that

$$-\frac{y'(t)}{\sqrt{2F(y(t))}} = 1 \quad \text{for all } 0 < t < 1. \tag{4.19}$$

An integration in (4.19) yields

$$\Psi(y(t)) = t \quad \text{for all } 0 < t < 1, \tag{4.20}$$

where

$$\Psi(t) = \int_t^c \frac{1}{\sqrt{2F(s)}} ds \quad 0 \leq t \leq c.$$

Notice that $\Psi : [0, c] \to [0, \Psi(0)]$ is bijective and (4.20) holds for all $0 \leq t \leq 1$. In this way, the unique solution y of (4.17) has the form

$$y(t) = \begin{cases} \Psi^{-1}(-t) & \text{if } -1 \leq t < 0, \\ \Psi^{-1}(t) & \text{if } 0 \leq t \leq 1. \end{cases}$$

4.3 A complete description for positive potentials

Motivated by the study of (4.17) we now consider the following more general boundary value problem:

$$\begin{cases} -\Delta u = p(x)g(u) + \lambda f(x, u) + \mu k(x) & \text{in } \Omega, \\ u > 0 & \text{in } \Omega, \\ u = 0 & \text{on } \partial\Omega, \end{cases} \tag{4.21}$$

where Ω is a smooth bounded domain in $\mathbb{R}^N$ $(N \geq 1)$ and λ, $\mu \geq 0$. We assume that $p : \overline{\Omega} \to (0, \infty)$ and $f : \overline{\Omega} \times [0, \infty) \to [0, \infty)$ are Hölder continuous functions such that $f > 0$ on $\overline{\Omega} \times (0, \infty)$. We also assume that f is nondecreasing with respect to the second variable and is sublinear. That is,

$(f1)$ the mapping $(0, \infty) \ni t \longmapsto \dfrac{f(x,t)}{t}$ is nonincreasing for all $x \in \overline{\Omega}$;

$(f2)$ $\displaystyle\lim_{t \searrow 0} \frac{f(x,t)}{t} = \infty$ and $\displaystyle\lim_{t \to \infty} \frac{f(x,t)}{t} = 0$ uniformly for all $x \in \overline{\Omega}$.

We suppose that $g \in C^1(0, \infty)$ is a nonnegative and decreasing function satisfying

$(g1)$ $\lim_{t \searrow 0} g(t) = \infty$;

$(g2)$ there exists $0 < \alpha < 1$ such that $g(t) = O(t^{-\alpha})$ as $s \searrow 0$.

Also we assume that $k : \Omega \to (0, \infty)$ is a Hölder continuous function that satisfies $\inf_{x \in \overline{\Omega}} k(x) > 0$ and

$(k1)$ there exists $0 < \beta < 1$ such that

$$k(x) = O(d(x)^{-\beta}) \quad \text{as } d(x) \searrow 0.$$

Obviously problem (4.17) is a particular case of (4.21), because $f(x, t) = t^p$ and $g(t) = t^{-\alpha}$, $0 < p$, $\alpha < 1$ satisfy $(f1) - (f2)$ and $(g1) - (g2)$ respectively.

As we have already argued in the beginning of this chapter, we cannot expect to find solutions in $C^2(\overline{\Omega})$. Define

$$\mathcal{E} = \{u \in C^2(\Omega) \cap C^{1-\gamma}(\overline{\Omega}) : \Delta u \in L^1(\Omega)\},$$

where $\gamma = \max\{\alpha, \beta\}$. Here, and in the rest of this chapter, we are looking for solutions of (4.21) in the class $\mathcal{E}$.

Remark 4.3.1 *Let $u \in C^2(\Omega) \cap C(\overline{\Omega})$ be a classical solution of (4.21). Then $\Delta u \in L^1(\Omega)$ if and only if $p(x)g(u) \in L^1(\Omega)$.*

The main result of this section is stated next.

Theorem 4.3.2 *Assume that $p > 0$ in $\overline{\Omega}$ and $(f1)-(f2)$, $(g1)-(g2)$, $(k1)$ hold. Then, for all λ, $\mu \geq 0$ problem (4.21) has a unique solution $u_{\lambda\mu} \in \mathcal{E}$. Moreover, $u_{\lambda\mu}$ is increasing with respect to λ and μ.*

For a fixed $\mu \geq 0$, the dependence on λ of the unique solution $u_\lambda = u_{\lambda\mu}$ is depicted in Figure 4.1.

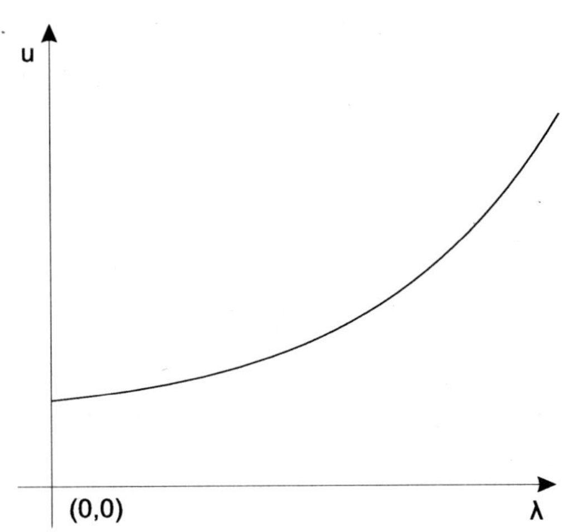

FIGURE 4.1. The dependence on $\lambda \geq 0$ in Theorem 4.3.2.

Proof For our convenience, we divide the proof into three steps.
Step 1: Existence. Fix $\lambda, \mu \geq 0$. Remark first that the result in Theorem 1.2.5 does not apply here directly, because the mapping

$$\Psi(x,t) = p(x)g(t) + \lambda f(x,t) + \mu k(x)$$

is not defined on $\partial\Omega \times (0,\infty)$. By Theorem 4.1.1, there exists a unique solution $v \in C^2(\Omega) \cap C(\overline{\Omega})$ of the problem

$$\begin{cases} -\Delta v = g(v) & \text{in } \Omega, \\ v > 0 & \text{in } \Omega, \\ v = 0 & \text{on } \partial\Omega. \end{cases}$$

Because $g \in L^1(0,1)$, from (4.6) we also have

$$m_1 d(x) \leq v(x) \leq m_2 d(x) \quad \text{in } \Omega, \tag{4.22}$$

for some $m_1, m_2 > 0$. Clearly, $\underline{u} := v$ is a subsolution of (4.21). To provide a supersolution, we need the following auxiliary result.

Lemma 4.3.3 *There exists a unique positive solution $w \in C^2(\Omega) \cap C(\overline{\Omega})$ of the problem*

$$\begin{cases} -\Delta w = 2 + k(x) & \text{in } \Omega, \\ w = 0 & \text{on } \partial\Omega. \end{cases} \tag{4.23}$$

Moreover, $w \le cd(x)$ in Ω, for some $c > 0$.

Proof Obviously $\underline{w} := 0$ is a subsolution of (4.23). Consider the problem

$$\begin{cases} -\Delta z = z^{-\beta} & \text{in } \Omega, \\ z > 0 & \text{in } \Omega, \\ z = 0 & \text{on } \partial\Omega. \end{cases}$$

Because $0 < \beta < 1$, by Theorem 4.1.1, the previous problem has a unique solution $z \in C^2(\Omega) \cap C(\overline{\Omega})$ such that

$$C_1 d(x) \le z(x) \le C_2 d(x) \quad \text{in } \Omega, \tag{4.24}$$

for some $C_1, C_2 > 0$. Hence, $-\Delta z \ge C_2^{-\beta} d(x)^{-\beta}$ in Ω. In view of the assumption $(k1)$ we can find $M > 1$ large enough such that $-\Delta(Mz) \ge 2 + k(x)$ in Ω. This means that $\overline{w} := Mz$ is a supersolution of (4.23). Hence, (4.23) has a solution $w \in C^2(\Omega) \cap C(\overline{\Omega})$ such that $\underline{w} \le w \le \overline{w}$ in $\overline{\Omega}$. By the strong maximum principle, w is the unique solution of (4.23). The fact that $w \le cd(x)$ in Ω follows from (4.24). This finishes the proof of Lemma 4.3.3. □

Now we can proceed to construct a supersolution of (4.21). Because f is sublinear, we have

$$\lim_{t\to\infty} \frac{f(x, t\|w\|_\infty)}{t} = 0 \quad \text{uniformly for } x \in \Omega.$$

Thus, there exists $M > 1$ such that

$$M > \lambda f(x, Mw) \quad \text{in } \Omega. \tag{4.25}$$

Furthermore, by taking $M > 1$ large enough, we can assume that

$$M > \max_{x\in\overline{\Omega}} f(x, Mv) \quad \text{and } M > \max\{1, \mu, \|p\|_\infty\}. \tag{4.26}$$

We claim that $\overline{u} := M(v + w)$ is a supersolution of (4.21) provided that M satisfies (4.25) and (4.26). Indeed, we have

$$\begin{aligned} -\Delta\overline{u} &= Mg(v) + 2M + Mk(x) \\ &\ge p(x)g(v) + \lambda(f(x, Mv) + f(x, Mw)) + \mu k(x) \\ &\ge p(x)g(M(v + w)) + \lambda f(x, M(v + w)) + \mu k(x) \\ &= p(x)g(\overline{u}) + \lambda f(x, \overline{u}) + \mu k(x) \quad \text{in } \Omega. \end{aligned}$$

Hence, $\overline{u}$ is a supersolution of (4.21) and $\underline{u} \le \overline{u}$ in Ω. By Theorem 1.2.3 it follows that problem (4.21) has at least one solution, $u_{\lambda\mu} \in C^2(\Omega) \cap C(\overline{\Omega})$, such that

$\underline{u} \le u_{\lambda\mu} \le \overline{u}$ in $\overline{\Omega}$. Using (4.22) and Lemma 4.3.3, there exist $c_1, c_2 > 0$ such that

$$c_1 d(x) \le u_{\lambda\mu}(x) \le c_2 d(x) \quad \text{in } \Omega. \tag{4.27}$$

Step 2: Uniqueness and dependence on λ and μ. To achieve the uniqueness, let us first remark that by (4.27) and $(g2)$ we have

$$p(x)g(u_{\lambda\mu}) \le p(x)g(c_1 d(x)) \le Cd(x)^{-\alpha} \quad \text{in } \Omega$$

for some positive constant $C > 0$. Thus $p(x)g(u_{\lambda\mu}) \in L^1(\Omega)$, and by Remark 4.3.1 we derive that $\Delta u_{\lambda\mu} \in L^1(\Omega)$. The uniqueness of the solution follows now from Theorem 1.3.17.

Let us obtain the dependence on λ and μ. Fix $\mu \ge 0$ and $\lambda_1 > \lambda_2 \ge 0$. Let $u_{\lambda_1\mu}$ and $u_{\lambda_2\mu}$ be the solutions of (4.21) obtained for the pairs (λ_1, μ) and (λ_2, μ), respectively. Then, by Theorem 1.3.17 for $\Phi_{\lambda_1\mu}(x,t) = p(x)g(t) + \lambda_1 f(x,t) + \mu k(x)$, we deduce $u_{\lambda_2\mu} \le u_{\lambda_1\mu}$ in $\overline{\Omega}$. By virtue of the strong maximum principle we find $u_{\lambda_2\mu} < u_{\lambda_1\mu}$ in Ω. In a similar way we obtain the dependence on μ.

Step 3: Regularity of solution. We have already proved that the unique solution of (4.21) verifies $u_{\lambda\mu} \in C^2(\Omega) \cap C(\overline{\Omega})$ and $\Delta u_{\lambda\mu} \in L^1(\Omega)$. It remains to show that $u_{\lambda\mu} \in C^{1,1-\gamma}(\overline{\Omega})$. To this aim, let $\mathcal{G}$ be the Green function associated with the Laplace operator in Ω. We shall use the following estimates, which is the work of Widman [195].

Lemma 4.3.4 *There exists a positive constant $c > 0$ such that for all $x, y \in \Omega$, $x \neq y$ we have*

(i) $|\mathcal{G}_x(x,y)| \le c\dfrac{\min\{|x-y|, d(y)\}}{|x-y|^N}$;

(ii) $|\mathcal{G}_{xx}(x,y)| \le c\dfrac{\min\{|x-y|, d(y)\}}{|x-y|^{N+1}}$.

Then

$$u_{\lambda\mu}(x) = -\int_\Omega \mathcal{G}(x,y)\Phi(y)dy \quad \text{for all } x \in \Omega,$$

where $\Phi(y) = a(y)g(u_{\lambda\mu}(y)) + \lambda f(y, u_{\lambda\mu}(y)) + \mu k(y)$. Hence

$$\nabla u_{\lambda\mu}(x) = -\int_\Omega \mathcal{G}_x(x,y)\Phi(y)dy \quad \text{for all } x \in \Omega. \tag{4.28}$$

A very useful tool in our approach is the following technical result.

Lemma 4.3.5 *There exists $c_0 = c_0(\Omega) > 0$ and $\delta_0 = \delta_0(\Omega) > 0$ such that for all $x_1, x_2 \in \Omega$, $0 < |x_1 - x_2| < \delta_0$ there exists a C^1 path $\xi : [0,1] \to \Omega$ with the properties*

(i) $\xi(0) = x_1$ *and* $\xi(1) = x_2$;

(ii) $|\xi'(t)| \le c_0|x_1 - x_2|$, *for all* $0 \le t \le 1$.

Fix $x_1, x_2 \in \Omega$, $0 < |x_1 - x_2| < \delta_0$ and let $\xi : [0,1] \to \Omega$ be the corresponding path in Lemma 4.3.5.

By (4.28) we have

$$
\begin{aligned}
|\nabla u_{\lambda\mu}(x_1) - \nabla u_{\lambda\mu}(x_2)| &\le \int_\Omega |\mathcal{G}_x(x_1,y) - \mathcal{G}_x(x_2,y)|\Phi(y)dy \\
&\le \underbrace{\int_{B_r(x_1)} |\mathcal{G}_x(x_1,y) - \mathcal{G}_x(x_2,y)|\Phi(y)dy}_{I} \\
&\quad + \underbrace{\int_{\Omega\setminus B_r(x_1)} |\mathcal{G}_x(x_1,y) - \mathcal{G}_x(x_2,y)|\Phi(y)dy}_{II} ,
\end{aligned}
\tag{4.29}
$$

where $r = (c_0 + 1)|x_1 - x_2|$ and c_0 is the constant appearing in Lemma 4.3.5.

Before evaluating I and II, let us remark that by $(g2)$, $(k1)$, and (4.27) there exists $c_1 > 0$ such that $\Phi(y) \le c_1 d(y)^{-\gamma}$ for all $y \in \Omega$. Then

$$
\begin{aligned}
I &\le c_1 \int_{B_r(x_1)} |\mathcal{G}_x(x_1,y) - \mathcal{G}_x(x_2,y)| d^{-\gamma}(y)dy \\
&\le c_1 \int_{B_r(x_1)} |\mathcal{G}_x(x_1,y)| d^{-\gamma}(y)dy + c_1 \int_{B_R(x_2)} |\mathcal{G}_x(x_2,y)| d^{-\gamma}(y)dy,
\end{aligned}
$$

where $R = r + |x_1 - x_2|$. Let $y \in B_r(x_1)$.
If $d(y) \ge |x_1 - y|$ then, by Lemma 4.3.4 (i), we have

$$
|\mathcal{G}_x(x_1,y)| d^{-\gamma}(y) \le c|x_1 - y|^{-N+1} d^{-\gamma}(y) \le c|x_1 - y|^{-N+1-\gamma}.
$$

If $d(y) < |x_1 - y|$ then, by Lemma 4.3.4 (i), we obtain

$$
|\mathcal{G}_x(x_1,y)| d^{-\gamma}(y) \le c|x_1 - y|^{-N} d^{1-\gamma}(y) \le c|x_1 - y|^{-N+1-\gamma}.
$$

Therefore, for all $y \in B_r(x_1)$ we have

$$
|\mathcal{G}_x(x_1,y)| d^{-\gamma}(y) \le c|x_1 - y|^{-N+1-\gamma},
$$

and similarly

$$
|\mathcal{G}_x(x_2,y)| d^{-\gamma}(y) \le c|x_2 - y|^{-N+1-\gamma} \qquad \text{for all } y \in B_R(x_2).
$$

Hence,

$$
\begin{aligned}
I &\le c_2 \int_{B_r(x_1)} |x_1 - y|^{-N+1-\gamma} dy + c_2 \int_{B_R(x_2)} |x_2 - y|^{-N+1-\gamma} dy \\
&\le c_3 \int_0^R t^{-\gamma} dt \le c_4 |x_1 - x_2|^{1-\gamma}.
\end{aligned}
\tag{4.30}
$$

To evaluate II, we first apply the mean value theorem. We have

$$\begin{aligned}
II &\leq c_1 \int_{\Omega\setminus B_r(x_1)} |\mathcal{G}_x(\xi(0),y) - \mathcal{G}_x(\xi(1),y)| d^{-\gamma}(y)dy \\
&\leq c_1 \int_{\Omega\setminus B_r(x_1)} \int_0^1 |\mathcal{G}_{xx}(\xi(t),y)||\xi'(t)| d^{-\gamma}(y)dtdy \\
&\leq c_5|x_1 - x_2| \int_{\Omega\setminus B_r(x_1)} \int_0^1 |\mathcal{G}_{xx}(\xi(t),y)| d^{-\gamma}(y)dtdy.
\end{aligned}$$

As shown earlier, by Lemma 4.3.4 (ii) we obtain

$$|\mathcal{G}_x(\xi(t),y)|d^{-\gamma}(y) \leq c|\xi(t) - y|^{-N-\gamma} \quad \text{for all } y \in B_r(x_1) \text{ and } 0 \leq t \leq 1.$$

Let $c_6 = 1/(1+c_0)$, where c_0 is the constant from Lemma 4.3.5 that depends only on Ω and not on x_1, x_2. Then for all $y \in \Omega \setminus B_r(x_1)$,

$$\begin{aligned}
|\xi(t) - y| &\geq |x_1 - y| - |\xi(t) - x_1| = |x_1 - y| - |\xi(t) - \xi(0)| \\
&\geq |x_1 - y| - t|\xi'(c_t)| \geq |x_1 - y| - c_0|x_1 - x_2| \\
&\geq c_6|x_1 - y|.
\end{aligned}$$

Combining the last two estimates we obtain

$$|\mathcal{G}_x(\xi(t),y)|d^{-\gamma}(y) \leq c_7|x_1 - y|^{-N-\gamma} \quad \text{for all } y \in B_r(x_1) \text{ and } 0 \leq t \leq 1.$$

Hence, we may write

$$\begin{aligned}
II &\leq c_7|x_1 - x_2| \int_{\Omega\setminus B_r(x_1)} |x_1 - y|^{-N-\gamma} dy \\
&\leq c_7|x_1 - x_2| \int_r^{\infty} t^{-1-\gamma} dt \leq c_8|x_1 - x_2| r^{-\gamma} \qquad (4.31) \\
&\leq c_9|x_1 - x_2|^{1-\gamma}.
\end{aligned}$$

The conclusion follows now from (4.29), (4.30), and (4.31).

This finishes the proof of Theorem 4.3.2. □

4.4 An example

Consider the problem

$$\begin{cases}
-\Delta u = u^{-\alpha} + f(x,u) & \text{in } \Omega, \\
u > 0 & \text{in } \Omega, \\
u = 0 & \text{on } \partial\Omega,
\end{cases} \qquad (4.32)$$

where $0 < \alpha < 1$ and f is a Hölder continuous function of class $C^{0,\gamma}$ $(0 < \gamma < 1)$, which satisfies $(f1)$ and $(f2)$ stated in the beginning of Section 4.3.

Proposition 4.4.1 *For all $0 < \alpha < 1$ problem (4.32) has a unique solution $u_\alpha \in C^2(\Omega) \cap C^{1,1-\alpha}(\overline{\Omega})$ with $\Delta u_\alpha \in L^1(\Omega)$. Furthermore we have*

$$\|u_\alpha - v\|_{C^1(\overline{\Omega})} \to 0 \quad \text{as } \alpha \searrow 0, \tag{4.33}$$

where v is the unique solution of the problem

$$\begin{cases} -\Delta v = 1 + f(x, v) & \text{in } \Omega, \\ v > 0 & \text{in } \Omega, \\ v = 0 & \text{on } \partial\Omega. \end{cases} \tag{4.34}$$

Proof The existence and uniqueness of solution $u_\alpha \in C^2(\Omega) \cap C^{1,1-\alpha}(\overline{\Omega})$ with $\Delta u_\alpha \in L^1(\Omega)$ follows directly from Theorem 4.3.2.

To prove (4.33) we first remark that for all $0 < \alpha < 1$ we have $u_\alpha \geq \xi$ in Ω, where $\xi \in C^2(\overline{\Omega})$ is the unique solution of

$$\begin{cases} -\Delta \xi = f(x, \xi) & \text{in } \Omega, \\ \xi > 0 & \text{in } \Omega, \\ \xi = 0 & \text{on } \partial\Omega. \end{cases}$$

Using the regularity of ξ we deduce that for all $0 < \alpha < 1$,

$$c_1 d(x) \leq \xi(x) \leq u_\alpha(x) \quad \text{in } \Omega, \tag{4.35}$$

where c_1 is a positive constant that does not depend on α.

Let us fix $q > p > N/(1-\gamma)$. We prove in the following that there exists $c_2 > 0$ not depending on α such that for all $0 < \alpha < 1/q$ there holds

$$u_\alpha(x) \leq c_2 d(x) \quad \text{in } \Omega. \tag{4.36}$$

For this purpose consider the problem

$$\begin{cases} -\Delta w = w^{-1/q} + 1 + f(x, w) & \text{in } \Omega, \\ w > 0 & \text{in } \Omega, \\ w = 0 & \text{on } \partial\Omega. \end{cases} \tag{4.37}$$

By Theorem 4.3.2, there exists a unique solution $w \in C^2(\Omega) \cap C^1(\overline{\Omega})$ of (4.37). Furthermore, there exists $c_2 > 0$ such that $w(c) \leq c_2 d(x)$ in Ω.

Let

$$F(x, t) = t^{-\alpha} + f(x, t), \quad (x, t) \in \overline{\Omega} \times (0, \infty).$$

Then F satisfies the hypotheses of Theorem 1.3.17. Note also that

$$t^{-\alpha} \leq t^{-1/q} + 1 \qquad \text{for all } t > 0 \text{ and } 0 < \alpha < 1/q.$$

Hence,

$$\Delta w + F(x, w) \leq 0 \leq \Delta u_\alpha + F(x, u_\alpha) \quad \text{in } \Omega.$$

Because $w, u_\alpha > 0$ in Ω, $\Delta u_\alpha \in L^1(\Omega)$ and $w = u_\alpha = 0$ on $\partial\Omega$, by Theorem 1.3.17 we obtain $u_\alpha \leq w$ in Ω for all $0 < \alpha < 1/q$, and (4.36) follows.

From (4.35) and (4.36) we deduce that the sequence $(F(x, u_\alpha(x)))_{0<\alpha<1/q}$ is bounded in $L^p(\Omega)$. By standard elliptic estimates it follows that $(u_\alpha)_{0<\alpha<1/q}$ is bounded in $W^{2,p}(\Omega)$. Because the embedding $W^{2,p}(\Omega) \hookrightarrow C^1(\overline{\Omega})$ is compact, we deduce that there exists $v \in C^1(\overline{\Omega})$ such that, up to a subsequence, we have

$$\|u_\alpha - v\|_{C^1(\overline{\Omega})} \to 0 \quad \text{as } \alpha \searrow 0.$$

Furthermore, by Schauder estimates, $(u_\alpha)_{0<\alpha<1/q}$ is bounded in $C^{2,\gamma}(\overline{\omega})$ for all $\omega \subset\subset \Omega$. Because $C^{2,\gamma}(\overline{\omega})$ is compactly embedded in $C^2(\overline{\omega})$, we deduce that, up to a subsequence, we have $u_\alpha \to v$ in $C^2(\overline{\omega})$ as $\alpha \searrow 0$. Because $\omega \subset\subset \Omega$ was arbitrary, passing to the limit with $\alpha \searrow 0$ in (4.32), it follows that $-\Delta v = 1 + f(x, v)$ in Ω. Moreover, from (4.35) and (4.36) we derive

$$c_1 d(x) \le v(x) \le c_2 d(x) \quad \text{in } \Omega.$$

Thus, we can extend $v = 0$ on $\partial\Omega$, which yields $v \in C(\partial\Omega)$. Hence, $v \in C^2(\Omega) \cap C(\overline{\Omega})$ is a classical solution of (4.34). In conclusion, any subsequence of $(u_\alpha)_{0<\alpha<1}$ has a subsequence that converges in the $C^1(\overline{\Omega})$ norm to the unique solution v of (4.34). Hence $u_\alpha \to v$ in $C^1(\overline{\Omega})$ as $\alpha \searrow 0$. This finishes the proof of Proposition 4.4.1. □

For $N = 1$, $\Omega = (-1, 1)$, and $f(x, t) = \sqrt{t}$, convergence (4.33) in Proposition 4.4.1 is depicted in Figure 4.2. The continuous line represents the solution v of problem (4.34) (which corresponds to the case $\alpha = 0$) whereas the dashed lines represent the solution u_α of problem (4.32) plotted for $\alpha = 0.1$ and $\alpha = 0.3$.

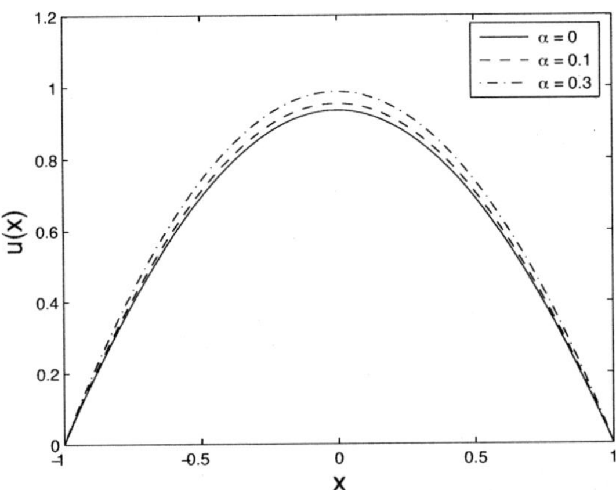

FIGURE 4.2. Convergence (4.33) in a one-dimensional case.

4.5 Bifurcation for negative potentials

This section is devoted to the case $p < 0$ in (4.21). For convenience, let us pass the singular nonlinearity $p(x)g(u)$ from the right-hand side to the left-hand side. Therefore, we shall be concerned with the problem

$$\begin{cases} -\Delta u + p(x)g(u) = \lambda f(x,u) + \mu k(x) & \text{in } \Omega, \\ u > 0 & \text{in } \Omega, \\ u = 0 & \text{on } \partial\Omega, \end{cases} \tag{4.38}$$

where f, g, and k satisfy $(f1)$–$(f2)$, $(g1)$–$(g2)$, and $(k1)$, respectively, and $p : \overline{\Omega} \to (0,\infty)$ is a Hölder continuous potential.

Theorem 4.5.1 *Assume that $p > 0$ in $\overline{\Omega}$ and conditions $(f1)$–$(f2)$, $(g1)$–$(g2)$, and $(k1)$ are fulfilled. Then, there exists $\lambda_* > 0$ and $\mu_* > 0$ such that*

- *problem (4.38) has at least one solution in $\mathcal{E}$ if $\lambda > \lambda_*$ or $\mu > \mu_*$;*
- *problem (4.38) has no solution in $\mathcal{E}$ if $0 \le \lambda < \lambda_*$ and $0 \le \mu < \mu_*$.*

Moreover, if $\lambda > \lambda_*$ or $\mu > \mu_*$, then (4.38) has a maximal solution in $\mathcal{E}$, which is increasing with respect to λ and μ.

The diagram of dependence on λ and μ in Theorem 4.5.1 is depicted in Figure 4.3.

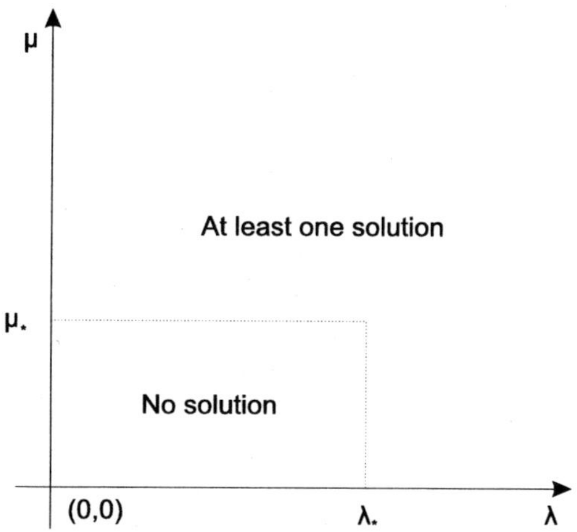

FIGURE 4.3. The dependence on λ and μ in Theorem 4.5.1.

Proof We split the proof into several steps.
Step 1: Existence of a solution for large λ. Let $\mu \ge 0$ be fixed. Repeating the arguments in the proof of Theorem 4.3.2 we find that for all $\lambda > 0$ there exists a unique solution $v_{\lambda\mu} \in \mathcal{E}$ of the problem

$$\begin{cases} -\Delta v = \lambda f(x,v) + \mu k(x) & \text{in } \Omega, \\ v > 0 & \text{in } \Omega, \\ v = 0 & \text{on } \partial\Omega. \end{cases} \tag{4.39}$$

Obviously, $\overline{u}_{\lambda\mu} := v_{\lambda\mu}$ is a supersolution of (4.38). The main point is to find a subsolution $\underline{u}_{\lambda\mu}$ of (4.38) such that $\underline{u}_{\lambda\mu} \leq v_{\lambda\mu}$ in Ω. For this purpose, let

$$\Phi : [0,\infty) \to [0,\infty), \quad \Phi(t) = \int_0^t \frac{1}{\sqrt{2\int_0^s g(\tau)d\tau}} ds.$$

Remark first that Φ is well defined, because $g \in L^1(0,1)$. Indeed, there exists $m > 0$ such that $g(s) \geq m$, for all $0 < s < 1$. This yields

$$\left(\int_0^s g(\tau)d\tau\right)^{-1/2} \leq (\sqrt{ms})^{-1} \quad \text{for all } 0 < s < 1,$$

which implies

$$\int_0^1 \left(\int_0^t g(s)ds\right)^{-1/2} dt < \infty. \tag{4.40}$$

We claim that Φ is bijective. Indeed, Φ is increasing and, if $M := g(1)$, then

$$\int_0^s g(\tau)d\tau \leq \int_0^1 g(\tau)d\tau + M(s-1) \quad \text{for all } s \geq 1.$$

Thus, there exists $c > 0$ such that

$$\int_0^s g(\tau)d\tau \leq Ms + c \quad \text{for all } s \geq 1.$$

It follows that

$$\Phi(t) \geq \int_1^t \frac{1}{\sqrt{2(Ms+c)}} ds \geq \frac{1}{M}(\sqrt{2(Mt+c)} - c_1) \quad \text{for all } t \geq 1.$$

This yields $\lim_{t\to\infty} \Phi(t) = \infty$, and the claim follows.

Let $h : [0,\infty) \to [0,\infty)$ be the inverse of Φ. Then h satisfies

$$\begin{cases} h > 0 & \text{in } (0,\infty), \\ h'(t) = \sqrt{2\int_0^{h(t)} g(s)ds} & \text{in } (0,\infty), \\ h''(t) = g(h(t)) & \text{in } (0,\infty), \\ h(0) = h'(0) = 0. \end{cases} \tag{4.41}$$

Hence, $h \in C^2(0,\infty) \cap C^1[0,\infty)$.

The key result for this part of the proof is the following lemma.

Lemma 4.5.2 *There exists a positive constant $M > 0$ (which depends on λ) such that $\underline{u}_{\lambda\mu} := Mh(\varphi_1)$ is a subsolution of (4.38) provided $\lambda > 0$ is large enough.*

Proof In view of the strong maximum principle there exist $\delta > 0$ and $\Omega_0 \subset\subset \Omega$ such that

$$|\nabla\varphi_1| > \delta \quad \text{in } \Omega \setminus \Omega_0. \tag{4.42}$$

We choose $M > 1$ such that

$$M|\nabla\varphi_1|^2 > 1 + \|p\|_\infty \quad \text{in } \Omega \setminus \Omega_0. \tag{4.43}$$

Moreover, because $-g(h(\varphi_1))+M\lambda_1\varphi_1 h'(\varphi_1) \to -\infty$ as $d(x) \searrow 0$, we can assume that

$$-g(h(\varphi_1)) + M\lambda_1\varphi_1 h'(\varphi_1) < 0 \quad \text{in } \Omega \setminus \Omega_0. \tag{4.44}$$

Now we are in position to show that $\underline{u}_{\lambda\mu} := Mh(\varphi_1)$ is a subsolution of (4.38) for large $\lambda > 0$. First we have

$$-\Delta\underline{u}_{\lambda\mu} = -Mg(h(\varphi_1))|\nabla\varphi_1|^2 + M\lambda_1\varphi_1 h'(\varphi_1) \quad \text{in } \Omega.$$

Because g is decreasing and $M > 1$, it follows that

$$\begin{aligned}-\Delta\underline{u}_{\lambda\mu} + p(x)g(\underline{u}_{\lambda\mu}) \leq & (\|p\|_\infty - M|\nabla\varphi_1|^2)g(h(\varphi_1)) \\ & + M\lambda_1\varphi_1 h'(\varphi_1) \quad \text{in } \Omega.\end{aligned} \tag{4.45}$$

From relations (4.43) through (4.45) we have

$$-\Delta\underline{u}_{\lambda\mu} + p(x)g(\underline{u}_{\lambda\mu}) \leq -g(h(\varphi_1)) + M\lambda_1\varphi_1 h'(\varphi_1) < 0 \quad \text{in } \Omega \setminus \Omega_0.$$

Hence,

$$-\Delta\underline{u}_{\lambda\mu} + p(x)g(\underline{u}_{\lambda\mu}) \leq \lambda f(x, \underline{u}_{\lambda\mu}) + \mu k(x) \quad \text{in } \Omega \setminus \Omega_0. \tag{4.46}$$

Furthermore, from (4.45) we have

$$-\Delta\underline{u}_{\lambda\mu} + p(x)g(\underline{u}_{\lambda\mu}) \leq \|p\|_\infty g(h(\varphi_1))|\nabla\varphi_1|^2 + M\lambda_1\varphi_1 h'(\varphi_1) \quad \text{in } \Omega. \tag{4.47}$$

Because $\varphi_1 > 0$ in $\overline{\Omega}_0$, we can choose $\lambda > 0$ such that

$$\lambda \min_{x\in\overline{\Omega}_0} f(x, h(\varphi_1)) \geq \max_{x\in\overline{\Omega}_0}\{\|p\|_\infty g(h(\varphi_1)) + M\lambda_1\varphi_1 h'(\varphi_1)\}. \tag{4.48}$$

Combining (4.47) and (4.48) we derive

$$-\Delta\underline{u}_{\lambda\mu} + p(x)g(\underline{u}_{\lambda\mu}) \leq \lambda f(x, \underline{u}_{\lambda\mu}) + \mu k(x) \quad \text{in } \Omega_0. \tag{4.49}$$

From (4.46) and (4.49) we note that $u_{\lambda\mu}$ is a subsolution of (4.38), and the proof of our lemma is now complete. □

To end the proof in this step, it remains to show that $\underline{u}_{\lambda\mu} \leq \overline{u}_{\lambda\mu}$ in Ω. This follows immediately by using Theorem 1.3.17 for the mapping $\Phi_{\lambda\mu}(x,t) = \lambda f(x,t) + \mu k(x)$, $(x,t) \in \Omega \times (0,\infty)$. Thus, problem (4.38) has at least one solution, $u_{\lambda\mu}$, such that $Mh(\varphi_1) \leq u_{\lambda\mu} \leq v_{\lambda\mu}$ in Ω. The regularity of $u_{\lambda\mu}$ and the fact that $u_{\lambda\mu} \in \mathcal{E}$ follows in the same manner as in the proof of Theorem 4.3.2. Note that from the previous inequality we derive the same estimates as in (4.27), which are necessary to achieve the regularity $u_{\lambda\mu} \in C^{1,1-\gamma}(\overline{\Omega})$.

Step 2: Existence of a solution for large μ. The proof is the same as that shown earlier. We have only to replace (4.48) with

$$\mu \min_{x\in\overline{\Omega}_0} k(x) \geq \max_{x\in\overline{\Omega}_0}\{\|p\|_\infty g(h(\varphi_1)) + M\lambda_1\varphi_1 h'(\varphi_1)\}.$$

Step 3: Nonexistence for small λ and μ. Let $m = \inf_{t>0}\{t\lambda_1 + p_* g(t)\}$, where $p_* = \min_{x\in\overline{\Omega}} p(x) > 0$. Because g is positive and decreasing on $(0,\infty)$, we have $m > 0$. Denote by $\zeta \in \mathcal{E}$ the unique solution of (4.39) for $\lambda = \mu = 1$. Because $\inf_{x\in\overline{\Omega}} k(x) > 0$, there exists $c > 0$ such that $f(x,\zeta) \leq ck(x)$ in Ω.

We claim that (4.38) has no solutions for λ and μ in the following range:

$$0 < \lambda, \mu < 1 \quad \text{and} \quad (\lambda c + \mu)\|k\|_1\|\varphi_1\|_\infty < m\|\varphi_1\|_1. \tag{4.50}$$

Suppose by contradiction that problem (4.38) has a solution $u_{\lambda\mu} \in \mathcal{E}$ for λ, μ that satisfies (4.50). In this case, ζ is a supersolution of (4.38) and, by Theorem 1.3.17, we deduce that $u_{\lambda\mu} \leq \zeta$ in Ω. We multiply by φ_1 in (4.38) and then we integrate over Ω. Therefore,

$$\int_\Omega (\lambda_1 u + p_* g(u))\varphi_1 dx \leq \int_\Omega (\lambda f(x,\zeta) + k(x))\varphi_1 dx.$$

This yields

$$m\|\varphi_1\|_1 \leq (\lambda c + \mu)\int_\Omega k(x)\varphi_1 dx \leq (\lambda c + \mu)\|\varphi_1\|_\infty\|k\|_1,$$

which contradicts (4.50). Hence, problem (4.38) has no solutions, provided λ and μ fulfill (4.50).

Step 4: Existence of a maximal solution for problem (4.38). We show that if problem (4.38) has a solution $u_{\lambda\mu} \in \mathcal{E}$, then it has a maximal solution that belongs to $\mathcal{E}$.

Let $\lambda, \mu > 0$ be such that (4.38) has a solution $u_{\lambda\mu} \in \mathcal{E}$. If $v_{\lambda\mu}$ is the corresponding solution of (4.39), in view of Theorem 1.3.17 we deduce $u_{\lambda\mu} \leq v_{\lambda\mu}$ in Ω. For $n \geq 1$ large enough, define

$$\Omega_n := \left\{x \in \Omega : d(x) > \frac{1}{n}\right\}.$$

Let $u_0 = v_{\lambda\mu}$ and u_n be the solution of

$$\begin{cases} -\Delta w + p(x)g(u_{n-1}) = \lambda f(x, u_{n-1}) + \mu h(x) & \text{in } \Omega_n, \\ w = u_{n-1} & \text{in } \Omega \setminus \Omega_n. \end{cases}$$

Using the fact that the mapping $\Phi_{\lambda\mu}(x,t) = \lambda f(x,t) + \mu k(x)$, $(x,t) \in \Omega \times (0,\infty)$ is nondecreasing with respect to the second variable, we obtain

$$u_{\lambda\mu} \le u_n \le u_{n-1} \le u_0 \quad \text{in } \Omega.$$

Define

$$U_{\lambda\mu}(x) := \lim_{n\to\infty} u_n(x) \quad \text{for all } x \in \overline{\Omega}.$$

By standard elliptic arguments it follows that $U_{\lambda\mu}$ is a solution of (4.38) and, obviously, $U_{\lambda\mu}$ is the maximal solution of (4.38). Moreover, with the same reasoning as in the proof of Theorem 4.3.2, we have $U_{\lambda\mu} \in \mathcal{E}$.

Step 5: Dependence on λ and μ. We first prove the dependence on λ of the maximal solution $U_{\lambda\mu} \in \mathcal{E}$ of (4.38). For this purpose, define

$$A := \{\lambda > 0 : \text{ for all } \mu \ge 0, \text{ problem (4.38) has at least one solution in } \mathcal{E}\}.$$

Let $\lambda_* := \inf A$. From the previous steps we have $A \neq \emptyset$ and $\lambda_* > 0$. Let $\lambda_1 \in A$, $\mu \ge 0$, and $U_{\lambda_1\mu}$ be the maximal solution of (4.38) for $\lambda = \lambda_1$. We prove that $(\lambda_1, \infty) \subset A$. Indeed, if $\lambda_2 > \lambda_1$, then $U_{\lambda_1\mu}$ is a subsolution of (4.38) with $\lambda = \lambda_2$. On the other hand, if $v_{\lambda_2\mu}$ is the solution of (4.39) with $\lambda = \lambda_2$, then

$$\Delta v_{\lambda_2\mu} + \Phi_{\lambda_2\mu}(x, v_{\lambda_2\mu}) \le 0 \le \Delta U_{\lambda_1\mu} + \Phi_{\lambda_2\mu}(x, U_{\lambda_1\mu}) \quad \text{in } \Omega,$$

$$v_{\lambda_2\mu}, U_{\lambda_1\mu} > 0 \quad \text{in } \Omega,$$

$$v_{\lambda_2\mu} \ge U_{\lambda_1\mu} \quad \text{on } \partial\Omega,$$

$$\Delta U_{\lambda_1\mu} \in L^1(\Omega).$$

By Theorem 1.3.17 we deduce $U_{\lambda_1\mu} \le v_{\lambda_2\mu}$ in $\overline{\Omega}$. Therefore, by the sub- and supersolution method, problem (4.38) has a solution $u_{\lambda_2\mu} \in \mathcal{E}$ such that

$$U_{\lambda_1\mu} \le u_{\lambda_2\mu} \le v_{\lambda_2\mu} \quad \text{in } \overline{\Omega}.$$

Hence, $\lambda_2 \in A$ and so $(\lambda_*, \infty) \subset A$. If $U_{\lambda_2\mu} \in \mathcal{E}$ is the maximal solution of (4.38) with $\lambda = \lambda_2$, the previous relation implies $U_{\lambda_1\mu} \le U_{\lambda_2\mu}$ in $\overline{\Omega}$ and, by virtue of the strong maximum principle, it follows that $U_{\lambda_1\mu} < U_{\lambda_2\mu}$ in Ω. Thus, $U_{\lambda\mu}$ is increasing with respect to λ.

To achieve dependence on μ we define

$$B := \{\mu > 0 : \text{ for all } \lambda \ge 0, \text{ problem (4.38) has at least one solution in } \mathcal{E}\}.$$

Let $\mu_* := \inf B$. The conclusion follows in the same manner as just demonstrated.

The proof of Theorem 4.5.1 is now complete. □

4.6 Existence for large values of parameters in the sign-changing case

In this section we study problem (4.21) assuming that the potential p changes sign in Ω. We shall obtain the existence of solutions in the class $\mathcal{E}$ for large values of parameters λ and μ. Let

$$p^* := \max_{x\in\overline{\Omega}} p(x) \quad \text{and} \quad p_* := \min_{x\in\overline{\Omega}} p(x).$$

Theorem 4.6.1 *Assume that $p^* > 0 > p_*$. Then there exists $\lambda_*, \mu_* > 0$ such that (4.21) has at least one solution $u_{\lambda\mu} \in \mathcal{E}$ if $\lambda > \lambda_*$ or $\mu > \mu_*$. Moreover, for $\lambda > \lambda_*$ or $\mu > \mu_*$, $u_{\lambda\mu}$ is increasing with respect to λ and μ.*

Proof Because $p_* < 0$, Theorem 4.5.1 implies that there exist $\lambda_*, \mu_* > 0$ such that the problem

$$\begin{cases} -\Delta u = p_* g(u) + \lambda f(x,u) + \mu k(x) & \text{in } \Omega, \\ u > 0 & \text{in } \Omega, \\ u = 0 & \text{on } \partial\Omega \end{cases}$$

has a maximal solution $U_{\lambda\mu} \in \mathcal{E}$, provided $\lambda > \lambda_*$ or $\mu > \mu_*$. Then $U_{\lambda\mu}$ is a subsolution of (4.21). Moreover, $U_{\lambda\mu}$ is increasing with respect to both λ and μ.

Consider the problem

$$\begin{cases} -\Delta v = p^* g(v) + \lambda f(x,v) + \mu k(x) & \text{in } \Omega, \\ v > 0 & \text{in } \Omega, \\ v = 0 & \text{on } \partial\Omega. \end{cases} \tag{4.51}$$

Because $p^* > 0$, by Theorem 4.3.2, problem (4.51) has a unique solution $v_{\lambda\mu} \in \mathcal{E}$. Obviously, $v_{\lambda\mu}$ is a supersolution of (4.21). Note also that $\Phi_{\lambda\mu}(x,t) = p^* g(t) + \lambda f(x,t) + \mu k(x)$, $(x,t) \in \Omega \times (0,\infty)$, fulfills the hypotheses of Theorem 1.3.17. According to this result, $U_{\lambda\mu} \le v_{\lambda\mu}$ in Ω. Hence, by Theorem 1.2.2, there exists a minimal solution $u_{\lambda\mu}$ of (4.21) with respect to the ordered pair $(U_{\lambda\mu}, v_{\lambda\mu})$. Furthermore, using Theorem 1.3.17 it is easy to see that $u_{\lambda\mu}$ is minimal in the class of solutions $u \in \mathcal{E}$ of (4.21) that fulfill $U_{\lambda\mu} \le u$ in Ω.

Now let us prove the dependence on λ. Fix $\lambda_1 > \lambda_2 > \lambda_*$, $\mu \ge 0$ and let $u_{\lambda_1\mu}, u_{\lambda_2\mu} \in \mathcal{E}$ be the solutions of (4.21) for $\lambda = \lambda_1$ and $\lambda = \lambda_2$, respectively, that we obtained earlier. Note that $u_{\lambda_1\mu}$ is a supersolution of (4.21) with $\lambda = \lambda_2$. Moreover, because $U_{\lambda\mu}$ is increasing with respect to $\lambda > \lambda_*$, we obtain

$$u_{\lambda_1\mu} \ge U_{\lambda_1\mu} \ge U_{\lambda_2\mu} \quad \text{in } \Omega.$$

Using the minimality of $u_{\lambda_2\mu}$, from the previous relation we deduce $u_{\lambda_1\mu} \ge u_{\lambda_2\mu}$ in Ω and, by the strong maximum principle, we deduce that $u_{\lambda_1\mu} > u_{\lambda_2\mu}$ in Ω. The dependence on μ follows in the same way. This completes the proof of Theorem 4.6.1. □

4.7 Singular elliptic problems in the whole space

In this section we extend the study of singular elliptic problems to the case where the domain Ω is the whole space $\mathbb{R}^N$. We shall be concerned with problems of the type

$$\begin{cases} -\Delta u = \Phi(x,u) & \text{in } \mathbb{R}^N, \\ u > 0 & \text{in } \mathbb{R}^N, \\ u(x) \to 0 & \text{as } |x| \to \infty, \end{cases} \tag{4.52}$$

where $\Phi : \mathbb{R}^N \times (0,\infty) \to [0,\infty)$, $N \geq 3$, is a Hölder continuous function having a singular behavior at the origin with respect to the second argument.

4.7.1 *Existence of entire solutions*

We are concerned here with the existence of classical solutions to (4.52)—that is, functions $u \in C^2(\mathbb{R}^N)$ (or even $u \in C^{2,\gamma}_{\text{loc}}(\mathbb{R}^N)$ for some $0 < \gamma < 1$) that verify (4.52) pointwise in $\mathbb{R}^N$. A solution of (4.52) is often called a *ground-state* solution. Roughly speaking, such a solution achieves a minimal level of energy as a result of the prescribed condition at infinity.

A natural way to construct solutions to (4.52) is to provide a monotone sequence of positive functions that are solutions of similar problems to (4.52), but in bounded domains. The main point is to supply a supersolution w of (4.52) that must satisfy $w(x) \to 0$ as $|x| \to \infty$. Then, standard elliptic arguments will imply that the pointwise limit of the previously mentioned sequence is, in fact, a genuine solution of (4.52). This procedure clearly fails in the absence of a comparison principle for problems like (4.52) in bounded domains.

Bearing these points in mind, we start out the study of existence by considering first the case where Φ in (4.52) is monotone with respect to its second variable. This particular situation will help us to construct in an easier way the monotone sequence with a limit that is the solution of (4.52). Let us consider the problem

$$\begin{cases} -\Delta u = p(x)g(u) & \text{in } \mathbb{R}^N, \\ u > 0 & \text{in } \mathbb{R}^N, \\ u(x) \to 0 & \text{as } |x| \to \infty, \end{cases} \tag{4.53}$$

where $g \in C^1(0,\infty)$ satisfies $g > 0$, $g' < 0$ in $(0,\infty)$ and $\lim_{t \searrow 0} g(t) = \infty$. We shall also assume that $p : \mathbb{R}^N \to (0,\infty)$ is a Hölder continuous function.

Let

$$\phi(r) := \max_{|x|=r} p(x) \quad \text{and} \quad \psi(r) := \min_{|x|=r} p(x), \quad r \geq 0.$$

Theorem 4.7.1 *The following properties are valid:*

(i) *If $\int_0^\infty r\psi(r)dr = \infty$, then problem (4.53) has no classical solutions.*

(ii) *If $\int_0^\infty r\phi(r)dr < \infty$, then problem (4.53) has a unique classical solution.*

Proof (i) Assume that (4.53) has a solution $u \in C^2(\mathbb{R}^N)$ and let U be the spherical average of u defined as

$$U(r) = \frac{1}{\omega_N r^{N-1}} \int_{|x|=r} u(x) d\sigma(x) \quad \text{for all } r \geq 0,$$

where ω_N is the surface area of the unit sphere in $\mathbb{R}^N$. Because u is a positive ground-state solution of (4.52), it follows that U is positive and $U(r) \to 0$ as $r \to \infty$. Let $m := \min_{x \in \mathbb{R}^N} g(u(x))$. Because $u(x) \to \infty$ as $|x| \to \infty$, it follows that $m > 0$ is finite. With the change of variable $x = ry$, we have

$$U(r) = \frac{1}{\omega_N} \int_{|y|=1} u(ry) d\sigma(y) \qquad \text{for all } r \geq 0$$

and

$$U'(r) = \frac{1}{\omega_N} \int_{|y|=1} \nabla u(ry) \cdot y d\sigma(y) \quad \text{for all } r \geq 0.$$

Hence,

$$U'(r) = \frac{1}{\omega_N} \int_{|y|=1} \frac{\partial u}{\partial r}(ry) d\sigma(y) = \frac{1}{\omega_N r^{N-1}} \int_{|x|=r} \frac{\partial u}{\partial r}(x) d\sigma(x).$$

That is,

$$U'(r) = \frac{1}{\omega_N r^{N-1}} \int_{|x|<r} \Delta u(x) dx \quad \text{for all } r \geq 0.$$

This yields

$$\begin{aligned} -\left(r^{N-1} U'(r)\right)' &= \frac{1}{\omega_N} \frac{d}{dr} \left(\int_{|x|<r} -\Delta u(x) dx \right) \\ &= \frac{1}{\omega_N} \int_{|x|=r} -\Delta u(x) d\sigma(x) \quad \text{for all } r \geq 0, \end{aligned}$$

which finally leads us to

$$-(r^{N-1} U'(r))' \geq m r^{N-1} \psi(r) \quad \text{for all } r \geq 0. \tag{4.54}$$

Integrating in (4.54) we find

$$U(0) - U(r) \geq m \int_0^r t^{1-N} \int_0^t s^{N-1} \psi(s) dt \quad \text{for all } r \geq 0. \tag{4.55}$$

Because $\int_0^\infty r\psi(r) dr = \infty$, according to Lemma 2.2.4 we deduce that the right-hand side in (4.55) tends to infinity as $r \to \infty$. Thus, passing to the limit with $r \to \infty$ in (4.55) we obtain a contradiction. Hence, problem (4.53) has no solution in this case.

(ii) We first apply Theorem 1.2.5 for $B_n := \{x \in \mathbb{R}^N : |x| < n\}$. Thus, for all $n \geq 1$ there exists $u_n \in C^{2,\gamma}(B_n) \cap C(\overline{B_n})$ $(0 < \gamma < 1)$ such that

$$\begin{cases} -\Delta u_n = p(x)g(u_n) & \text{in } B_n, \\ u_n > 0 & \text{in } B_n, \\ u_n = 0 & \text{on } \partial B_n. \end{cases}$$

We extend u_n by zero outside of B_n. Because g is decreasing, we deduce

$$u_1 \leq u_2 \leq \cdots \leq u_n \leq u_{n+1} \leq \cdots \quad \text{in } \mathbb{R}^N.$$

We next focus on finding a supersolution w of (4.53). To this aim, define

$$\zeta(r) := r^{1-N} \int_0^r t^{N-1}\phi(t)dt \quad \text{for all } r \geq 0. \tag{4.56}$$

By Lemma 2.2.4, and that fact that $\int_0^\infty r\phi(r)dr < \infty$, we find $\int_0^\infty r\zeta(r)dr < \infty$. Consider now

$$\xi(x) = \int_{|x|}^\infty \zeta(t)dt \quad \text{for all } x \in \mathbb{R}^N.$$

Then ξ satisfies

$$\begin{cases} -\Delta \xi = \phi(|x|) & \text{in } \mathbb{R}^N, \\ \xi > 0 & \text{in } \mathbb{R}^N, \\ \xi(x) \to 0 & \text{as } |x| \to \infty. \end{cases}$$

Because the mapping $[0,\infty) \ni t \longmapsto \int_0^t ds/g(s) \in [0,\infty)$ is bijective, we can implicitly define $w : \mathbb{R}^N \to (0,\infty)$ by

$$\int_0^{w(x)} \frac{dt}{g(t)} = \xi(x) \quad \text{for all } x \in \mathbb{R}^N.$$

It is easy to see that $w \in C^2(\mathbb{R}^N)$, $w > 0$ in $\mathbb{R}^N$, and $w(x) \to 0$ as $|x| \to \infty$. Furthermore, we have

$$-\Delta \xi = \frac{g'(w)}{g(w)^2}|\nabla w|^2 - \frac{\Delta w}{g(w)} = \phi(|x|) \quad \text{in } \mathbb{R}^N.$$

Because $g' < 0$, it follows that

$$-\Delta w \geq \phi(|x|)g(w) \geq p(x)g(w) \quad \text{in } \mathbb{R}^N. \tag{4.57}$$

Therefore, w is a supersolution of (4.53). Now, it is easy to deduce that $u_n \leq w$ in $\mathbb{R}^N$. By standard elliptic arguments we find that $u(x) := \lim_{n\to\infty} u_n(x)$, $x \in \mathbb{R}^N$, satisfies $u \in C^{2,\gamma}_{\text{loc}}(\mathbb{R}^N)$ for some $0 < \gamma < 1$, and $-\Delta u = p(x)g(u)$ in $\mathbb{R}^N$. Because $u_n \leq w$ in $\mathbb{R}^N$, we obtain that $u \leq w$ in $\mathbb{R}^N$, which yields $u(x) \to 0$ as $|x| \to \infty$. Hence, u is a solution of (4.53).

Finally, if u_1 and u_2 are two solutions, by the maximum principle and the fact that $u_1 - u_2$ tends to zero at infinity we derive $u_1 \equiv u_2$—that is, problem (4.53) has a unique solution. This concludes the proof. □

Next we study the existence of classical solutions for the problem

$$\begin{cases} -\Delta u = p(x)(g(u) + f(u)) & \text{in } \mathbb{R}^N, \\ \quad u > 0 & \text{in } \mathbb{R}^N, \\ u(x) \to 0 & \text{as } |x| \to \infty, \end{cases} \tag{4.58}$$

where g and p are as considered earlier and $f : [0, \infty) \to [0, \infty)$ satisfies the sublinear conditions $(f1)$ and $(f2)$ introduced in the beginning of this chapter. As a result of the lack of monotonicity, the uniqueness is not so obvious as in the study of problem (4.53). This matter will be discussed in the last part of this section for radially symmetric solutions.

Theorem 4.7.2 *Assume that $\int_0^\infty r\phi(r)dr < \infty$. Then problem (4.58) has at least one classical solution.*

Proof We use the same approach as in Theorem 4.7.1 (ii). By Theorem 1.2.5, for all $n \geq 1$ there exists $u_n \in C^{2,\gamma}(B_n) \cap C(\overline{B_n})$ such that

$$\begin{cases} -\Delta u_n = p(x)\Psi(u_n) & \text{in } B_n, \\ \quad u_n > 0 & \text{in } B_n, \\ \quad u_n = 0 & \text{on } \partial B_n, \end{cases}$$

where $\Psi(t) = g(t) + f(t)$, $t > 0$. We have

$$\Delta u_{n+1} + p(x)\Psi(u_{n+1}) \leq 0 \leq \Delta u_n + p(x)\Psi(u_n) \quad \text{in } B_n,$$

$$u_n = 0 < u_{n+1} \quad \text{on } \partial B_n.$$

Notice that $\Delta u_{n+1} \in L^1(B_n)$, because u_{n+1} is bounded away from zero in B_n. Thus, by Theorem 1.3.17 it follows that $u_n \leq u_{n+1}$ in B_n. We extend u_n by zero outside of B_n. Then $0 \leq u_1 \leq \cdots \leq u_n \leq u_{n+1} \leq \cdots$ in $\mathbb{R}^N$. It remains to find an upper bound for $(u_n)_{n\geq 1}$. By Theorem 4.7.1 with g replaced with $g + 1$, there exists $v \in C^2(\mathbb{R}^N)$ such that

$$\begin{cases} -\Delta v = p(x)(g(v) + 1) & \text{in } \mathbb{R}^N, \\ \quad v > 0 & \text{in } \mathbb{R}^N, \\ v(x) \to 0 & \text{as } |x| \to \infty. \end{cases} \tag{4.59}$$

Because f is sublinear and v is bounded in $\mathbb{R}^N$, we can find $M > 0$ such that $M > f(Mv)$ in $\mathbb{R}^N$. Hence, $w := Mv$ satisfies

$$\begin{cases} -\Delta w = p(x)(g(w) + f(w)) & \text{in } \mathbb{R}^N, \\ \quad w > 0 & \text{in } \mathbb{R}^N, \\ w(x) \to 0 & \text{as } |x| \to \infty. \end{cases}$$

As seen earlier, we find $u_n \leq w$ in $\mathbb{R}^N$ and $u(x) := \lim_{n\to\infty} u_n(x)$, for all $x \in \mathbb{R}^N$, is the solution of (4.58). This finishes the proof. □

4.7.2 *Uniqueness of radially symmetric solutions*

We discuss in the sequel the uniqueness of a radially symmetric solution to (4.53). More precisely, we shall be concerned with the uniqueness of classical solutions to the problem

$$\begin{cases} -u''(r) - \dfrac{N-1}{r}u'(r) = p(r)\Psi(u(r)) & \text{in } [0,\infty), \\ u > 0 & \text{in } [0,\infty), \\ u'(0) = 0, & \\ u(r) \to 0 & \text{as } r \to \infty, \end{cases} \tag{4.60}$$

where $p : [0,\infty) \to [0,\infty)$ and $\Psi : (0,\infty) \to (0,\infty)$ are Hölder continuous functions such that

$(A1)$ the mapping $(0,\infty) \ni t \longmapsto \dfrac{\Psi(t)}{t}$ is decreasing;

$(A2)$ there exists $t_0 > 0$ such that Ψ is decreasing on $(0,t_0)$ and increasing on $[t_0,\infty)$;

$(A3)$ there exists $\lim\limits_{t\searrow 0} t^{\alpha}\Psi(t) \in (0,\infty)$ for some $\alpha > 0$.

Notice that these hypotheses are quite natural (see Figure 4.4). Typical examples of nonlinearities encountered so far, such as $\Psi(t) = t^{-\alpha} + t^{q}$, $\alpha > 0$, $0 < q < 1$, or $\Psi(t) = t^{-\alpha} + \ln(1+t)$, $\alpha > 0$, satisfy $(A1)$ through $(A3)$. Obviously, any solution u of (4.60) provides a radially symmetric solution $\widetilde{u}(x) = u(|x|)$, $x \in \mathbb{R}^N$, of problem (4.52) with $\Phi(x,t) = p(|x|)\Psi(t)$.

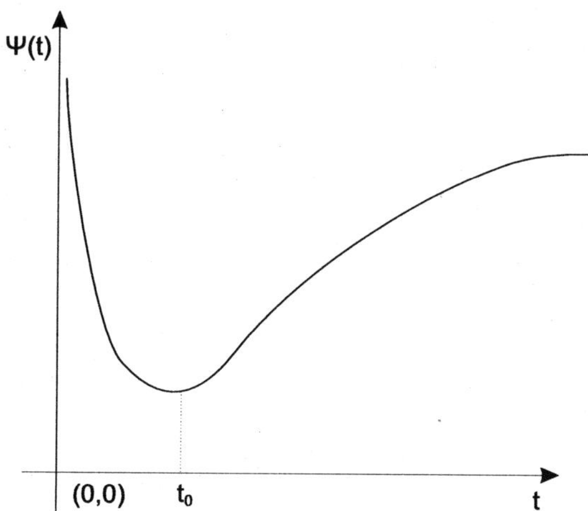

FIGURE 4.4. The shape of function Ψ.

Theorem 4.7.3 *Assume that* $(A1)$ *through* $(A3)$ *hold. Then, problem (4.60) has at most one solution.*

Proof Assume that there exist $u, v \in C^2[0,\infty)$ two distinct solutions of (4.60). By virtue of Theorem 1.3.17 we easily deduce the following useful result.

Lemma 4.7.4 *If there exists* $R > 0$ *such that* $u(R) = v(R)$, *then* $u \equiv v$ *in* $[0, R)$.

By virtue of Lemma 4.7.4, one of the following three situations may occur:

1. $u > v$ in $[0,\infty)$.
2. $u(0) = v(0)$ and $u > v$ in $(0,\infty)$.
3. There exists $R > 0$ such that $u \equiv v$ in $[0, R)$ and $u > v$ in (R,∞).

We shall discuss these three situations separately.

CASE 1: $u > v$ in $[0,\infty)$. Let us notice that u and v verify

$$-(r^{N-1}u'(r))' = r^{N-1}p(r)\Psi(u(r)), \tag{4.61}$$

$$-(r^{N-1}v'(r))' = r^{N-1}p(r)\Psi(v(r)), \tag{4.62}$$

for all $r \geq 0$. We multiply the first equality by v', the second one by u', and then subtract. We obtain

$$-\Big(r^{N-1}(u'v - uv')\Big)' = r^{N-1}p(r)uv\Big(\frac{\Psi(u)}{u} - \frac{\Psi(v)}{v}\Big) \quad \text{for all } r \geq 0.$$

An integration over $[0, r]$, $r > 0$ yields

$$r^{N-1}(u'v - uv') = \int_0^r t^{N-1}p(t)uv\Big(\frac{\Psi(u)}{u} - \frac{\Psi(v)}{v}\Big)dt \quad \text{for all } r \geq 0.$$

Therefore, $u'v - uv' > 0$ in $[0,\infty)$, which implies

$$\frac{u}{v} \text{ is increasing and } \quad \frac{u}{v} > \frac{u'}{v'} \quad \text{in } (0,\infty). \tag{4.63}$$

Hence, there exists $\ell := \lim_{r\to\infty} u(r)/v(r)$. Because $u(0) > v(0)$, it follows that $1 < \ell \leq \infty$.

From (4.61) and (4.62), for all $r \geq 0$, we have

$$u'(r) = -r^{1-N}\int_0^r t^{N-1}p(t)\Psi(u)dt, \tag{4.64}$$

$$v'(r) = -r^{1-N}\int_0^r t^{N-1}p(t)\Psi(v)dt. \tag{4.65}$$

Assume first that $\int_0^\infty t^{N-1}p(t)\Psi(u)dt = \infty$. Because $u > v$ in $[0,\infty)$, and both u and v tend to zero as $r \to \infty$, by $(A2)$ we find $\Psi(v(r)) > \Psi(u(r))$ for

r large enough. This yields $\int_0^\infty t^{N-1}p(t)\Psi(v)dt = \infty$. Then, by l'Hospital's rule we find

$$1 < \ell = \lim_{r\to\infty} \frac{u(r)}{v(r)} = \lim_{r\to\infty} \frac{u'(r)}{v'(r)} = \lim_{r\to\infty} \frac{\int_0^r t^{N-1}p(t)\Psi(u)dt}{\int_0^r t^{N-1}p(t)\Psi(v)dt} = \lim_{r\to\infty} \frac{\Psi(u(r))}{\Psi(v(r))}.$$

By $(A3)$ we have

$$\lim_{r\to\infty} \frac{\Psi(u(r))}{\Psi(v(r))} = \lim_{r\to\infty} \left(\frac{u(r)}{v(r)}\right)^{-\alpha} \frac{u^\alpha\Psi(u)}{v^\alpha\Psi(v)} = \ell^{-\alpha} < 1,$$

which is a contradiction.

Let us assume now that $\int_0^\infty t^{N-1}p(t)\Psi(u)dt < \infty$. If $\int_0^\infty t^{N-1}p(t)\Psi(v)dt = \infty$, as before we have

$$1 < \ell = \lim_{r\to\infty} \frac{u(r)}{v(r)} = \lim_{r\to\infty} \frac{u'(r)}{v'(r)} = \lim_{r\to\infty} \frac{\int_0^r t^{N-1}p(t)\Psi(u)dt}{\int_0^r t^{N-1}p(t)\Psi(v)dt} = 0,$$

which is again a contradiction.

It remains to analyze the case when both integrals $\int_0^\infty t^{N-1}p(t)\Psi(u)dt$ and $\int_0^\infty t^{N-1}p(t)\Psi(v)dt$ are finite.

Lemma 4.7.5 *There exists a unique $a > 0$ such that*

$$\begin{aligned} &\Psi(u(r)) > \Psi(v(r)) \quad \textit{for all } 0 \le r < a,\\ &\Psi(u(a)) = \Psi(v(a)),\\ &\Psi(u(r)) < \Psi(v(r)) \quad \textit{for all } r > a. \end{aligned}$$

Proof Let us first remark that $\Psi(u(0)) > \Psi(v(0))$. If we would have the converse inequality, then, by $(A2)$ and $u(0) > v(0)$, we obtain $v(0) \le t_0$. It follows that for all $r \ge 0$, $\Psi(v(r)) > \Psi(u(r))$, which yields

$$\int_0^\infty t^{N-1}p(t)\Psi(u)dt < \int_0^\infty t^{N-1}p(t)\Psi(v)dt.$$

Again by l'Hospital's rule we find

$$1 < \ell = \lim_{r\to\infty} \frac{u(r)}{v(r)} = \lim_{r\to\infty} \frac{u'(r)}{v'(r)} = \frac{\int_0^\infty t^{N-1}p(t)\Psi(u)dt}{\int_0^\infty t^{N-1}p(t)\Psi(v)dt} < 1,$$

which is a contradiction. Hence, $\Psi(u(0)) > \Psi(v(0))$. Furthermore, because $u > v$ and $u(t) \to 0$, $v(r) \to 0$ as $r \to \infty$, for $r > 0$ large enough we obtain $\Psi(u(r)) < \Psi(v(r))$. Thus, by continuity arguments, we can find $a > 0$ such that $\Psi(u(a)) = \Psi(v(a))$.

Assume that there exists two real numbers $a_1 > a_2 > 0$ such that $\Psi(u(a_i)) = \Psi(v(a_i))$, $i = 1, 2$. Then we must have $u(a_i) > t_0 > v(a_i)$, $i = 1, 2$. Because $a_1 > a_2$, we deduce $u(a_1) < u(a_2)$, and by $(A2)$ we further obtain

$$\Psi(v(a_1)) = \Psi(u(a_1)) < \Psi(u(a_2)) = \Psi(v(a_2)).$$

Because $v(a_i) < t_0$, $i = 1, 2$, the last equality implies $v(a_1) > v(a_2)$, which in turn implies $a_1 < a_2$, but this is clearly a contradiction. Now, the rest of the proof follows from the uniqueness of the number $a > 0$ with $\Psi(u(a)) = \Psi(v(a))$. □

By l'Hospital's rule we have found

$$\ell = \frac{\int_0^\infty t^{N-1} p(t) \Psi(u) dt}{\int_0^\infty t^{N-1} p(t) \Psi(v) dt}.$$

Thus, for $0 < \varepsilon < 1$ there exists $r_0 > a$ large enough such that

$$\int_0^r t^{N-1} p(t) \Psi(u) dt > (\ell - \varepsilon) \int_0^r t^{N-1} p(t) \Psi(v) dt \quad \text{for all } r \geq r_0.$$

By Lemma 4.7.5 we have $\Psi(u(t)) < \Psi(v(t))$, for all $t > a$, which yields

$$\int_0^a t^{N-1} p(t) \Psi(u) dt + \int_a^r t^{N-1} p(t) \Psi(v) dt > (\ell - \varepsilon) \int_0^r t^{N-1} p(t) \Psi(v) dt,$$

for all $r \geq r_0$. This leads us to

$$\int_0^a t^{N-1} p(t) (\Psi(u) - \Psi(v)) dt > (\ell - 1 - \varepsilon) \int_0^r t^{N-1} p(t) \Psi(v) dt \quad \text{for all } r \geq r_0.$$

Passing to the limit with $r \to \infty$, and because $0 < \varepsilon < 1$ was arbitrarily chosen, we obtain

$$\int_0^a t^{N-1} p(t) (\Psi(u) - \Psi(v)) dt > (\ell - 1) \int_0^\infty t^{N-1} p(t) \Psi(v) dt.$$

That is,

$$a^{N-1}(v'(a) - u'(a)) > (\ell - 1) \int_0^\infty t^{N-1} p(t) \Psi(v) dt. \tag{4.66}$$

On the other hand, from (4.63) we have $u'/v' < u/v < \ell$. This yields

$$v'(a) - u'(a) = v'(a) \left(1 - \frac{u'(a)}{v'(a)} \right) < (1 - \ell) v'(a).$$

Hence,

$$
\begin{aligned}
a^{N-1}(v'(a) - u'(a)) &< (1-\ell)a^{N-1}v'(a) \\
&= (\ell - 1)\int_0^a t^{N-1}p(t)\Psi(v)dt \\
&< (\ell - 1)\int_0^\infty t^{N-1}p(t)\Psi(v)dt.
\end{aligned} \tag{4.67}
$$

From (4.66) and (4.67) we obtain the desired contradiction. This concludes the proof in Case 1.

CASE 2: $u(0) = v(0)$ and $u > v$ in $(0, \infty)$. Let $w = u - v$. Then w verifies

$$
\begin{cases}
-(r^{N-1}w')' = r^{N-1}p(r)(\Psi(u) - \Psi(v)) & \text{in } [0,\infty), \\
w > 0 & \text{in } [0,\infty), \\
w(0) = w'(0) = 0, & \\
w(r) \to 0 & \text{as } r \to \infty.
\end{cases} \tag{4.68}
$$

Integrating in (4.68) we have

$$
w(r) = -\int_0^r t^{1-N}\int_0^t s^{N-1}p(s)(\Psi(u) - \Psi(v))dsdt \quad \text{for all } r \geq 0. \tag{4.69}
$$

We claim that $u(0) = v(0) \leq t_0$. If $u(0) = v(0) > t_0$, then $u(t) > v(t) > t_0$ for $t > 0$ small enough, and by the hypothesis $(A2)$ on Ψ we deduce $\Psi(u) > \Psi(v)$ in a small neighborhood of the origin. But using (4.69) this yields $w(t) < 0$ for some $t > 0$, which is a contradiction. Hence, $u(0) = v(0) \leq t_0$. Then

$$
v(r) < u(r) \leq u(0) \leq t_0 \quad \text{for all } r \geq 0,
$$

which implies $\Psi(u) - \Psi(v) < 0$ in $(0, \infty)$. Now, to obtain a contradiction we look at (4.69). The left-hand side tends to zero as $r \to \infty$ whereas the right-hand side tends to a positive quantity as $r \to \infty$. This is clearly a contradiction.

CASE 3: There exists $R > 0$ such that $u \equiv v$ in $[0, R)$ and $u > v$ in (R, ∞). We proceed exactly in the same manner as we did in Case 2 for $\widetilde{u}(r) = u(r + R)$ and $\widetilde{v}(r) = v(r + R)$, $r \geq 0$. This finishes the proof of Theorem 4.7.3. □

4.8 Comments and historical notes

Singular elliptic problems arise in various branches of mathematics (see [38], [56], [134], and [160] for more details). To our best knowledge, the first study in this direction is from Fulks and Maybee [81] and have been intensively studied after the pioneering work by Crandall, Rabinowitz and Tartar [57].

In their work [57], a similar problem to (4.3) has been studied for more general differential operators. The existence of a classical solution to (4.3) has been obtained by considering the perturbed problem

$$\begin{cases} -\Delta u = g(u+\varepsilon),\ 0<\varepsilon<1 & \text{in } \Omega, \\ u>0 & \text{in } \Omega, \\ u=0 & \text{on } \partial\Omega. \end{cases} \tag{4.70}$$

Then (4.70) has a unique solution $u_\varepsilon \in C^2(\overline{\Omega})$, which is decreasing with respect to ε. If $u(x) := \lim_{\varepsilon \searrow 0} u_\varepsilon(x)$, $x \in \overline{\Omega}$, then $u \in C^2(\Omega) \cap C(\overline{\Omega})$ is the unique solution of problem (4.3).

In the case of pure powers in nonlinearities and $\mu = 0$, problem (4.21) was studied by Shi and Yao [182] and was then generalized by Ghergu and Rădulescu in [83]. For more results concerning singular elliptic equations in bounded domains we refer to [40], [42], [48], [55], [61], [67], [84], [85], [86], [96], [101], [127], [174], [201], and [202] as well as to recent surveys [102], [175].

If $\alpha < 0$, problem (4.12) is called the Lane–Emden–Fowler equation and arises in the boundary-layer theory of viscous fluids (see Fowler [78] and Wong [197]).

As we have already mentioned, as a result of the meaning of the unknowns (concentrations, populations), the positive solutions are relevant in most cases. But nonnegative solutions may also arise in some situations. In fact, the presence of the singular term $g(u)$, which is not locally Lipschitz near the origin, may give rise to *dead core* solutions—that is, solutions identically zero on subdomains of Ω with positive measure.

For more results on dead core solutions involving singular nonlinearities, we refer the reader to Dávila and Montenegro [59], where the following problem was considered:

$$\begin{cases} -\Delta u = \chi_{\{u>0\}}(-u^{-\alpha} + \lambda u^p) & \text{in } \Omega, \\ u=0 & \text{on } \partial\Omega, \end{cases} \tag{4.71}$$

with $0 < \alpha, p < 1$. It is proved in [59] that for all $\lambda > 0$, problem (4.71) has a maximal solution u_λ. Moreover, there exists λ^* such that for all $\lambda > \lambda^*$ the maximal solution u_λ is positive and stable.

Another direction regarding the study of singular elliptic problems refers to sign-changing solutions for such problems and was considered by McKenna and Reichel [137]. In this respect, the precise asymptotic behavior of a solution at the boundary where it changes sign plays an essential role in the study of existence.

The approach in obtaining the $C^{1,\gamma}(\overline{\Omega})$ regularity of the classical solution to (4.21) follows the method in Gui and Lin [99] in the case $\lambda = \mu = 0$, $g(t) = t^{-\alpha}$, and $p(x)$ behaves like $d(x)^\sigma$, $0 < \sigma < 1$ near $\partial\Omega$. The method in [99] can be extended to more general nonlinearities than those we dealt with in this chapter. As a conclusion, $C^{1,\gamma}(\overline{\Omega})$ $(0 < \gamma < 1)$ is the best regularity of solutions we can expect for these kinds of problems. We have seen in Theorem 4.1.2 that if g decays fast in the neighborhood of the origin, the solutions may not be in $C^1(\overline{\Omega})$ or, even more unexpected, they do not belong to the usual Sobolev space $H_0^1(\Omega)$. In other words, it may happen that a classical solution is not a weak solution. We will come back to this issue in Chapter 7.

The growth condition $(g2)$ on g is a natural assumption. This allowed us to show that if problem (4.38) has a classical solution $u \in C^2(\Omega) \cap C(\overline{\Omega})$, then $u \in \mathcal{E}$. We shall see in Chapter 9 that condition $g \in L^1(\Omega)$ is necessary to have the existence of a classical solution.

Condition (4.40) is often called the *Keller–Osserman condition* around the origin. As proved by Bénilan, Brezis, and Crandall [23], condition (4.40) is equivalent to the *property of compact support.* That is, for every $k \in L^1(\mathbb{R}^N)$ with compact support, there exists a unique $u \in W^{1,1}(\mathbb{R}^N)$ with compact support such that $\Delta u \in L^1(\mathbb{R}^N)$ and

$$-\Delta u + g(u) = k(x) \qquad \text{almost everywhere in } \mathbb{R}^N.$$

Theorem 4.5.1 leaves open the question of multiplicity. This is clearly a delicate issue even in simpler cases. In this sense we refer the reader to Ouyang, Shi, and Yao [156], who studied the existence of radially symmetric solutions of the problem

$$\begin{cases} -\Delta u = \lambda(u^p - u^{-\alpha}) & \text{in } B, \\ u > 0 & \text{in } B, \\ u = 0 & \text{on } \partial B, \end{cases} \tag{4.72}$$

where B is the unit ball in $\mathbb{R}^N$, $(N \geq 1)$, $0 < \alpha, p < 1$ and $\lambda > 0$. Using a bifurcation theorem of Crandall and Rabinowitz, it has been shown in [156] that there exists $\lambda_1 > \lambda_0 > 0$ such that problem (4.72) has no solutions for $\lambda < \lambda_0$, one solution for $\lambda = \lambda_0$ or $\lambda > \lambda_1$, and two solutions for $\lambda_1 \geq \lambda > \lambda_0$.

Finally, the results obtained for singular elliptic problems in bounded domains allowed us to extend the study to the case when the domain is the whole space. The first results in this sense concern problem (4.53) and have been obtained by Kusano and Swanson [122] and Edelson [74]. Condition $\int_0^\infty r\phi(r)dr < \infty$ in the statement of Theorem 4.7.1 was first supplied by Lair and Shaker [124] (see also [125] also [200]), where the existence of a ground-state solution of (4.53) is obtained under a weaker assumption on the potential p, namely, p satisfies the assumption $(p1)$ in Section 2.2. That is, p is nonnegative and any zero of p is surrounded by a region where $p > 0$. In the case $\Psi(x,t) = p(x)(t^{-\alpha} + t^p)$, $0 < p$, $\alpha < 1$, the uniqueness of a radially symmetric solution to problem (4.60) has been obtained in [199].

5

BIFURCATION AND ASYMPTOTIC ANALYSIS: THE MONOTONE CASE

> A mathematician, like a painter or poet, is a maker of patterns. If his patterns are more permanent than theirs, it is because they are made with ideas.
>
> Godfrey H. Hardy (1877–1947)

In the current chapter and in the following one, we are concerned with bifurcation problems of the type

$$\begin{cases} -\Delta u = \lambda f(u) & \text{in } \Omega, \\ u > 0 & \text{in } \Omega, \\ u = 0 & \text{on } \partial\Omega, \end{cases} \tag{5.1}$$

where Ω is a smooth bounded domain in $\mathbb{R}^N$ $(N \geq 1)$, λ is a positive parameter, and $f : [0, \infty) \to (0, \infty)$ is a continuous function. We study the existence and the qualitative properties of solutions $u \in C^2(\Omega) \cap C(\overline{\Omega})$ (provided such solutions exist), as well as their asymptotic behavior with respect to λ.

To give an elementary example, let us consider the class of functions f with linear or superlinear growth—that is,

$$\text{there exists } a > 0 \text{ such that } f(t) \geq at, \text{ for all } t \geq 0. \tag{5.2}$$

We observe that if f satisfies condition (5.2), then there exists $0 < \lambda_0 \leq \Lambda_0 < \infty$ such that

(i) problem (5.1) has at least one solution if $\lambda \in (0, \lambda_0)$;

(ii) problem (5.1) has no solution if $\lambda > \Lambda_0$.

Of course, natural questions are the following:

- Is $\lambda_0 = \Lambda_0$?
- If not, what happens if $\lambda \in (\lambda_0, \Lambda_0)$?

Assertion (i) follows by the implicit function theorem (Theorem B.1 in Appendix B). To argue (ii), let φ_1 denote the positive eigenfunction corresponding to the first eigenvalue λ_1 of $(-\Delta)$ in $H_0^1(\Omega)$. Thus, by multiplication with φ_1 in (5.1) and integration, we find

$$\lambda_1 \int_\Omega u\varphi_1 dx = \lambda \int_\Omega f(u)\varphi_1 dx \geq \lambda a \int_\Omega u\varphi_1 dx\,.$$

This relation shows that problem (5.1) has no solution if $\lambda > \lambda_1/a$.

An important feature in our analysis concerns the notion of *stability*, which is viewed in terms of a related linearized problem. More precisely, to any solution (u, λ) of the nonlinear problem (5.1), we associate the linear eigenvalue problem

$$\begin{cases} -\Delta w - \lambda f'(u)w = \mu w & \text{in } \Omega, \\ w = 0 & \text{on } \partial\Omega. \end{cases} \tag{5.3}$$

Problem (5.3) has a sequence of real eigenvalues $\mu_1 < \mu_2 \leq \cdots \leq \mu_n \cdots \to \infty$, and μ_1 is a simple eigenvalue with a corresponding eigenfunction ψ_1 such that $\psi_1 > 0$ in Ω. If $\mu_1 > 0$, then solution u of problem (5.1) is said to be *stable*; if $\mu_1 < 0$, then u is *unstable*; and if $\mu_1 = 0$, then the solution u is *neutrally stable.* Usually, a neutrally stable solution is referred as an *unstable solution.* If $\mu_1 = 0$, then solution u is *degenerate*; otherwise, it is *nondegenerate.* If $\mu_1 < 0$, then the number of negative eigenvalues μ_k (counting their multiplicity) of problem (5.3) is the *Morse index* of solution u.

5.1 Introduction

Consider the bifurcation problem

$$\begin{cases} -\Delta u = \lambda f(u) & \text{in } \Omega, \\ u = 0 & \text{on } \partial\Omega, \end{cases} \tag{5.4}$$

where $\Omega \subset \mathbb{R}^N$ $(N \geq 1)$ is a smooth bounded domain and $\lambda \geq 0$. We assume that $f : [0, \infty) \to (0, \infty)$ is a C^1 convex function such that $f'(0) > 0$ and f is asymptotically linear. That is,

$$\lim_{t\to\infty} \frac{f(t)}{t} = a \in (0, \infty). \tag{5.5}$$

A first existence and bifurcation result concerning problem (5.4) is given by Amann's bifurcation theorem in Section 5.2 in a more general framework.

Notice that because f is convex, by (5.5) we have $\lim_{t\to\infty} f'(t) = a$, and the mapping $g : [0, \infty) \to [0, \infty)$, $g(t) = f(t) - at$ is nonincreasing. Hence, there exists

$$\ell := \lim_{t\to\infty} g(t) \in [-\infty, \infty).$$

The sign of ℓ plays a crucial role in the analysis of (5.4). The current chapter is concerned with the case $\ell \geq 0$. Thus, we study here the *monotone case* on f

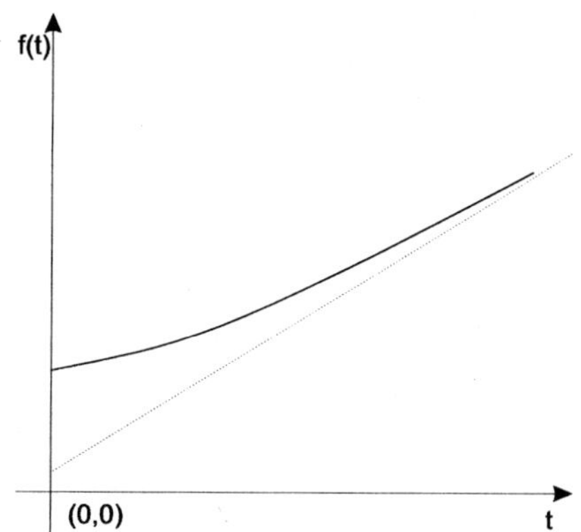

FIGURE 5.1. The monotone case on f.

because the mapping $t \longmapsto f(t)/t$ is nonincreasing, so that $a = \inf_{t>0} f(t)/t$ (see Figure 5.1).

In turn, if $\ell < 0$ then $m := \inf_{t>0} f(t)/t > a$, and both numbers a and m will play an important role in the study of (5.4). This fact will be emphasized in Chapter 6.

For $\alpha \in L^\infty(\Omega)$ we denote by $\lambda_1(\alpha)$ and $\varphi_1(\alpha)$ the first eigenvalue and the first eigenfunction, respectively, of the operator $-\Delta - \alpha$ in $H_0^1(\Omega)$. By means of the variational characterization of the first eigenvalue of a linear operator, $\lambda_1(\alpha)$ is defined as

$$\lambda_1(\alpha) = \inf_{u \in H_0^1(\Omega), \|u\|_2 = 1} \int_\Omega (|\nabla u|^2 - \alpha(x) u^2) dx. \tag{5.6}$$

Definition 5.1.1 *A solution u of (5.4) is called* stable *if $\lambda_1(\lambda f'(u)) > 0$, and, is* unstable *otherwise.*

5.2 A general bifurcation result

The main result in this section is the following.

Theorem 5.2.1 (Amann's bifurcation theorem) *Assume that $f : \mathbb{R} \to \mathbb{R}$ is a C^2 function that is convex, positive, and such that $f'(0) > 0$. Then there exists $\lambda^* \in (0, \infty)$ such that the following holds:*

(i) *If $\lambda > \lambda^*$, then problem (5.4) has no classical solutions.*

(ii) *If $0 \le \lambda < \lambda^*$, then problem (5.4) has a minimal solution $u_\lambda \in C^2(\overline{\Omega})$ such that*

  (ii1) *$u_\lambda > 0$ in Ω;*

  (ii2) *the mapping $(0, \lambda^*) \ni \lambda \longmapsto u_\lambda(x)$ is increasing, for every $x \in \Omega$;*

  (ii3) *u_λ is the unique stable solution of the problem (5.4) for all $0 < \lambda < \lambda^*$.*

Proof We divide the proof of Theorem 5.2.1 into several steps.

Step 1: Definition of λ^*. To apply the implicit function theorem (see Theorem B.1) to our problem (5.4), define $F : C_0^{2,\alpha}(\overline{\Omega}) \times \mathbb{R} \to C^{0,\alpha}(\overline{\Omega})$ by

$$F(u, \lambda) = -\Delta u - \lambda f(u), \quad (u, \lambda) \in C^{2,\alpha}(\overline{\Omega}) \times \mathbb{R},$$

where $0 < \alpha < 1$. It is clear that F verifies all the assumptions of the implicit function theorem. Hence, there exists a maximal neighborhood of the origin I and a unique map $u = u_\lambda$ that is the solution of problem (5.4) for all $\lambda \in I$ and such that the linearized operator $-\Delta - \lambda f'(u_\lambda)$ is bijective. In other words, for every $\lambda \in I$, problem (5.4) admits a stable solution, which is given by the implicit function theorem. Let $\lambda^* := \sup I \leq \infty$.

Step 2: The mapping $(0, \lambda^*) \ni \lambda \longmapsto u_\lambda(x)$ *is increasing and* $u_\lambda > 0$ *in* Ω. For $x \in \overline{\Omega}$, denote by $v_\lambda(x)$ the derivative with respect to λ of the mapping $(0, \lambda^*) \ni \lambda \longmapsto u_\lambda(x)$. We differentiate in (5.4) with respect to λ and we obtain

$$\begin{cases} (-\Delta - \lambda f'(u_\lambda))v_\lambda = \lambda f(u_\lambda) & \text{in } \Omega, \\ v_\lambda = 0 & \text{on } \partial\Omega. \end{cases}$$

Note also that the operator $-\Delta - \lambda f'(u_\lambda)$ is coercive. Therefore, by Stampacchia's maximum principle (see Theorem 1.3.10), either $v_\lambda \equiv 0$ in Ω or $v_\lambda > 0$ in Ω. The first variant is not convenient, because it would imply that $f(u_\lambda) = 0$, which is impossible, by our initial hypotheses. Hence, $v_\lambda > 0$ in Ω, which then yields $u_\lambda > 0$ in Ω.

Step 3: u_λ *is stable.* Set $\psi(\lambda) := \lambda_1(\lambda f'(u_\lambda))$. Thus, $\psi(0) = \lambda_1 > 0$. Furthermore, by the implicit function theorem the linearized operator $-\Delta - \lambda f'(u_\lambda)$ is bijective, which yields $\psi(\lambda) \neq 0$ for all $\lambda < \lambda^*$. Now, by the continuity of the mapping $\lambda \longmapsto \lambda f'(u_\lambda)$, it follows that ψ is continuous, which further implies that $\psi > 0$ on $[0, \lambda^*)$.

Step 4: $\lambda^* < \infty$. Because f is convex, it follows that $f'(u_\lambda) \geq f'(0) > 0$, which leads to $\lambda_1(\lambda f'(0)) \geq \lambda_1(\lambda f'(u_\lambda)) > 0$ for every $\lambda \in I$. Notice that $\lambda_1(\lambda f'(0)) = \lambda_1 - \lambda f'(0)$. This yields $\lambda_1 - \lambda f'(0) > 0$ for any $\lambda < \lambda^*$—that is, $\lambda^* \leq \lambda_1/f'(0) < \infty$.

Step 5: Problem (5.4) has no classical solutions provided $\lambda > \lambda^*$.

Assume that there exists $\mu > \lambda^*$ and a corresponding solution v to problem (5.4). We first prove that for every $\lambda < \lambda^*$ we have $u_\lambda < v$ in Ω. Indeed, by the convexity of f it follows that

$$\begin{aligned} -\Delta(v - u_\lambda) &= \mu f(v) - \lambda f(u_\lambda) \\ &\geq \lambda(f(v) - f(u_\lambda)) \\ &\geq \lambda f'(u_\lambda)(v - u_\lambda) \quad \text{in } \Omega. \end{aligned}$$

Hence,

$$(-\Delta - \lambda f'(u_\lambda))(v - u_\lambda) \geq 0 \quad \text{in } \Omega. \tag{5.7}$$

Because the operator $-\Delta - \lambda f'(u_\lambda)$ is coercive, by Stampacchia's maximum principle, we deduce $v \geq u_\lambda$ in Ω for all $\lambda < \lambda^*$. Therefore, u_λ is bounded in $L^\infty(\Omega)$ by v and there exists $u^* = \lim_{\lambda \nearrow \lambda^*} u_\lambda$ in $\overline{\Omega}$ and $u^* = 0$ on $\partial\Omega$.

We claim that $\lambda_1(\lambda^* f'(u^*)) = 0$. We already know that $\lambda_1(\lambda^* f'(u^*)) \geq 0$. Assume by contradiction that $\lambda_1(\lambda^* f'(u^*)) > 0$. In other words, the operator $-\Delta - \lambda^* f'(u^*)$ is coercive. By applying the implicit function theorem to $F(u,\lambda) = -\Delta u - \lambda f(u)$ at the point (u^*, λ^*), there exists a curve of solutions of the problem (5.4) passing through (u^*, λ^*), which contradicts the maximality of λ^*. Hence, $\lambda_1(\lambda^* f'(u^*)) = 0$, which implies the existence of $\varphi_1 > 0$ in Ω such that $\varphi_1 = 0$ on $\partial\Omega$ and

$$-\Delta\varphi_1 - \lambda^* f'(u^*)\varphi_1 = 0 \quad \text{in } \Omega. \tag{5.8}$$

Passing to the limit with $\lambda \nearrow \lambda^*$ in (5.7), we find

$$(-\Delta - \lambda^* f'(u^*))(v - u^*) \geq 0 \quad \text{in } \Omega.$$

Multiplying this inequality by φ_1 and integrating over Ω, we obtain

$$-\int_\Omega (v - u^*)\Delta\varphi_1 dx - \lambda^* \int_\Omega f'(u^*)(v - u^*)\varphi_1 dx \geq 0.$$

According to (5.8), the previous relation is in fact an equality, which implies that $-\Delta(v - u^*) = \lambda^* f'(u^*)(v - u^*)$ in Ω. It follows that $\mu f(v) = \lambda^* f(u^*)$ in Ω. But $\mu > \lambda^*$ and $f(v) \geq f(u^*)$ so that $f(v) = 0$, which is impossible. Hence, problem (5.4) has no solutions for $\lambda > \lambda^*$.

Step 6: u_λ is the minimal and the unique stable solution of (5.4). Fix an arbitrary $\lambda < \lambda^*$ and let v be another solution of problem (5.4). We have

$$-\Delta(v - u_\lambda) = \lambda f(v) - \lambda f(u_\lambda) \geq \lambda f'(u_\lambda)(v - u_\lambda) \quad \text{in } \Omega.$$

Again, by Stampacchia's maximum principle applied to the coercive operator $-\Delta - \lambda f'(u_\lambda)$, we find that $v \geq u_\lambda$ in Ω.

Now let w be another stable solution of (5.4) for some $\lambda < \lambda^*$. With the same reasoning as applied earlier to the coercive operator $-\Delta - \lambda f'(w)$, we derive that $u_\lambda \geq w$ in Ω and, finally, $u_\lambda \equiv w$. This concludes the proof. □

5.3 Existence and bifurcation results

By Amann's bifurcation theorem we have the following information about problem (5.4):

(i) There exists $\lambda^* \in (0, \infty)$ such that (5.4) has solution when $\lambda \in (0, \lambda^*)$, and no solution exists if $\lambda \in (\lambda^*, \infty)$.

(ii) For $\lambda \in (0, \lambda^*)$, among the solutions of (5.4) there exists a minimal one, say u_λ.

(iii) The mapping $\lambda \longmapsto u_\lambda$ is a C^1 convex and increasing function.

(iv) u_λ can be characterized as the only solution u of (5.4) such that the operator $-\Delta - \lambda f'(u)$ is coercive; that is, u_λ is the only stable solution of (5.4).

Assuming now (5.5), we aim to discuss in an unified way some natural questions raised by (5.4), namely

- what can be said when $\lambda = \lambda^*$?
- what is the behavior of u_λ when λ approaches λ^*?
- are these results still valid if f is unbounded around the origin?
- is there other solution of (5.4) excepting u_λ?
- if so, what is their behavior?

In the rest of this chapter we try to answer these questions. The approach we present in the following is the result of the work by Mironescu and Rădulescu [142], [143]. We are first concerned with the *monotone case*, corresponding to

$$\ell := \lim_{t\to\infty} (f(t) - at) \geq 0,$$

where $a := \lim_{t\to\infty} f(t)/t \in (0, \infty)$.

Theorem 5.3.1 *Assume $\ell \geq 0$. Then we have*

(i) $\lambda^* = \lambda_1/a$;
(ii) *The problem (5.4) has no solution for $\lambda = \lambda^*$;*
(iii) u_λ *is the unique solution of (5.4) for all $0 < \lambda < \lambda^*$;*
(iv) $\lim_{\lambda\nearrow\lambda^*} u_\lambda = \infty$ *uniformly on compact subsets of Ω.*

The bifurcation diagram in the monotone case is depicted in Figure 5.2.

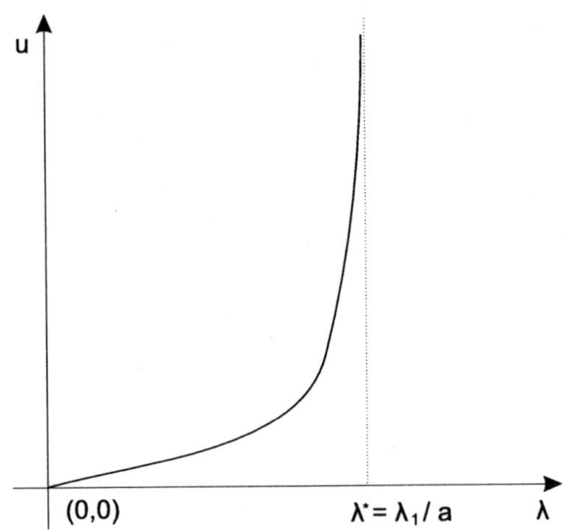

FIGURE 5.2. The bifurcation diagram in Theorem 5.3.1.

Proof (i)–(ii) If $0 < \lambda < \lambda_1/a$, then λf satisfies the hypotheses of Theorem 1.2.5. Hence, problem (5.4) has at least one solution for all $0 < \lambda < \lambda_1/a$, so $\lambda^* \geq \lambda_1/a$.

We claim that problem (5.4) has no classical solutions for $\lambda \geq \lambda_1/a$. Assume that there exists $\lambda \geq \lambda_1/a$ such that (5.4) has a solution u. Because the mapping $s \longmapsto f(t)/t$ is nonincreasing, it follows that $\lambda f(u) \geq \lambda_1 u$ in Ω. Then, multiplying by φ_1 in (5.4) and integrating by parts, we have

$$\lambda_1 \int_\Omega u\varphi_1 dx = \lambda \int_\Omega f(u)\varphi_1 dx \geq \lambda_1 \int_\Omega u\varphi_1 dx.$$

This yields $\lambda = \lambda_1/a$ and $f(u) = au$; that is, $f(t) = at$ for all $0 \leq t \leq \max_{x\in\overline{\Omega}} u(x)$. But this is a contradiction because $f(0) > 0$. Therefore $\lambda^* = \lambda_1/a$ and (5.4) has no solutions for $\lambda = \lambda^*$.

(iii) As we have argued in the previous section, the fact that f is convex and fulfills (5.5) implies that $g : [0,\infty) \to [0,\infty)$, $g(t) = f(t) - at$ is nonincreasing. Then u_λ satisfies

$$-\Delta u_\lambda - \lambda a u_\lambda = \lambda g(u_\lambda) \quad \text{in } \Omega.$$

Let v_λ be another solution of (5.4). According to the minimality of u_λ, we have $u_\lambda \leq v_\lambda$ in Ω. Thus, $w := v_\lambda - u_\lambda \geq 0$ satisfies $-\Delta w - \lambda a w \leq 0$ in Ω. Because $a\lambda < \lambda_1$, the strong maximum principle yields $w = 0$ in Ω and the uniqueness follows.

(iv) Because of the special character of our problem, we will be able to prove that, in certain cases, L^2 boundedness implies H_0^1 boundedness!

Assume by contradiction that $(u_\lambda)_{0<\lambda<\lambda^*}$ does not converge to ∞ on compact subsets of Ω as $\lambda \nearrow \lambda^*$ and define

$$u^*(x) = \lim_{\lambda \nearrow \lambda^*} u_\lambda(x) \qquad \text{for all } x \in \Omega.$$

According to our assumption, $u^* \not\equiv \infty$. Because u_λ is increasing with respect to λ, it follows that u^* is lower semicontinuous. Let $x_0 \in \Omega$ be such that $u^*(x_0) \neq \infty$ and denote by $\widetilde{u}_\lambda$ the spherical average of u_λ with respect to x_0—that is,

$$\widetilde{u}_\lambda(r) = \frac{1}{\omega_N} \int_{|y|=1} u_\lambda(x_0 + ry) d\sigma(y) \quad 0 \leq r < \operatorname{dist}(x_0, \partial\Omega).$$

Because u_λ is a superharmonic function we have

$$\begin{aligned}
\widetilde{u}_\lambda(r) &= \frac{1}{\omega_N} \int_{|y|=1} \nabla u_\lambda(x_0 + ry) \cdot y \, d\sigma(y) \\
&= \frac{1}{\omega_N} \int_{|y|=1} \frac{\partial u_\lambda}{\partial n}(x_0 + ry) \, d\sigma(y) \\
&= \frac{1}{\omega_N} \int_{|y|=1} \Delta u_\lambda(x_0 + ry) \, d\sigma(y) \\
&\leq 0.
\end{aligned}$$

Hence, $\widetilde{u}_\lambda$ is decreasing in $[0, R)$, which implies

$$u_\lambda(x_0) = \widetilde{u}_\lambda(x_0) \geq \frac{1}{\omega_N} \int_{|y|=1} u_\lambda(x_0 + ry)d\sigma(y) \quad \text{for all } 0 \leq r < \operatorname{dist}(x_0, \partial\Omega).$$

Integrating in the previous relation we obtain

$$\int_{B_R(0)} u_\lambda(x)dx \leq R\omega_N u_\lambda(x_0) \leq Cu^*(x_0) < \infty.$$

By Fatou's lemma we derive $u^* \in L^1(B_R(0))$. Let

$$X := \{x \in \Omega : u^* \text{ is integrable in a neighborhood of } x\}.$$

From the previous arguments we deduce that X is open and nonempty. Furthermore, because u^* is lower semicontinuous, it follows that X is also closed. Because Ω is connected we derive that $X = \Omega$—that is, $u^* \in L^1_{\text{loc}}(\Omega)$. This implies that $(u_\lambda)_{0<\lambda<\lambda^*}$ is bounded in $L^1_{\text{loc}}(\Omega)$.

We next prove that $(u_\lambda)_{\lambda<\lambda^*}$ is bounded in $L^2(\Omega)$. We argue by contradiction. Suppose that $(u_\lambda)_{\lambda<\lambda^*}$ is not bounded in $L^2(\Omega)$. Thus, passing eventually at a subsequence we have $u_\lambda = k(\lambda)w_\lambda$, where

$$k(\lambda) = \|u_\lambda\|_2 \to \infty \text{ as } \lambda \nearrow \lambda^* \text{ and } w_\lambda \in L^2(\Omega), \ \ \|w_\lambda\|_2 = 1. \tag{5.9}$$

Notice that $f(t) \leq at + b$ for all $t \geq 0$, where $b = f(0)$. This implies

$$\frac{\lambda f(u_\lambda)}{k(\lambda)} \to 0 \quad \text{in } L^1_{\text{loc}}(\Omega) \text{ as } \lambda \nearrow \lambda^*.$$

That is,

$$-\Delta w_\lambda \to 0 \quad \text{in } L^1_{\text{loc}}(\Omega) \text{ as } \lambda \nearrow \lambda^*. \tag{5.10}$$

By Green's first identity, for all $\phi \in C_0^\infty(\Omega)$ we have

$$\int_\Omega \nabla w_\lambda \cdot \nabla\phi \, dx = -\int_\Omega \phi \, \Delta w_\lambda dx = -\int_{\text{Supp}\,\phi} \phi \, \Delta w_\lambda dx. \tag{5.11}$$

Using (5.10) we obtain

$$\begin{aligned} \left| \int_{\text{Supp}\,\phi} \phi \, \Delta w_\lambda \, dx \right| &\leq \int_{\text{Supp}\,\phi} |\phi||\Delta w_\lambda| \, dx \\ &\leq \|\phi\|_\infty \int_{\text{Supp}\,\phi} |\Delta w_\lambda| \, dx \to 0 \quad \text{as } \lambda \nearrow \lambda^*. \end{aligned} \tag{5.12}$$

Combining (5.11) and (5.12) we derive that for all $\phi \in C_0^\infty(\Omega)$ we have

$$\int_\Omega \nabla w_\lambda \cdot \nabla\phi \, dx \to 0 \quad \text{as } \lambda \nearrow \lambda^*. \tag{5.13}$$

By definition, the sequence $(w_\lambda)_{0<\lambda<\lambda^*}$ is bounded in $L^2(\Omega)$. We claim that $(w_\lambda)_{0<\lambda<\lambda^*}$ is bounded in $H_0^1(\Omega)$. Indeed, using Hölder's inequality, we have

$$\begin{aligned}\int_\Omega |\nabla w_\lambda|^2 dx &= -\int_\Omega w_\lambda \Delta w_\lambda dx = \frac{\lambda}{k(\lambda)}\int_\Omega w_\lambda f(u_\lambda)dx\\ &\le \frac{\lambda^*}{k(\lambda)}\int_\Omega w_\lambda(au_\lambda+b)dx\\ &= \lambda^* a\int_\Omega w_\lambda^2 dx + \frac{\lambda^* b}{k(\lambda)}\int_\Omega w_\lambda dx\\ &\le \lambda^* a + c_1|\Omega|^{1/2},\end{aligned}$$

where $c_1 > 0$ does not depend on λ. From the previous estimates it is easy to see that $(w_\lambda)_{0<\lambda<\lambda^*}$ is bounded in $H_0^1(\Omega)$, so the claim follows. Thus, there exists $w \in H_0^1(\Omega)$ such that up to a subsequence and as $\lambda \nearrow \lambda^*$ we have

$$\begin{aligned} w_\lambda &\rightharpoonup w \quad \text{weakly in } H_0^1(\Omega),\\ w_\lambda &\to w \quad \text{strongly in } L^2(\Omega).\end{aligned} \tag{5.14}$$

On the one hand, by (5.9) and (5.14), we derive that $\|w\|_2 = 1$. Furthermore, using (5.13) and (5.14), we infer that

$$\int_\Omega \nabla w\cdot\nabla\phi\, dx = 0 \quad \text{for all } \phi\in C_0^\infty(\Omega).$$

Because $w \in H_0^1(\Omega)$, using the previous relation and the definition of $H_0^1(\Omega)$, we find $w = 0$. This contradiction shows that $(u_\lambda)_{0<\lambda<\lambda^*}$ is bounded in $L^2(\Omega)$. As noted earlier for $(w_\lambda)_{0<\lambda<\lambda^*}$, we derive that $(u_\lambda)_{0<\lambda<\lambda^*}$ is bounded in $H_0^1(\Omega)$. Hence, there exists $u^* \in H_0^1(\Omega)$ such that, up to a subsequence and as $\lambda \nearrow \lambda^*$, there holds

$$\begin{aligned} u_\lambda &\rightharpoonup u^* \quad \text{weakly in } H_0^1(\Omega),\\ u_\lambda &\to u^* \quad \text{strongly in } L^2(\Omega),\\ u_\lambda &\to u^* \quad \text{almost everywhere in } \Omega.\end{aligned} \tag{5.15}$$

Now we obtain the desired contradiction in the same manner as in the proof of (i)–(ii). We first multiply by φ_1 in (5.4) and then we integrate over Ω. We obtain

$$\lambda_1\int_\Omega u_\lambda\varphi_1 dx = \lambda\int_\Omega f(u_\lambda)\varphi_1 dx \ge \lambda a\int_\Omega u_\lambda\varphi_1 dx \quad \text{for all } 0<\lambda<\lambda^*. \tag{5.16}$$

Passing to the limit in (5.16) with $\lambda \nearrow \lambda^*$, by virtue of Lebesgue's theorem on dominated convergence we find

$$\lambda_1\int_\Omega u^*\varphi_1 = \frac{\lambda_1}{a}\int_\Omega f(u^*)\varphi_1 dx.$$

Hence, $f(u^*) = au^*$, which contradicts the fact that $f(0) > 0$. This shows that $\lim_{\lambda\nearrow\lambda^*} u_\lambda = \infty$ uniformly on compact subsets of Ω, and the proof is now complete. □

Remark 5.3.2 *The results in Theorem 5.3.1 hold in a little more general context —namely, the mapping $(0, \infty) \ni t \longmapsto f(t)/t$ is nonincreasing, f satisfies (5.5), and $f(0) > 0$. By Theorem 1.2.5 and the strong maximum principle we may show that (i)-(iv) in Theorem 5.3.1 hold. However, the convexity assumption on f provides a complete description of the unique solution u_λ to problem (5.4). We will see in Chapter 6 that the hypothesis f convex is strongly needed when the issue of multiplicity arises.*

Example 5.1 Let us consider the one-dimensional problem

$$\begin{cases} -u'' = \lambda\sqrt{u^2 + u + 1} & \text{in } (0, \pi), \\ u(0) = u(\pi) = 0. \end{cases} \tag{5.17}$$

Clearly, $f(t) = \sqrt{t^2 + t + 1}$, $t \geq 0$, obeys the monotone case described in this chapter. In this case $\lambda_1 = 1$, and by Theorem 5.3.1 we have that (5.17) has a solution if and only if $0 \leq \lambda < 1$. Moreover, for all $0 \leq \lambda < 1$, problem (5.17) has a unique solution u_λ, which is stable, and $\lim_{\lambda \nearrow 1} u_\lambda = \infty$ uniformly on compact subsets of $(0, \pi)$.

Using the collocation method we have computed the solution u_λ of (5.17) when λ approaches λ_1. The solution is plotted in Figure 5.3.

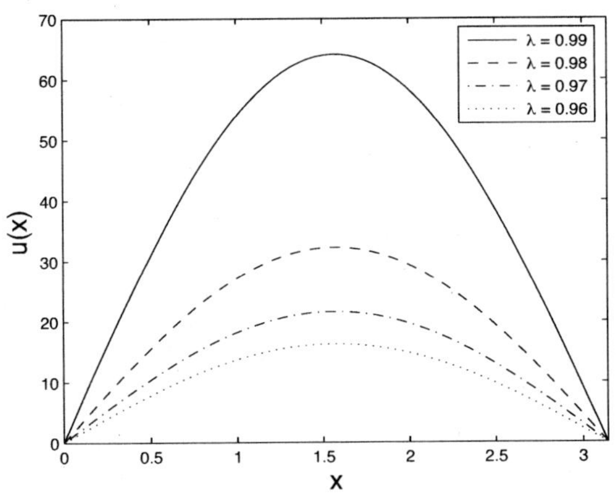

FIGURE 5.3. The unique solution of problem (5.17).

5.4 Asymptotic behavior of the solution with respect to parameters

In this section we discuss the asymptotic behavior of the unique solution u_λ of (5.4) with respect to λ. We have seen that $\lim_{\lambda \nearrow \lambda^*} u_\lambda = \infty$ uniformly on compact subsets of Ω.

We consider the normalized sequence $(w_\lambda)_{0<\lambda<\lambda^*}$ of $(u_\lambda)_{0<\lambda<\lambda^*}$ in $L^2(\Omega)$. That is,

$$u_\lambda = k(\lambda)w_\lambda \quad \text{with } \|w_\lambda\|_2 = 1.$$

Because $\lim_{\lambda\nearrow\lambda^*} u_\lambda = \infty$ uniformly on each compact subset of Ω, we deduce that $k(\lambda) \to \infty$ as $\lambda \nearrow \lambda^*$. From the proof of Theorem 5.3.1 we have a uniform bound for $(w_\lambda)_{0<\lambda<\lambda^*}$ in $H_0^1(\Omega)$ as $\lambda \nearrow \lambda^*$. Assuming that the first eigenvalue φ_1 of $(-\Delta)$ in $H_0^1(\Omega)$ is normalized in $L^2(\Omega)$, we obtain the following result.

Theorem 5.4.1 *We have $w_\lambda \to \varphi_1$ in $C^1(\overline{\Omega})$ as $\lambda \nearrow \lambda^*$.*

Proof Because $(w_\lambda)_{0<\lambda<\lambda^*}$ is bounded in $H_0^1(\Omega)$ as $\lambda \nearrow \lambda^*$, there exists $w \in H_0^1(\Omega)$ such that, up to a subsequence, we have as $\lambda \nearrow \lambda^*$:

$$\begin{aligned} w_\lambda &\rightharpoonup w \quad \text{weakly in } H_0^1(\Omega),\\ w_\lambda &\to w \quad \text{strongly in } L^2(\Omega),\\ w_\lambda &\to w \quad \text{almost everywhere in } \Omega. \end{aligned} \tag{5.18}$$

Then $-\Delta w(\lambda) \to -\Delta w$ in $\mathcal{D}'(\Omega)$. On the other hand, we have

$$-\Delta w_\lambda = \lambda \frac{f(u_\lambda)}{k(\lambda)} \quad \text{in } \Omega. \tag{5.19}$$

We claim that $f(u_\lambda)/k(\lambda) \to aw$ almost everywhere in Ω as $\lambda \nearrow \lambda^*$.

If $x \in \Omega$ is such that $w(x) > 0$, then, by (5.5), we obtain

$$\lim_{\lambda\nearrow\lambda^*} \frac{f(k(\lambda)w_\lambda(x))}{k(\lambda)} = \lim_{\lambda\nearrow\lambda^*} \frac{f(k(\lambda)w_\lambda(x))}{k(\lambda)w_\lambda(x)} w_\lambda(x) = aw(x).$$

If $w(x) = 0$, let $\varepsilon > 0$ and $0 < \lambda_0 < \lambda^*$ be such that $w_\lambda(x) < \varepsilon$ for all $\lambda_0 < \lambda < \lambda^*$. Because $f(t) \le at + b$ for all $t \ge 0$ (with $b = f(0)$) we deduce

$$\frac{f(k(\lambda)w_\lambda(x))}{k(\lambda)} \le a\varepsilon + \frac{b}{k(\lambda)} \quad \text{for all } \lambda_0 < \lambda < \lambda^*.$$

This yields $f(w_\lambda(x))/k(\lambda) \to 0$ as $\lambda \nearrow \lambda^*$, which proves the claim.

Passing to the limit with $\lambda \nearrow \lambda^*$ in (5.19) we obtain

$$-\Delta w = \lambda_1 w, \ w \in H_0^1(\Omega), \ w \ge 0, \ \|w\|_2 = 1.$$

But this means exactly that $w = \varphi_1$. Moreover, $(w_\lambda)_{0<\lambda<\lambda^*}$ is relatively compact in $C^1(\overline{\Omega})$. This can be done by showing that the sequence is bounded in $C^{1,1/2}(\overline{\Omega})$, which follows from the fact that $(w_\lambda)_{0<\lambda<\lambda^*}$ is bounded in $H_0^1(\Omega)$ and a bootstrap argument (note that a uniform bound for w_λ in some $L^p(\Omega)$, $1 < p < \infty$ provides a uniform bound for $-\Delta w_\lambda$). This ends the proof. □

Theorem 5.4.2 *The sequence $(\varphi_1/w_\lambda)_{0<\lambda<\lambda^*}$ is uniformly bounded when λ approaches λ^*.*

Proof The strong maximum principle implies that $\partial w_\lambda/\partial n < 0$ on $\partial\Omega$. Therefore, φ_1/w_λ can be extended to a continuous function on $\overline{\Omega}$ by setting

$$\frac{\varphi_1}{w_\lambda}(x) = \frac{\partial\varphi_1(x)/\partial n}{\partial w_\lambda(x)/\partial n} \qquad \text{for all } x \in \partial\Omega.$$

Using the regularity of $\partial\Omega$ we can find $\delta > 0$ such that if

$$A_\delta := \{x \in \mathbb{R}^N : d(x) < \delta\},$$

then

(i) for each $x \in A_\delta$ there is a unique $x_0 \in \partial\Omega$ such that $d(x) = |x - x_0|$;
(ii) if $\pi : A_\delta \to \partial\Omega$ is defined by $\pi(x) = x_0$, then $\pi \in C^1(A_\delta)$;
(iii) if $|x-\pi(x)| = \varepsilon$ then $x = \pi(x)-\varepsilon n(\pi(x))$ or $x = \pi(x)+\varepsilon n(\pi(x))$, according to the case $x \in \Omega$ or $x \notin \Omega$;
(iv) if $x \in \Omega$ then $[x, \pi(x)) \subset \Omega$.

Let $\Omega_\delta = \Omega \cap A_\delta$. Because $w_\lambda \to \varphi_1$ uniformly in $\overline{\Omega}$, it is easy to see that φ_1/w_λ is uniformly bounded in $\Omega \setminus \Omega_\delta$. If $x \in \Omega_\delta$, let $x_0 = \pi(x)$. Then

$$\frac{\varphi_1(x)}{w_\lambda(x)} = \frac{\varphi_1(x) - \varphi_1(x_0)}{w_\lambda(x) - w_\lambda(x_0)} = \frac{\partial\varphi_1(x_0 + \tau(x - x_0))/\partial n(x_0)}{\partial w_\lambda(x_0 + \tau(x - x_0))/\partial n(x_0)}$$

for some $\tau \in (0, 1)$. Taking a smaller δ if necessary, we may assume that

$$\frac{\partial w_\lambda(x)}{\partial\nu(\pi(x))} < 0 \qquad \text{on } \overline{\Omega}_\delta.$$

Thus, the quotient φ_1/w_λ is uniformly bounded in $\overline{\Omega}_\delta$ for λ near λ^*. This concludes the proof. □

Theorem 5.4.3 *We have*

$$\lim_{\lambda\nearrow\lambda^*} (\lambda^* - \lambda)\|u_\lambda\|_2 = \frac{\lambda_1\ell}{a}\|\varphi_1\|_1.$$

Proof Let ℓ_0 be a limit point of $(\lambda^* - \lambda)\|u_\lambda\|_2$ as $\lambda \nearrow \lambda^*$. Multiplying by φ_1 in (5.4) and then integrating by parts we obtain

$$\lambda_1 \int_\Omega u_\lambda\varphi_1 dx = \lambda \int_\Omega f(u_\lambda)\varphi_1 dx,$$

which may be rewritten in the form

$$\int_\Omega \varphi_1(\lambda_1 - a\lambda)u_\lambda dx = \lambda \int_\Omega (f(u_\lambda) - au_\lambda)\varphi_1 dx.$$

Hence,

$$\int_\Omega \varphi_1(\lambda_1 - a\lambda)\|u_\lambda\|_2 w_\lambda dx = \int_\Omega \lambda\varphi_1(f(u_\lambda) - au_\lambda)dx. \tag{5.20}$$

Note that the right-hand side integrand in (5.20) converges dominated to $\lambda^*\ell\varphi_1$ as $\lambda \nearrow \lambda^*$. Also the left-hand side integrand tends to $\ell_0\varphi_1^2$. Now, passing to the limit in (5.20) we find $\ell_0 = (\lambda_1\ell/a)\|\varphi_1\|_1$. This finishes the proof. □

5.5 Examples

Theorem 5.4.3 describes the growth rate of $(\lambda^* - \lambda)\|u_\lambda\|_2$ as $\lambda \nearrow \lambda^*$. If $\ell = 0$ the convergence depends heavily on f, and from Theorem 5.4.3 we deduce only that $\|u_\lambda\|_2$ grows slower than $(\lambda^* - \lambda)^{-1}$. In the curent section we illustrate this aspect by the following two examples.

5.5.1 *First example*

Proposition 5.5.1 *Let $f(t) = t + 1/(t+2)$, $t \geq 0$. Then, the unique solution u_λ of (5.4) satisfies*

$$\lim_{\lambda \nearrow \lambda_1} \sqrt{\lambda_1 - \lambda}\|u_\lambda\|_2 = \sqrt{\lambda_1|\Omega|}.$$

Proof With the usual decomposition $u_\lambda = k(\lambda)w_\lambda$ and dividing (5.20) by $\sqrt{\lambda_1 - \lambda}$ we have

$$\int_\Omega \sqrt{\lambda_1 - \lambda}k(\lambda)w_\lambda\varphi_1 dx = \int_\Omega \frac{\lambda\varphi_1}{\sqrt{\lambda_1 - \lambda}(k(\lambda)w_\lambda + 2)}dx. \tag{5.21}$$

We first infer that $\liminf_{\lambda \nearrow \lambda_1} \sqrt{\lambda_1 - \lambda}k(\lambda) > 0$. Otherwise, passing to a subsequence we may suppose that $\sqrt{\lambda_1 - \lambda}k(\lambda) \to 0$. Then

$$\sqrt{\lambda_1 - \lambda}k(\lambda)w_\lambda\varphi_1 \to 0 \quad \text{uniformly in } \overline{\Omega}$$

and

$$\sqrt{\lambda_1 - \lambda}(k(\lambda)w_\lambda + 2) \to 0 \quad \text{uniformly in } \overline{\Omega},$$

which contradicts (5.21).

We also claim that $\limsup_{\lambda \nearrow \lambda_1} \sqrt{\lambda_1 - \lambda}k(\lambda) < \infty$. Supposing the contrary, and passing to a subsequence, we have $\sqrt{\lambda_1 - \lambda}k(\lambda) \to \infty$ as $\lambda \nearrow \lambda_1$. Then the left-hand side in (5.21) tends to ∞ whereas the right-hand side remains bounded because by Theorem 5.4.2, the quotient φ_1/w_λ is uniformly bounded in $\overline{\Omega}$.

Let $c \in (0, \infty)$ be a limit point of $\sqrt{\lambda_1 - \lambda}k(\lambda)$ as $\lambda \nearrow \lambda_1$. Then the left-hand side of (5.21) tends to c, whereas the right-hand side integrand converges dominated to λ_1/c. Hence, $c = \lambda_1|\Omega|/c$, which finishes the proof. □

Remark 5.5.2 *A similar computation can be made for $f(t) = \sqrt{t^2 + 1}$, $t > 0$.*

5.5.2 *Second example*

If $f(t) - at$ decays to ∞ faster than $1/t$, then the behavior becomes more complicated. This is illustrated by the following example.

Proposition 5.5.3 *Let $f(t) = t + (t+1)^{-2}$. Then $\|u_\lambda\|_2$ tends to ∞ as $\lambda \nearrow \lambda_1$ like no power of $(\lambda_1 - \lambda)$. More precisely, we have*

(i) $\lim_{\lambda \nearrow \lambda_1} (\lambda_1 - \lambda)^\alpha \|u_\lambda\|_2 = \infty$ *if* $\alpha \leq 1/3$;

(ii) $\lim_{\lambda \nearrow \lambda_1} (\lambda_1 - \lambda)^\alpha \|u_\lambda\|_2 = 0$ *if* $\alpha > 1/3$.

Proof We first need the following auxiliary result.

Lemma 5.5.4 *There exists $\varepsilon_0 > 0$ and two positive constants $c_1, c_2 > 0$ such that*

$$c_1 |\ln \varepsilon| \leq \int_{\{\varphi_1 > \varepsilon\}} \frac{1}{\varphi_1} dx \leq c_2 |\ln \varepsilon| \quad \textit{for all } 0 < \varepsilon < \varepsilon_0. \tag{5.22}$$

Proof Consider $c > 0$ such that

$$c d(x) \leq \varphi_1 \leq \frac{1}{c} d(x) \quad \text{in } \Omega. \tag{5.23}$$

Let $\delta > 0$ and A_δ be as in the proof of Theorem 5.4.2. Define

$$\Phi : A_\delta \to \partial\Omega \times (-\delta, \delta) \quad \text{and} \quad \Psi : \partial\Omega \times (-\delta, \delta) \to A_\delta$$

by

$$\Phi(x) = (\pi(x), \langle x - \pi(x), n(\pi(x)) \rangle) \quad \text{and} \quad \Psi(x_0, \varepsilon) = x_0 + \varepsilon n(x_0).$$

Then Φ, Ψ are smooth and $\Psi = \Phi^{-1}$. Replacing δ if necessary, by a smaller number, we may assume that there exist $c_1, c_2 > 0$ such that

$$0 < c_1 \leq |J(\Psi)| \leq c_2 \quad \text{in } \partial\Omega \times (-\delta, \delta), \tag{5.24}$$

where $J(\Psi)$ is the Jacobian of Ψ. Let

$$\varepsilon_0 := \min\{\inf_{\Omega \setminus \Omega_\delta} \varphi_1, c\delta\}.$$

Now, for $0 < \varepsilon < \varepsilon_0$ we decompose

$$\int_{\{\varphi_1 > \varepsilon\}} \frac{1}{\varphi_1} dx = \int_{\{\varphi_1 \geq \varepsilon_0\}} \frac{1}{\varphi_1} dx + \int_{\{\varepsilon < \varphi_1 < \varepsilon_0\}} \frac{1}{\varphi_1} dx. \tag{5.25}$$

Note that by (5.23) we have

$$\{\varepsilon / c < d(x) < c\varepsilon_0\} \subset \{\varepsilon < \varphi_1 < \varepsilon_0\} \subset \{c\varepsilon < d(x) < \varepsilon_0 / c\}.$$

Hence,

$$\begin{aligned} c\int_{\{\varepsilon/c<d(x)<c\varepsilon_0\}}\frac{1}{d(x)}dx &\le \int_{\{\varepsilon<\varphi_1<\varepsilon_0\}}\frac{1}{\varphi_1}dx \\ &\le \frac{1}{c}\int_{\{c\varepsilon<d(x)<\varepsilon_0/c\}}\frac{1}{d(x)}dx. \end{aligned} \tag{5.26}$$

With the change of coordinates $x = \Psi(x_0, \eta)$ and by virtue of (5.24) we obtain

$$\begin{aligned} \int_{\{\varepsilon/c<d(x)<c\varepsilon_0\}}\frac{1}{d(x)}dx &= \int_{\partial\Omega\times(\frac{\varepsilon}{c},c\varepsilon_0)}\frac{|J(\Psi)|}{\eta}d\sigma(x_0)d\eta \\ &\ge c_1|\partial\Omega|\ln\frac{c^2\varepsilon_0}{\varepsilon}. \end{aligned} \tag{5.27}$$

Similarly

$$\begin{aligned} \int_{\{c\varepsilon<d(x)<\frac{\varepsilon_0}{c}\}}\frac{1}{d(x)}dx &= \int_{\partial\Omega\times(c\varepsilon,\frac{\varepsilon_0}{c})}\frac{|J(\Psi)|}{\eta}d\sigma(x_0)d\eta \\ &\le c_2|\partial\Omega|\ln\frac{\varepsilon_0}{c^2\varepsilon}. \end{aligned} \tag{5.28}$$

Now the claim follows from (5.25) through (5.28). □

Let us come back to the proof of Proposition 5.5.3.

(i) It is enough to show that

$$\lim_{\lambda\nearrow\lambda_1}(\lambda_1-\lambda)^{1/3}\|u_\lambda\|_2 = \infty.$$

Supposing to the contrary, there exists $c \in [0, \infty)$ such that up to a subsequence we have $(\lambda_1 - \lambda)^{1/3}k(\lambda) \to c$ as $\lambda \nearrow \lambda^*$, where $k(\lambda) = \|u_\lambda\|_2$.

Dividing by $(\lambda_1 - \lambda)^{2/3}$ in (5.20) we obtain

$$\int_\Omega(\lambda_1-\lambda)^{1/3}\varphi_1 k(\lambda)w_\lambda dx = \lambda\int_\Omega\frac{\varphi_1}{(\lambda_1-\lambda)^{2/3}(k(\lambda)w_\lambda+1)^2}dx. \tag{5.29}$$

If $c = 0$ then the left-hand side in (5.29) converges to 0, whereas the right-hand side tends to ∞. Hence, $c \in (0, \infty)$. Let $\varepsilon > 0$ be fixed. Because $(\lambda_1 - \lambda)^{2/3}(k(\lambda)w_\lambda + 1)^2 \to c^2\varphi_1^2$ as $\lambda \nearrow \lambda_1$, we can find $0 < \lambda_0 < \lambda_1$ such that

$$(\lambda_1-\lambda)^{2/3}(k(\lambda)w_\lambda+1)^2 < c^2(\varphi_1^2+\varepsilon) \quad \text{for all } \lambda_0<\lambda<\lambda_1.$$

Then (5.29) yields

$$\int_\Omega(\lambda_1-\lambda)^{1/3}\varphi_1 k(\lambda)w_\lambda dx \ge \frac{\lambda}{c^2}\int_\Omega\frac{\varphi_1}{\varphi_1^2+\varepsilon}dx \quad \text{for all } \lambda_0<\lambda<\lambda_1.$$

Because the limit of the left-hand side is c, we deduce

$$c \ge \frac{\lambda^*}{c^2}\int_\Omega\frac{\varphi_1}{\varphi_1^2+\varepsilon}dx \quad \text{for all } \varepsilon>0.$$

Letting $\varepsilon \searrow 0$, by Lemma 5.5.4 we obtain $c = \infty$, the desired contradiction.

(ii) We argue again by contradiction. Thus, there exists $\alpha > 1/3$ such that up to a subsequence we have $(\lambda_1 - \lambda)^\alpha k(\lambda) \to c$ as $\lambda \nearrow \lambda_1$ and $c > 0$.

Dividing by $(\lambda_1 - \lambda)^{1-\alpha}$ in (5.20) we obtain

$$\int_\Omega (\lambda_1 - \lambda)^\alpha \varphi_1 k(\lambda) dx = \lambda \int_\Omega \frac{\varphi_1}{(\lambda_1 - \lambda)^{2\alpha-\beta}(k(\lambda)w_\lambda + 1)^2} dx. \tag{5.30}$$

The limit of the left-hand side is $c \in (0, \infty]$. To estimate the right-hand side, we have

$$\begin{aligned} 0 &< \int_{\{\varphi_1 < \lambda_1 - \lambda\}} \frac{\varphi_1}{(\lambda_1 - \lambda)^{1-\alpha}(k(\lambda)w_\lambda + 1)^2} dx \\ &\le \int_\Omega \frac{\lambda_1 - \lambda}{(\lambda_1 - \lambda)^{1-\alpha}} dx = (\lambda_1 - \lambda)^\alpha |\Omega| \to 0 \quad \text{as } \lambda \nearrow \lambda^*. \end{aligned}$$

By Lemma 5.5.4 we also obtain

$$\begin{aligned} 0 &< \int_{\{\varphi_1 \ge \lambda_1 - \lambda\}} \frac{\varphi_1}{(\lambda_1 - \lambda)^{1-\alpha}(k(\lambda)w_\lambda + 1)^2} dx \\ &\le \frac{C(\lambda_1 - \lambda)^{3\alpha - 1}}{c^2} \int_{\{\varphi_1 \ge \lambda_1 - \lambda\}} \frac{1}{\varphi_1} dx \to 0 \quad \text{as } \lambda \nearrow \lambda^*, \end{aligned}$$

where $C = \sup_{0<\lambda<\lambda^*} \max_{\overline{\Omega}} (\varphi_1^2 / w_\lambda^2) < \infty$ (as shown in Theorem 5.4.2). From these estimates, we deduce that the right-hand side in (5.30) converges to zero, which is a contradiction. This ends the proof. □

5.6 The case of singular nonlinearities

Our aim in this section is to supply a similar result to that in Section 3.3 in case f is singular at the origin. Assume that f satisfies the following:

$(f1)$ The mapping $(0, \infty) \ni t \longmapsto \frac{f(t)}{t}$ is decreasing and there exists

$$\lim_{t\to\infty} \frac{f(t)}{t} = a \in (0, \infty).$$

$(f2)$ $\lim_{s \searrow 0} f(s) = \infty$ and there exists $0 < \alpha < 1$ such that

$$f(t) = O(t^{-\alpha}) \quad \text{as } t \searrow 0.$$

Thus, we are dealing with the monotone case, but in its nonsmooth variant. Note that, because of the assumptions $(f1)$ and $(f2)$, the nonlinearity f is no longer monotone. Also notice that condition $(f1)$ allows us to relax the convexity requirement on f. Furthermore, in view of $(f1)$ we have $f(t) - at \ge 0$ for all $t > 0$ and we allow $\ell := \lim_{t\to\infty}(f(t) - at)$ to be ∞. As may be easily verified by the reader, the mapping $f(t) = 1/\sqrt{t} + \sqrt{t} + t$, $t > 0$ satisfies $(f1)$ and $(f2)$, is not convex, and $\ell = \infty$ (see Figure 5.4). The convexity assumption of f is needed in the framework of Amann's theorem and provides the stability of the minimal solution. In our context we prove the uniqueness of a classical solution whereas the stability of the solution will be achieved in a more general setting in Chapter 8.

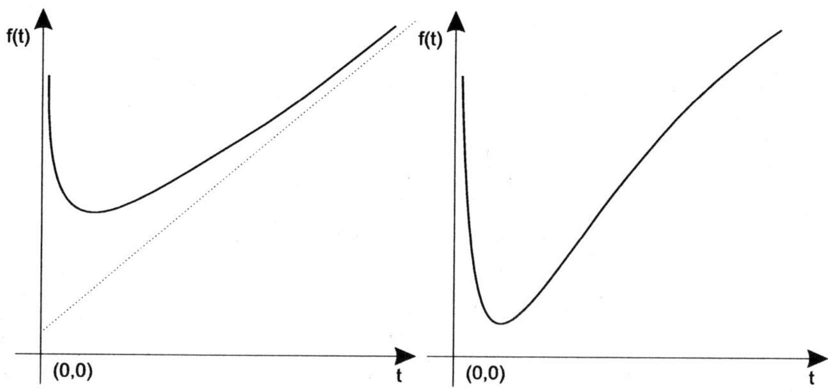

FIGURE 5.4. The singular monotone case on f with $\ell < \infty$ respectively $\ell = \infty$.

Theorem 5.6.1 *Assume that condition* $(f1)$ *and* $(f2)$ *are fulfilled.*

(i) *If* $\lambda \geq \lambda_1/a$, *then problem (5.4) has no classical solutions.*

(ii) *If* $0 < \lambda < \lambda_1/a$, *then problem (5.4) has a unique solution* u_λ *such that*

  (ii1) $u_\lambda \in C^2(\Omega) \cap C^{1,1-\alpha}(\overline{\Omega})$;

  (ii2) $\lim_{\lambda \nearrow \lambda^*} u_\lambda = \infty$ *uniformly on compact subsets of* Ω.

Proof Existence and nonexistence of a classical solution as well as the uniqueness follows in the same manner as in the proof of Theorem 5.3.1. The regularity of the solution in (ii1) is similar to that obtained in Theorem 4.3.2. For the proof of (ii2) we claim that the arguments used in Theorem 5.3.1 (iv) still work here. Indeed, we have only to show that the sequence $w_\lambda = u_\lambda/\|u_\lambda\|_2$ $(0 < \lambda < \lambda_1/a)$ is bounded in $H_0^1(\Omega)$. Using the assumptions $(f1)$ and $(f2)$, there exists $b, c > 0$ such that $f(t) \leq at + bt^{-\alpha} + c$ for all $t > 0$. Then, by Hölder's inequality we may write

$$
\begin{aligned}
\int_\Omega |\nabla w_\lambda|^2 &= -\int_\Omega w_\lambda \Delta w_\lambda = \frac{\lambda}{k(\lambda)} \int_\Omega w_\lambda f(u_\lambda) \\
&\leq \frac{\lambda^*}{k(\lambda)} \int_\Omega w_\lambda (a u_\lambda + b u_\lambda^{-\alpha} + c) \\
&= \lambda^* a \int_\Omega w_\lambda^2 + \frac{\lambda^* b}{k(\lambda)^{1+\alpha}} \int_\Omega w_\lambda^{1-\alpha} + \frac{\lambda^* c}{k(\lambda)} \int_\Omega w_\lambda \\
&\leq \lambda^* a + \frac{\lambda^* b}{k(\lambda)^{1+\alpha}} |\Omega|^{(1+\alpha)/2} + \frac{\lambda^* c}{k(\lambda)} |\Omega|^{1/2}.
\end{aligned}
$$

From now on, we follow the proof of Theorem 5.3.1 line to line. This completes the proof. □

Finally, let us remark that the same result holds for bifurcation problems of the type

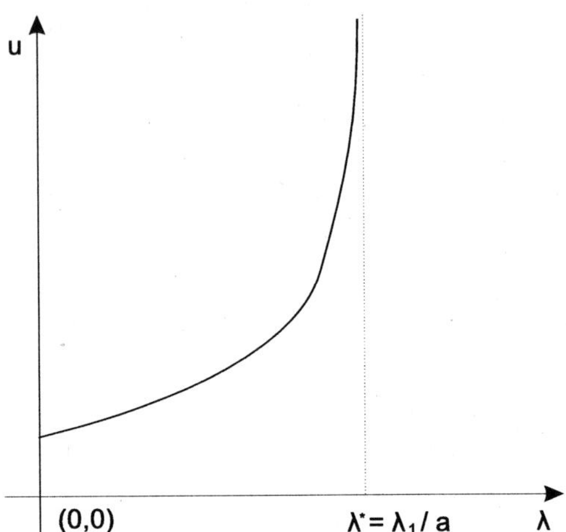

FIGURE 5.5. The bifurcation diagram for the singular problem (5.31).

$$\begin{cases} -\Delta u = g(u) + \lambda f(u) & \text{in } \Omega, \\ u > 0 & \text{in } \Omega, \\ u = 0 & \text{on } \partial\Omega, \end{cases} \tag{5.31}$$

where g is a singular nonlinearity that satisfies $(f2)$ and f is a C^1 function having a sublinear growth. The bifurcation diagram is depicted in Figure 5.5.

5.7 Comments and historical notes

The constant λ^* in the statement of Theorem 5.2.1 is also called *extremal value* (or the *Frank–Kamenetskii constant* in the combustion literature).

The class of functions satisfying assumption (5.2) includes the family of convex mappings $f : [0, \infty) \to (0, \infty)$, which are increasing in a neighborhood of ∞. The important example corresponding to $f(t) = e^t$ is related to the celebrated Liouville–Gelfand problem

$$\begin{cases} -\Delta u = \lambda e^u & \text{in } \Omega, \\ u = 0 & \text{on } \partial\Omega. \end{cases} \tag{5.32}$$

This problem was initially studied by Liouville [132] if $N = 1$ and, subsequently, by Gelfand [82]. The solutions of (5.32) arise as steady-state solutions of the nonlinear evolution problem (called the *solid fuel ignition model*)

$$\begin{cases} v_t = \Delta v + \lambda(1 - \varepsilon v)^m e^{v/(1+\varepsilon v)} & \text{in } \Omega, \\ v = 0 & \text{on } \partial\Omega, \end{cases}$$

within the approximation $\varepsilon \ll 1$, where λ is known as the Frank–Kamenetskii parameter (see [80]), v is a dimensionless temperature, and ε^{-1} is the activation energy.

6

BIFURCATION AND ASYMPTOTIC ANALYSIS: THE NONMONOTONE CASE

> We are servants rather than masters in mathematics.
>
> Charles Hermite (1822–1901)

6.1 Introduction

The *monotone case* in problem (5.4) corresponding to an increasing, positive, and convex nonlinear term f satisfying $f(t) \geq at$ for all $t \geq 0$, where $a := \lim_{t\to\infty} f(t)/t \in (0,\infty)$, was studied in Theorem 5.3.1. The proof of this theorem relies essentially on the maximum principle. The complementary *nonmonotone case* is more difficult and more rich in information. This different setting is analyzed in this chapter by means of combined arguments, including variational methods like the mountain pass theorem of Ambrosetti and Rabinowitz.

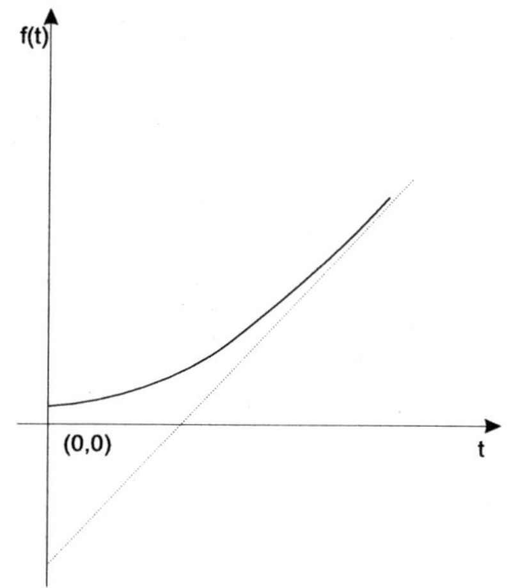

FIGURE 6.1. The nonmonotone case on f.

More precisely, we study the bifurcation problem (5.4) if the nonlinear term f obeys the *nonmonotone case* (see Figure 6.1)—that is, $f : [0,\infty) \to (0,\infty)$ is a C^1 convex function such that $f'(0) > 0$, there exists $a := \lim_{t\to\infty} f(t)/t \in (0,\infty)$, and

$$\ell := \lim_{t\to\infty} (f(t) - at) \in [-\infty, 0). \tag{6.1}$$

The basic hypothesis $\ell < 0$ will change radically the study of (5.4). At the same time, new and different methods are needed to deal with the multiplicity of solutions to (5.4).

6.2 Auxiliary results

Lemma 6.2.1 *Let $\alpha \in L^\infty(\Omega)$ and $w \in H_0^1(\Omega) \setminus \{0\}$, $w \geq 0$, be such that $\lambda_1(\alpha) \leq 0$ and $-\Delta w \geq \alpha w$ in Ω. Then the following properties are valid:*

(i) $\lambda_1(\alpha) = 0$.

(ii) $-\Delta w = \alpha w$ *in* Ω.

(iii) $w > 0$ *in* Ω.

Proof Let us multiply the inequality $-\Delta w \geq \alpha w$ by $\varphi_1(\alpha)$ and then integrate by parts. We obtain

$$\int_\Omega \alpha\varphi_1(\alpha)w dx + \lambda_1(\alpha)\int_\Omega \varphi_1(\alpha)w dx \geq \int_\Omega \alpha\varphi_1(\alpha)w dx.$$

This implies that $\lambda_1(\alpha) = 0$ and $-\Delta w = \alpha w$ in Ω. Because $w \geq 0$ and $w \not\equiv 0$, we deduce $w = c\varphi_1(\alpha)$ for some $c > 0$, which concludes the proof. □

It is obvious that λf satisfies the hypotheses of Theorem 1.2.5 for all $0 < \lambda < \lambda_1/a$. This yields $\lambda^* \geq \lambda_1/a$.

Lemma 6.2.2 *The following properties hold true:*

(i) *If problem (5.4) has a solution for $\lambda = \lambda^*$, then this solution is necessarily unstable.*

(ii) *Problem (5.4) has at most one solution when $\lambda = \lambda^*$.*

(iii) *u_λ is the only solution u of problem (5.4) such that $\lambda_1(\lambda f'(u)) \geq 0$.*

Proof (i) Suppose that (5.4) with $\lambda = \lambda^*$ has a solution u^* with $\lambda_1(\lambda^* f'(u^*)) > 0$. Thus, by the implicit function theorem applied to the mapping

$$G : C_0^{2,1/2}(\overline{\Omega}) \times \mathbb{R} \to C^{0,1/2}(\overline{\Omega}), \quad G(u,\lambda) = -\Delta u - \lambda f(u),$$

it follows that problem (5.4) has a solution for λ in a neighborhood of λ^*, contradicting by this the definition of λ^*.

(ii) Let u be a solution of (5.4) for $\lambda = \lambda^*$. Then u is a supersolution of (5.4) for all $0 < \lambda < \lambda^*$. Using the minimality of u_λ, it follows that $u \geq u_\lambda$ in Ω for all $0 < \lambda < \lambda^*$. This shows that u_λ (which increases with λ) converges to a certain u^* in $L^1(\Omega)$.

Let us multiply by u_λ in (5.4) and integrate by parts. We obtain

$$\int_\Omega |\nabla u_\lambda|^2 dx = \lambda \int_\Omega f(u_\lambda)u_\lambda dx \leq \lambda^* \int_\Omega uf(u)dx.$$

Hence, $(u_\lambda)_{0<\lambda<\lambda^*}$ is bounded in $H_0^1(\Omega)$.

We claim that $u_\lambda \rightharpoonup u^*$ in $H_0^1(\Omega)$ as $\lambda \nearrow \lambda^*$. Indeed, if v is a limit point of $(u_\lambda)_{0<\lambda<\lambda^*}$ as $\lambda \nearrow \lambda^*$, then, up to a subsequence, $u_\lambda \to v$ almost everywhere in Ω. But $u_\lambda \to u^*$ a.e in Ω. Hence, $u^* \in H_0^1(\Omega)$. The proof will be concluded when we show that $u = u^*$. Let $w = u - u^* \geq 0$. Then

$$-\Delta w = \lambda^*(f(u) - f(u^*)) \geq \lambda^* f'(u^*)w \quad \text{in } \Omega. \tag{6.2}$$

Because $\lambda_1(\lambda^* f'(u^*)) \leq 0$, by Lemma 6.2.1 it follows that either $w \equiv 0$ or

$$w > 0, \quad \lambda_1(\lambda^* f'(u^*)) = 0, \;\text{ and } \; -\Delta w = \lambda^* f'(u^*)w \;\text{ in } \Omega.$$

Assuming $w \not\equiv 0$, by (6.2) we deduce that f is linear in all the intervals of the form $[u^*(x), u(x)]$, $x \in \Omega$. It is easy to see that this forces f to be linear in $[0, \max_{\overline{\Omega}} u]$. Let $\alpha, \beta > 0$ be such that $f(u) = \alpha u + \beta$ and $f(u^*) = \alpha u^* + \beta$. We have

$$0 = \lambda_1(\lambda^* f'(u^*)) = \lambda_1(\lambda^* \alpha) = \lambda_1 - \lambda^* \alpha,$$

that is, $\lambda^* = \lambda_1/\alpha$. Hence, u satisfies

$$\begin{cases} -\Delta u = \lambda_1 u + \dfrac{\lambda_1 \beta}{\alpha} & \text{in } \Omega, \\ u > 0 & \text{in } \Omega, \\ u > 0 & \text{on } \partial\Omega, \end{cases}$$

which is a contradiction. Thus $u^* = u$.

(iii) Suppose that problem (5.4) has a solution $u \neq u_\lambda$ with $\lambda_1(\lambda f'(u)) \geq 0$. By the strong maximum principle we derive $u > u_\lambda$ in Ω. Let $w = u - u_\lambda > 0$. Then

$$-\Delta w = \lambda(f(u) - f(u_\lambda)) \leq \lambda f'(u)w \quad \text{in } \Omega. \tag{6.3}$$

Next, we multiply by $\varphi := \varphi_1(\lambda f'(u))$ in (6.3) and we integrate by parts. We find

$$\lambda \int_\Omega f'(u)\varphi w dx + \lambda_1(\lambda f'(u)) \int_\Omega \varphi w dx \leq \lambda \int_\Omega f'(u)\varphi w dx.$$

Thus, $\lambda_1(\lambda f'(u)) = 0$ and in (6.3) we have equality—that is, f is linear in $[0, \max_{\overline{\Omega}} u]$. Let $\alpha, \beta > 0$ be such that $f(u) = \alpha u + \beta$, $f(u_\lambda) = \alpha u_\lambda + \beta$. Then

$$0 = \lambda_1(\lambda f'(u)) = \lambda_1(\lambda f'(u_\lambda)),$$

which is a contradiction. This finishes the proof. □

Lemma 6.2.3 *The following conditions are equivalent:*

(i) $\lambda^* > \lambda_1/a$.

(ii) *Problem (5.4) has exactly one solution for* $\lambda = \lambda^*$.

(iii) *The sequence* $(u_\lambda)_{0<\lambda<\lambda^*}$ *converges uniformly in* $\overline{\Omega}$ *as* $\lambda \nearrow \lambda^*$ *to some* u^*, *which is the unique solution of (5.4) for* $\lambda = \lambda^*$.

Proof (i)⇒(ii) Assume that (5.4) has no solution for $\lambda = \lambda^*$. We claim that $u_\lambda \to \infty$ as $\lambda \nearrow \lambda^*$ uniformly on compact subsets of Ω. Supposing the contrary, with the same arguments as in the proof of Theorem 5.3.1 (iv) it follows that $(u_\lambda)_{0<\lambda<\lambda^*}$ is bounded in $H_0^1(\Omega)$. Then, up to a subsequence $(u_\lambda)_{0<\lambda<\lambda^*}$ is weakly convergent in $H_0^1(\Omega)$ and its weak limit is a solution of (5.4), which contradicts our assumption. Hence, $u_\lambda \to \infty$ as $\lambda \nearrow \lambda^*$ uniformly on compact subsets of Ω.

Let

$$u_\lambda = k(\lambda) w_\lambda \text{ with } \|w_\lambda\|_2 = 1, \quad \text{for all } 0 < \lambda < \lambda^*.$$

Because (5.4) has no solution for $\lambda = \lambda^*$, we easily derive $k(\lambda) \to \infty$ as $\lambda \nearrow \lambda^*$. Again we use the arguments in the proof of Theorem 5.3.1 (iv) to obtain that $(w_\lambda)_{0<\lambda<\lambda^*}$ is bounded in $H_0^1(\Omega)$ as $\lambda \nearrow \lambda^*$. Hence, there exists $w \in H_0^1(\Omega)$ such that up to a subsequence and as $\lambda \nearrow \lambda^*$ we have

$$\begin{aligned} w_\lambda &\rightharpoonup w && \text{weakly in } H_0^1(\Omega), \\ w_\lambda &\to w && \text{strongly in } L^2(\Omega), \\ w_\lambda &\to w && \text{almost everywhere in } \Omega. \end{aligned} \tag{6.4}$$

Then $-\Delta w_\lambda \to -\Delta w$ in $\mathcal{D}'(\Omega)$. On the other hand, we have

$$-\Delta w_\lambda = \lambda \frac{f(u_\lambda)}{k(\lambda)} \quad \text{in } \Omega. \tag{6.5}$$

As in the proof of Theorem 5.3.1, we further deduce

$$\frac{1}{k(\lambda)} f(u_\lambda) \to aw \quad \text{in } L^2(\Omega) \qquad \text{as } \lambda \nearrow \lambda^*.$$

Passing to the limit in (6.5) we obtain

$$-\Delta w = \lambda^* w, \ w \in H_0^1(\Omega), \ w \geq 0, \ \|w\|_2 = 1.$$

This yields $w = \varphi_1$ and $\lambda^* = \lambda_1/a$, which is a contradiction. Hence (5.4) has at least a solution, say u^*, when $\lambda = \lambda^*$. By Lemma 6.2.2 (ii), it follows that u^* is the unique solution of (5.4) for $\lambda = \lambda^*$.

(ii)⇒(iii) We have seen in Lemma 6.2.2 (ii) that if (5.4) has a solution u^* for $\lambda = \lambda^*$, then $u_\lambda \to u^*$ almost everywhere in Ω as $\lambda \nearrow \lambda^*$.

It suffices to prove that u_λ has a limit in $C(\overline{\Omega})$ as $\lambda \nearrow \lambda^*$. Even less, it is enough to prove that $(u_\lambda)_{0<\lambda<\lambda^*}$ is relatively compact in $C(\overline{\Omega})$. This follows by Arzelà–Ascoli's theorem and standard elliptic estimates, which implies that $(u_\lambda)_{0<\lambda<\lambda^*}$ is bounded in $C^{0,1/2}(\overline{\Omega})$.

(iii)⇒(i) Assume that $\lambda^* = \lambda_1/a$. According to Lemma 6.2.2 (i) we have $\lambda_1(\lambda^* f'(u^*)) \leq 0$. Furthermore, $\lambda_1 \leq \lambda^* f'(u^*) \leq a\lambda^* = \lambda_1$. Then f is linear on $[0, \max_{\overline{\Omega}} u^*]$ and, as in the proof of Lemma 6.2.2 (ii), we obtain a contradiction. This concludes the proof. □

The following result asserts that the limit of a sequence of unstable solutions corresponding to problem (5.4) is also unstable.

Lemma 6.2.4 *Let u_n be a solution of (5.4) for some $\lambda = \lambda_n > 0$ such that $\lambda_1(\lambda_n f'(u_n)) \le 0$ and as $n \to \infty$ we have $\lambda_n \to \lambda$, $u_n \rightharpoonup u$ in $H_0^1(\Omega)$. Then u is an unstable solution of problem (5.4)—that is, $\lambda_1(\lambda f'(u)) \le 0$.*

Proof The fact that u is a solution of (5.4) follows by standard elliptic arguments. To prove that u is unstable, let us first notice that by the minmax characterization (5.6) of the first eigenvalue, condition $\lambda_1(\alpha) \le 0$ for some $\alpha \in L^\infty(\Omega)$ is equivalent to the existence of $\psi \in H_0^1(\Omega)$ such that $\|\psi\|_2 = 1$ and

$$\int_\Omega |\nabla \psi|^2 dx \le \int_\Omega \alpha \psi^2 dx.$$

Therefore, for all $n \ge 1$ there exists $\psi_n \in H_0^1(\Omega)$, $\|\psi_n\|_2 = 1$ such that

$$\int_\Omega |\nabla \psi_n|^2 dx \le \int_\Omega \lambda_n f'(u_n) \psi_n^2 dx. \tag{6.6}$$

Because $f' \le a$, from (6.6) we derive that $(\psi_n)_{n\ge1}$ is bounded in $H_0^1(\Omega)$. Let $\psi \in H_0^1(\Omega)$ be such that, up to a subsequence, $\psi_n \rightharpoonup \psi$ in $H_0^1(\Omega)$ and $\psi_n \to \psi$ in $L^2(\Omega)$. Furthermore, up to a subsequence, $(\psi_n)_{n\ge1}$ is dominated in $L^2(\Omega)$. This yields $\|\psi\|_2 = 1$ and

$$\int_\Omega \lambda_n f'(u_n) \psi_n^2 dx \to \lambda \int_\Omega f'(u) \psi^2 dx \quad \text{as } n \to \infty. \tag{6.7}$$

By (6.6) and Fatou's lemma we obtain

$$\int_\Omega |\nabla \psi|^2 dx \le \liminf_{n\to\infty} \int_\Omega |\nabla \psi_n|^2 dx \le \int_\Omega \lambda f'(u) \psi^2 dx.$$

We have thus obtained the existence of $\psi \in H_0^1(\Omega)$ such that $\|\psi\|_2 = 1$ and $\int_\Omega |\nabla \psi|^2 dx \le \int_\Omega \lambda f'(u) \psi^2 dx$—that is, u is unstable. This finishes the proof of our lemma. □

6.3 Existence and bifurcation results in the nonmonotone case

In the framework of the curent chapter, existence and bifurcation results are completely different to those presented in Theorem 5.3.1 in the monotone case. Furthermore, the uniqueness no longer holds in a neighborhood of the bifurcation point λ^*. This will be proved by means of the mountain pass theorem of Ambrosetti and Rabinowitz (see Theorem D.1.3).

Let $m := \min_{t>0} f(t)/t$. As a result of (6.1) we have $m > a$.

Theorem 6.3.1 *Assume that f verifies (5.5) and (6.1). Then the following hold:*

(i) $\lambda^* \in (\lambda_1/a, \lambda_1/m)$.

(ii) *Problem (5.4) has exactly one solution* u^* *for* $\lambda = \lambda^*$;.
(iii) $\lim_{\lambda \nearrow \lambda^*} u_\lambda = u^*$ *uniformly in* $\overline{\Omega}$.
(iv) u_λ *is the unique solution of (5.4) for all* $0 < \lambda \leq \lambda_1/a$.
(v) *For all* $\lambda_1/a < \lambda \leq \lambda^*$, *problem (5.4) has at least an unstable solution* v_λ.

For any solution $v_\lambda \not\equiv u_\lambda$ *we have*

(vi) $\lim_{\lambda \searrow \lambda_1/a} v_\lambda = \infty$ *uniformly on compact subsets of* Ω;
(vii) $\lim_{\lambda \nearrow \lambda^*} v_\lambda = u^*$ *uniformly in* $\overline{\Omega}$.

The bifurcation diagram in the nonmonotone case is depicted in Figure 6.2.

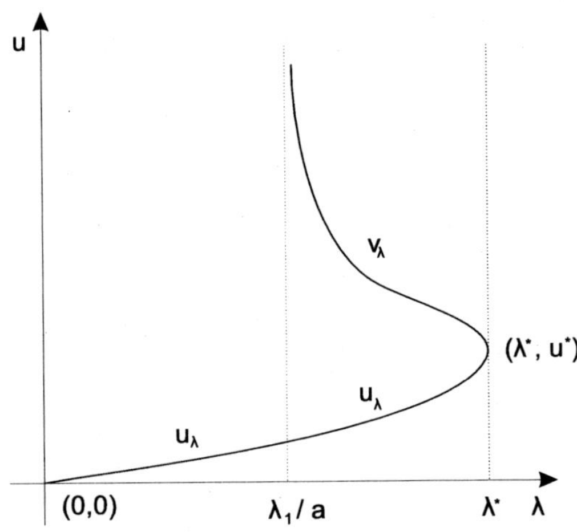

FIGURE 6.2. The bifurcation diagram in Theorem 6.3.1.

Proof We first prove that $\lambda^* \leq \lambda_1/m$. To this aim, it suffices to show that problem (5.4) has no solution for $\lambda = \lambda_1/m$. Suppose the contrary. Let u be a solution of (5.4) for $\lambda = \lambda_1/m$. Then, multiplying in (5.4) by φ_1 and integrating by parts we find

$$\lambda_1 \int_\Omega \varphi_1 u dx = \lambda \int_\Omega \varphi_1 f(u) dx. \tag{6.8}$$

Furthermore,

$$\lambda_1 \int_\Omega \varphi_1 u dx = \frac{\lambda_1}{m} \int_\Omega \varphi_1 f(u) \geq \lambda_1 \int_\Omega \varphi_1 u dx,$$

which forces $f(u) = mu$. This clearly contradicts the fact that $f(0) > 0$.

The remaining parts of (i), (ii), and (iii) are equivalent in view of Lemma 6.2.3. In this sense, it is enough to prove that $\lambda^* > \lambda_1/a$. Assume by contradiction that $\lambda^* \leq \lambda_1/a$. By Lemma 6.2.3 we derive that $\lambda^* = \lambda_1/a$. Then, problem (5.4) has no solution for $\lambda = \lambda^*$. Indeed, if u^* would be such a solution, then by Lemma

6.2.2, u^* is necessarily unstable. On the other hand, because $f'(u^*) \le a$ we obtain $0 \le \lambda_1(\lambda^* f'(u^*)) \le \lambda_1(\lambda^* a) = 0$. Hence, $\lambda_1(\lambda^* f'(u^*)) = 0$—that is, $f'(u^*) = a$. We now proceed as in the proof of Lemma 6.2.2 (ii) to obtain a contradiction. Hence, (5.4) has no solution for $\lambda = \lambda_1/a$. With the same arguments as in the proof of Lemma 6.2.3, we derive $\lim_{\lambda \nearrow \lambda^*} u_\lambda = \infty$ uniformly on compact subsets of Ω.

From (6.8) we have

$$\begin{aligned} 0 &= \int_\Omega \varphi_1 \left[\lambda_1 u_\lambda - \lambda f(u_\lambda)\right] dx \\ &= \int_\Omega \varphi_1 \left[(\lambda_1 - a\lambda)u_\lambda - \lambda(f(u_\lambda) - a u_\lambda))\right] dx \\ &\ge -\lambda \int_\Omega \varphi_1 \left[f(u_\lambda) - a u_\lambda\right] dx. \end{aligned} \tag{6.9}$$

Because $\lim_{\lambda \nearrow \lambda^*} u_\lambda(x) = \infty$ for all $x \in \Omega$, it follows that $f(u_\lambda(x)) - a u_\lambda(x) \to \ell$ as $\lambda \nearrow \lambda^*$, for all $x \in \Omega$. Passing to the limit in (6.9) we obtain the contradictory inequality $0 \ge -\ell \lambda^* \int_\Omega \varphi_1 > 0$.

We have proved that $\lambda_1/a < \lambda^* \le \lambda_1/m$, and by Lemma 6.2.3, we have that problem (5.4) has a solution when $\lambda = \lambda^*$. This shows that $\lambda^* < \lambda_1/m$.

(iv) Taking into account Lemma 6.2.2 (iii), it is enough to prove that for all $0 < \lambda \le \lambda_1/a$, any solution u of (5.4) verifies $\lambda_1(\lambda f'(u)) \ge 0$. Because $f'(u) \le a$ we obtain $\lambda_1(\lambda f'(u)) \ge \lambda_1(\lambda a) \ge 0$ for all $0 < \lambda \le \lambda_1/a$.

(v) We want to find solutions of (5.4) different from u_λ, which are critical points other than u_λ, of the energy functional

$$J : H_0^1(\Omega) \to \mathbb{R}, \quad J(u) = \frac{1}{2}\int_\Omega |\nabla u|^2 dx - \int_\Omega F(u) dx,$$

where $F(t) = \lambda \int_0^t f(s) ds$. In the following, for each $\lambda_1/a < \lambda < \lambda^*$ we will take u_λ as u_0 in the mountain pass theorem of Ambrosetti–Rabinowitz. Clearly, $J \in C^1(H_0^1(\Omega), \mathbb{R})$ and u_0 is a local minimum for J.

To apply the Ambrosetti–Rabinowitz theorem, we transform u_0 into a local strict minimum by modifying J. Let $\varepsilon \in (0,1)$ and define

$$J_\varepsilon : H_0^1(\Omega) \to \mathbb{R}, \quad J_\varepsilon(u) = J(u) + \frac{\varepsilon}{2}\int_\Omega |\nabla(u - u_0)|^2 dx.$$

Then $J_\varepsilon \in C^1(H_0^1(\Omega), \mathbb{R})$ and for all $u, v \in H_0^1(\Omega)$ we have

$$J'_\varepsilon(u)v = \int_\Omega \nabla u \cdot \nabla v dx - \lambda \int_\Omega f(u) v dx + \varepsilon \int_\Omega \nabla(u - u_0) \cdot \nabla v dx.$$

Moreover, u_0 is a local strict minimum for J_ε provided $\varepsilon > 0$ is small enough.

Lemma 6.3.2 *Let $\varepsilon_0 = (\lambda a - \lambda_1)/(2\lambda_1)$. Then there exists $v_0 \in H_0^1(\Omega)$ such that*

$$J_\varepsilon(v_0) < J_\varepsilon(u_0) \quad \textit{for all } 0 \le \varepsilon \le \varepsilon_0.$$

Proof Because J_ε is nonincreasing with respect to ε, it suffices to prove that

$$\lim_{t\to\infty} J_{\varepsilon_0}(t\varphi_1) = -\infty.$$

Notice that

$$\begin{aligned} J_\varepsilon(t\varphi_1) = & \frac{\lambda_1}{2}t^2 + \frac{\varepsilon_0}{2}\lambda_1 t^2 - \varepsilon_0\lambda_1 t\int_\Omega \varphi_1 u_0 dx \\ & + \frac{\varepsilon_0}{2}\int_\Omega |\nabla u_0|^2 dx - \int_\Omega F(t\varphi_1)dx. \end{aligned} \tag{6.10}$$

Let $\alpha = (3a\lambda + \lambda_1)/(4\lambda)$. Because $\alpha < a$, there exists $\beta \in \mathbb{R}$ such that $f(t) \geq \alpha t + \beta$ for all $t \geq 0$, which implies that $F(t) \geq \alpha\lambda/2t^2 + \beta\lambda t$ when $t \geq 0$. Then (6.10) shows that

$$\limsup_{t\to\infty} \frac{1}{t^2} J_{\varepsilon_0}(t\varphi_1) \leq \frac{\lambda_1 + \varepsilon_0\lambda_1 - \lambda\alpha}{2} < 0,$$

because of the choice of α. Hence, $v_0 := t\varphi_1$ satisfies the conclusion of lemma for large $t > 0$. □

The following result states that the Palais–Smale condition (see Appendix D) on J_ε is satisfied uniformly in ε.

Lemma 6.3.3 *Let $(u_n)_{n\geq 1} \subset H_0^1(\Omega)$ and $0 < \varepsilon_n \leq \varepsilon_0$ be such that*

$$(J_{\varepsilon_n}(u_n))_n \text{ is bounded in } \mathbb{R} \quad \text{and} \quad J'_{\varepsilon_n}(u_n) \to 0 \text{ in } H^{-1}(\Omega). \tag{6.11}$$

Then $(u_n)_{n\geq 1}$ is relatively compact in $H_0^1(\Omega)$.

Proof It suffices to prove that $(u_n)_{n\geq 1}$ contains a bounded subsequence in $H_0^1(\Omega)$. Indeed, suppose we have already proved this claim. Then, there exist $0 \leq \varepsilon \leq \varepsilon_0$ and $u \in H_0^1(\Omega)$ such that, up to a subsequence and as $n \to \infty$ we have $\varepsilon_n \to \varepsilon$ and

$$\begin{aligned} u_n &\rightharpoonup u && \text{weakly in } H_0^1(\Omega), \\ u_n &\to u && \text{strongly in } L^2(\Omega), \\ u_n &\to u && \text{almost everywhere in } \Omega. \end{aligned}$$

Because $J'_{\varepsilon_n}(u_n) \to 0$ in $H^{-1}(\Omega)$ we deduce

$$-\Delta u_n - \lambda f(u_n) - \varepsilon_n\Delta(u_n - u_0) \to 0 \quad \text{in } \mathcal{D}'(\Omega) \text{ as } n \to \infty. \tag{6.12}$$

Note that $f' \leq a$ implies

$$|f(u_n) - f(u)| \leq a|u_n - u| \quad \text{in } \Omega,$$

which yields $f(u_n) \to f(u)$ in $L^2(\Omega)$ as $n \to \infty$. Thus, from (6.12) we have

$$-\Delta u - \lambda f(u) - \varepsilon\Delta(u - u_0) = 0. \tag{6.13}$$

Multiplying (6.13) by u and integrating over Ω we obtain

$$(1+\varepsilon)\int_\Omega |\nabla u|^2 dx - \lambda \int_\Omega u f(u) dx - \varepsilon\lambda \int_\Omega u f(u_0) dx = 0. \tag{6.14}$$

From (6.11) and the boundedness of $(u_n)_{n\geq 1}$ we also deduce

$$J'_\varepsilon(u_n)u_n \to 0 \quad \text{as } n \to \infty.$$

Therefore, as $n \to \infty$ we have

$$(1+\varepsilon_n)\int_\Omega |\nabla u_n|^2 dx - \lambda \int_\Omega u_n f(u_n) dx - \varepsilon_n\lambda \int_\Omega u_n f(u_0) dx \to 0. \tag{6.15}$$

Combining now (6.14) with (6.15) we obtain $u_n \to u$ in $H_0^1(\Omega)$ as $n \to \infty$.

We now establish that $(u_n)_{n\geq 1}$ is, up to a subsequence, bounded in $H_0^1(\Omega)$. To this aim, it suffices to show that passing to a subsequence $(u_n)_{n\geq 1}$ is bounded in $L^2(\Omega)$. Then, by virtue of (6.11) we deduce that $(u_n)_{n\geq 1}$ is bounded in $H_0^1(\Omega)$.

Let $u_n = k_n w_n$ with $\|w_n\|_2 = 1$, $k_n > 0$ and assume by contradiction that up to a subsequence we have $k_n \to \infty$ as $n \to \infty$. We also may assume that $\varepsilon_n \to \varepsilon$. By (6.11) we derive

$$\frac{J_{\varepsilon_n}(u_n)}{k_n^2} \to 0 \quad \text{as } n \to \infty.$$

That is,

$$\frac{1}{2}\int_\Omega |\nabla w_n|^2 dx - \frac{1}{k_n^2}\int_\Omega F(u_n) dx + \frac{\varepsilon_n}{2}\int_\Omega \left|\nabla\left(w_n - \frac{u_0}{k_n}\right)\right|^2 dx \to 0 \tag{6.16}$$

as $n \to \infty$. Notice that

$$\begin{aligned}\frac{\varepsilon_n}{2}\int_\Omega \left|\nabla\left(w_n - \frac{u_0}{k_n}\right)\right|^2 dx &= \frac{\varepsilon_n}{2}\int_\Omega |\nabla w_n|^2 dx \\ &\quad + \frac{\varepsilon_n}{2k_n^2}\int_\Omega |\nabla u_0|^2 dx - \frac{\varepsilon_n\lambda}{k_n}\int_\Omega w_n f(u_0) dx,\end{aligned}$$

so that (6.16) produces

$$\lim_{n\to\infty}\left\{\frac{1+\varepsilon_n}{2}\int_\Omega |\nabla w_n|^2 dx - \frac{1}{k_n^2}\int_\Omega F(u_n) dx\right\} = 0. \tag{6.17}$$

Because $F(u_n) = F(k_n w_n) \leq \lambda a k_n^2 w_n^2/2 + \lambda b k_n w_n$ (where $b = f(0)$), from (6.17) we deduce that $(w_n)_{n\geq 1}$ is bounded in $H_0^1(\Omega)$. Let $w \in H_0^1(\Omega)$ be such that up to a subsequence and as $n \to \infty$ we have

$$\begin{aligned} w_n &\rightharpoonup w \quad \text{weakly in } H_0^1(\Omega), \\ w_n &\to w \quad \text{strongly in } L^2(\Omega), \\ w_n &\to w \quad \text{almost everywhere in } \Omega. \end{aligned}$$

We claim that

$$-(1+\varepsilon)\Delta w = \lambda a w. \tag{6.18}$$

Indeed, dividing (6.11) by k_n we have

$$\lim_{n\to\infty}\left\{(1+\varepsilon_n)\int_\Omega \nabla w_n\cdot\nabla v dx - \lambda\int_\Omega \frac{f(u_n)}{k_n} v dx\right\} = 0, \tag{6.19}$$

for each $v\in H_0^1(\Omega)$. This implies

$$(1+\varepsilon)\int_\Omega \nabla w\cdot\nabla v dx = \lambda \lim_{n\to\infty}\int_\Omega \frac{f(u_n)}{k_n} v dx \quad \text{for all } v\in H_0^1(\Omega). \tag{6.20}$$

With the same arguments as in the proof of Theorem 5.4.1, we derive that $(f(u_n)/k_n)_{n\geq 1}$ converges almost everywhere to aw in Ω. Moreover, because f is asymptotically linear and up to a subsequence $(w_n)_{n\geq 1}$ is dominated in $L^2(\Omega)$, we obtain $f(u_n)/k_n \to aw$ in $L^2(\Omega)$ as $n\to\infty$. Now, from (6.20) we obtain (6.18). Furthermore, from (6.18) we deduce that $w=\varphi_1$ and $\lambda a/(1+\varepsilon)=\lambda_1$, which contradicts the choice of ε_0. This completes the proof. □

Let us come back to the proof of (v) in Theorem 6.3.1. By the Ambrosetti–Rabinowitz theorem, for all $0\leq\varepsilon\leq\varepsilon_0$ there exists $v_\varepsilon\in H_0^1(\Omega)$ such that

$$-\Delta v_\varepsilon = \frac{\lambda}{1+\varepsilon} f(v_\varepsilon) + \frac{\lambda\varepsilon}{1+\varepsilon} f(u_0) \quad \text{in } \Omega. \tag{6.21}$$

Set

$$c_\varepsilon = J_\varepsilon(v_\varepsilon).$$

Because J_ε increases with ε, we have $c_0\leq c_\varepsilon\leq c_{\varepsilon_0}$—that is, $(J_\varepsilon(v_\varepsilon))_{\varepsilon>0}$ is bounded. Furthermore, by Lemma 6.3.3 there exists $v\in H_0^1(\Omega)$ such that, up to a subsequence, $v_\varepsilon\to v$ in $H_0^1(\Omega)$ as $\varepsilon\to 0$. Now (6.21) implies $-\Delta v=\lambda f(v)$ in Ω and, by standard regularity arguments, v is a classical solution of (5.4). To conclude the proof of (v) it remains only to show that $v\neq u_0=u_\lambda$. This will be achieved by Theorem 5.2.1 and the fact that v is an unstable solution of (5.4).

It can be readily seen that u_0 is the minimal solution of (6.21). Therefore, by Theorem 5.2.1 applied to the mapping

$$\mathbb{R}\ni t\longmapsto \frac{\lambda}{1+\varepsilon} f(t) + \frac{\lambda\varepsilon}{1+\varepsilon} f(u_0),$$

it follows that v_ε is unstable in the sense that

$$\lambda_1\left(\frac{\lambda}{1+\varepsilon} f'(v_\varepsilon)\right)\leq 0.$$

By virtue of Lemma 6.2.4 we derive that $\lambda_1(\lambda f'(v))\leq 0$ and thus $u_0\neq v$. This completes the proof of (v).

(vi) Suppose the contrary. With the same arguments as in the proof of Theorem 5.3.1 there exist $(\lambda_n)_{n\geq 1} \subset (0,\infty)$ such that $\lambda_n \searrow \lambda_1/a$ and $v_n \in C^2(\overline{\Omega}) \cap C(\Omega)$ a solution of (5.4) for $\lambda = \lambda_n$ which is unstable and the sequence $(v_n)_{n\geq 1}$ is bounded in $L^1_{\text{loc}}(\Omega)$. We first claim that $(v_n)_{n\geq 1}$ is unbounded in $H^1_0(\Omega)$. Otherwise, let $w \in H^1_0(\Omega)$ be such that, up to a subsequence, $v_n \rightharpoonup w$ weakly in $H^1_0(\Omega)$ and strongly in $L^2(\Omega)$. Then

$$-\Delta v_n \to -\Delta w \quad \text{in } \mathcal{D}'(\Omega) \text{ as } n \to \infty$$

and

$$f(v_n) \to f(w) \quad \text{in } L^2(\Omega) \text{ as } n \to \infty,$$

which yields $-\Delta w = \lambda_1 f(w)/a$. Thus, $w \in C^2(\Omega) \cap C(\overline{\Omega})$ is a solution of (5.4) for $\lambda = \lambda_1/a$. From Lemma 6.2.4 we also deduce that $\lambda_1(\lambda_1 f'(w)/a) \leq 0$—that is, $w \neq u_{\lambda_1/a}$, which contradicts (iv) in the statement of our theorem.

Because of the nature of our problem, the fact that $(v_n)_{n\geq 1}$ is unbounded in $L^2(\Omega)$ implies that $(v_n)_{n\geq 1}$ is also unbounded in $H^1_0(\Omega)$. Let $v_n = k_n w_n$ with $\|w_n\|_2 = 1$, $k_n > 0$ and, up to a subsequence, $k_n \to \infty$ as $n \to \infty$. This yields

$$-\Delta w_n = \frac{\lambda_n}{k_n} f(v_n) \to 0 \quad \text{in } L^1_{\text{loc}}(\Omega) \text{ as } n \to \infty.$$

Also remark that the previous convergence holds in the distribution space $\mathcal{D}'(\Omega)$, and $(w_n)_{n\geq 1}$ is bounded in $H^1_0(\Omega)$ with an already provided argument. If w is a limit point of $(w_n)_{n\geq 1}$ in $H^1_0(\Omega)$, we obtain $-\Delta w = 0$ in Ω and $\|w\|_2 = 1$, the desired contradiction.

(vii) As before, it is enough to prove that $(v_\lambda)_{\lambda<\lambda^*}$ is bounded in $L^2(\Omega)$ as $\lambda \nearrow \lambda^*$, and then to use the uniqueness of u^*.

Supposing the contrary, there exist $(\lambda_n)_{\lambda<\lambda^*}$, $\lambda_n \nearrow \lambda^*$ and v_n a solution of (5.4) for $\lambda = \lambda_n$ such that $\|v_n\|_2 \to \infty$ as $n \to \infty$. If we write again $v_n = k_n w_n$, then

$$-\Delta w_n = \frac{\lambda_n}{k_n} f(u_n) \quad \text{in } \Omega. \tag{6.22}$$

The fact that the right-hand side of (6.22) is bounded in $L^2(\Omega)$ implies that $(w_n)_{n\geq 1}$ is bounded in $H^1_0(\Omega)$. Let w be such that, up to a subsequence, $w_n \to w$ weakly in $H^1_0(\Omega)$ and strongly in $L^2(\Omega)$. A computation already done shows that

$$-\Delta w = \lambda^* a w,\ w \geq 0 \text{ in } \Omega \text{ and } \|w\|_2 = 1,$$

which forces $w \equiv \varphi_1$ and $\lambda^* = \lambda_1/a$. This is clearly a contradiction because $\lambda^* > \lambda_1/a$. This concludes the proof of Theorem 6.3.1. □

With the same arguments as in Section 5.4 we may describe the asymptotic behavior of the unstable solutions as $\lambda \searrow \lambda_1/a$.

Theorem 6.3.4 *Let $f : [0,\infty) \to (0,\infty)$ be a C^1 convex function that satisfies hypotheses (5.5), (6.1), and $f'(0) > 0$.*

(i) *For any unstable solution* v_λ *of (5.4) we have*

$$\lim_{\lambda \searrow \lambda_1/a} (\lambda_1 - a\lambda)\|v_\lambda\|_2 = \ell.$$

(ii) *If* $v_\lambda = k(\lambda)w_\lambda$ *with* $\|w_\lambda\|_2 = 1$ *we have* $w_\lambda \to \varphi_1$ *in* $C^1(\overline{\Omega})$ *as* $\lambda \searrow \lambda_1/a$, *and the quotient* φ_1/w_λ *is uniformly bounded when* λ *approaches* λ_1/a.

6.4 An example

Example 6.1 Let $f(t) = t + 2 - \sqrt{t+1}$, $t > 0$. Then

$$\lim_{\lambda \searrow \lambda_1} (\lambda - \lambda_1)^2 \|v_\lambda\|_2 = \left(\lambda_1 \int_\Omega \varphi_1^{3/2} dx\right)^2.$$

Proof Multiplying by φ_1 in (5.4) and integrating over Ω we obtain

$$(\lambda_1 - \lambda) \int_\Omega \varphi_1 v_\lambda dx = \lambda \int_\Omega (2 - \sqrt{v_\lambda + 1})\varphi_1 dx.$$

Adding $\lambda \int_\Omega \sqrt{v_\lambda}\varphi_1 dx$ to both sides of the previous equality and then multiplying by $\lambda - \lambda_1$ we obtain

$$\begin{aligned} \int_\Omega \varphi_1(\lambda - \lambda_1)\sqrt{v_\lambda}(\lambda - (\lambda - \lambda_1)\sqrt{v_\lambda})dx &= 2\lambda(\lambda - \lambda_1) \int_\Omega \varphi_1 dx \\ &\quad - \lambda \int_\Omega \varphi_1(\lambda - \lambda_1)(\sqrt{v_\lambda + 1} - \sqrt{v_\lambda})dx. \end{aligned} \tag{6.23}$$

Let $v_\lambda = k(\lambda)w_\lambda$, where $k(\lambda)$, w_λ are as usual.

We first prove that $\limsup_{\lambda \searrow \lambda_1} (\lambda - \lambda_1)^2 k(\lambda) < \infty$. Supposing the contrary, up to a subsequence, we have $(\lambda - \lambda_1)^2 k(\lambda) \to \infty$ as $\lambda \searrow \lambda_1$. Then the right-hand side of (6.23) tends to zero whereas the left-hand side goes to $-\infty$ as $\lambda \searrow \lambda_1$, which is a contradiction.

Furthermore, from (6.23) we have

$$\int_\Omega \varphi_1 \sqrt{v_\lambda}(\lambda - (\lambda - \lambda_1)\sqrt{v_\lambda})dx \le 2\lambda \int_\Omega \varphi_1 dx. \tag{6.24}$$

If $\liminf_{\lambda \searrow \lambda_1} (\lambda - \lambda_1)^2 k(\lambda) = 0$ then the left-hand side of (6.24) goes to ∞, which is again a contradiction. Now let $c \in (0, \infty)$ be a limit point of $(\lambda - \lambda_1)^2 k(\lambda)$ as $\lambda \searrow \lambda_1$. From (6.23) we deduce $c = (\lambda_1 \int_\Omega \varphi_1^{3/2} dx)^2$ and the proof is now complete. □

6.5 Comments and historical notes

We have seen in Theorem 6.3.1 that λ^* is a returning point from which unstable solutions v_λ begin to emanate, which exist for any $\lambda \in (\lambda_1/a, \lambda^*)$. It has been proved that any such solution v_λ has the same behavior, in the sense that

$$\lim_{\lambda \searrow \lambda_1/a} v_\lambda = \infty \qquad \text{uniformly on compact subsets of } \Omega$$

and

$$\lim_{\lambda \nearrow \lambda^*} v_\lambda = u^* \qquad \text{uniformly in } \overline{\Omega},$$

where u^* is the (unstable) solution corresponding to $\lambda = \lambda^*$.

An interesting *open problem* is to establish whether there is a unique unstable solution v_λ, for any $\lambda \in (\lambda_1/a, \lambda^*)$.

Returning points exist in many bifurcation problems. An interesting phenomenon is related to the Liouville–Gelfand equation (5.32) corresponding to $f(t) = e^t$, provided $\Omega \subset \mathbb{R}^N$ is a *ball* (see, for instance, [111]). In such a case the bifurcation diagram strongly depends on N, as depicted in Figure 6.3.

These results can be described as follows:

CASE 1: $1 \leq N \leq 2$. Then there exists a unique solution for $\lambda = \lambda^*$ and exactly two solutions for any $\lambda \in (0, \lambda^*)$. Moreover, the unstable solution v_λ tends to ∞ as $\lambda \searrow 0$, uniformly on compact subsets of Ω. This is a result of the work by Liouville [132] (if $N = 1$) and Bratu [29] (if $N = 2$).

CASE 2: $2 < N < 10$. Then there exists a continuum of solutions that oscillate around the line $\lambda = 2(N-2)$. This result was found by Gelfand [82].

CASE 3: $N \geq 10$. Then $\lambda^* = 2(N-2)$ and there is a unique solution for any $\lambda \in (0, 2(N-2))$. This result is a work by Joseph and Lundgren [112].

If $N \leq 9$ then the *extremal solution* u^* corresponding to λ^* is smooth. If $N \geq 10$ then u^* is unbounded and, in this case, $u^*(x) = \ln\left(1/|x|^2\right)$ provided Ω is the unit ball in $\mathbb{R}^N$. In this situation, u^* fails to be a classical solution but it is the unique weak solution of the Liouville–Gelfand problem in the limiting case $\lambda = \lambda^*$.

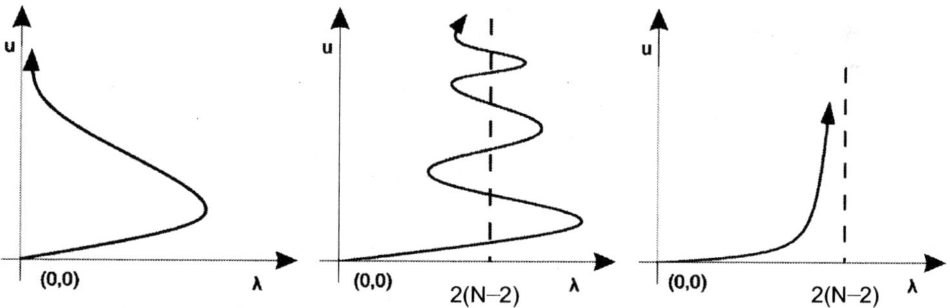

FIGURE 6.3. The bifurcation diagram for the Liouville–Gelfand problem.

7

SUPERLINEAR PERTURBATIONS OF SINGULAR ELLIPTIC PROBLEMS

> The art of doing mathematics consists in finding that special case which contains all the germs of generality.
>
> David Hilbert (1862–1943)

7.1 Introduction

In this chapter we are interested in the study of singular elliptic problems in the presence of smooth nonlinearities having a superlinear growth at infinity. Our analysis, which includes existence, regularity, bifurcation, and asymptotic behavior of solutions with respect to the parameters, will concern the model problem

$$\begin{cases} -\Delta u = u^{-\alpha} + \lambda u^p & \text{in } \Omega, \\ u > 0 & \text{in } \Omega, \\ u = 0 & \text{on } \partial\Omega \end{cases} \tag{7.1}$$

in a smooth bounded domain $\Omega \subset \mathbb{R}^N$, where $\lambda > 0$, $0 < \alpha < 1$.

As a result of the continuous (or even compact) embedding of $H_0^1(\Omega)$ into the standard $L^p(\Omega)$ spaces, we will be concerned with the case $N \geq 3$ and $1 < p \leq (N+2)/(N-2)$.

We have already seen in Theorem 4.3.2 that if $0 < p < 1$, then (7.1) has a unique solution for all $\lambda \geq 0$. Furthermore, if $p = 1$ then problem (7.1) has solutions if and only if $\lambda < \lambda_1$, and in this case the solution is unique. Uniqueness of the solution has been obtained so far via Theorem 1.3.17. In our setting the assumption $(H1)$ in Theorem 1.3.17 is not fulfilled, so that the issue of multiplicity of the solution to (7.1) is raised.

The first task in this chapter is to investigate the existence of *weak solutions* to problem (7.1) in the following sense.

Definition 7.1.1 *We say that $u \in H_0^1(\Omega)$ is a weak solution of (7.1) if $u > 0$ in Ω and*

$$\int_\Omega \nabla u \nabla \phi dx = \int_\Omega (u^{-\alpha} + \lambda u^p)\phi dx \quad \textit{for all } \phi \in H_0^1(\Omega). \tag{7.2}$$

In contrast to the case $0 < p \leq 1$, a different approach is needed here because the hypotheses of Theorems 1.2.5 and 1.3.17 are not satisfied. However, we are able to show that for small values of λ, problem (7.1) has at least two weak solutions, and no solutions exist if λ is large. As usual in this case, the existence is proved by considering the *energy functional*

$$J_\lambda(u) = \frac{1}{2}\int_\Omega |\nabla u|^2 dx - \frac{1}{1-\alpha}\int_\Omega u^{1-\alpha}dx - \frac{\lambda}{p+1}\int_\Omega u^{p+1}dx, \quad u \in H_0^1(\Omega).$$

The main difficulty in the study of (7.1) here consists not only of the fact that J_λ is not differentiable, but also of the presence of the unbounded term $u^{-\alpha}$ combined with the presence of the superlinear term u^p. The existence of the first solution is obtained by the sub- and supersolution method in the weak sense presented in the next section. Moreover, a direct analysis in the H^1 neighborhood of the solution reveals the fact that this solution is also a local minimizer with respect to the H^1 topology. The second solution is then obtained by Ekeland's variational principle. More precisely, the following result holds true.

Theorem 7.1.2 *Assume that $0 < \alpha < 1$ and $1 < p \leq (N+2)/(N-2)$. There exists $\lambda^* > 0$ such that*

(i) *for all $0 < \lambda < \lambda^*$, problem (7.1) has at least two weak solutions u_λ and v_λ such that $u_\lambda < v_\lambda$ in Ω;*
(ii) *for $\lambda = \lambda^*$, problem (7.1) has at least one weak solution u_λ;*
(iii) *for all $\lambda > \lambda^*$, problem (7.1) has no weak solutions.*

If α is small enough, then any weak solution of (7.1) is in fact a classical solution. This will be proved in Section 7.6. In Section 7.7, the problem of asymptotic behavior as $p \searrow 1$ will be addressed both for the bifurcation point λ^* and for solutions of (7.1). This matter will point out new and interesting features of problem (7.1). For instance, we shall see that if $\alpha < 1/N$, then (7.1) possesses a minimal solution (which is in fact u_λ in the statement of Theorem 7.1.2), and any other solution of (7.1) blows up in the L^∞ norm as $p \searrow 1$.

7.2 The weak sub- and supersolution method

In this section we present an existence method for weak solutions that is similar to that described in Theorem 1.2.2 for classical ones. To begin with, we introduce the concept of sub- and supersolution in the weak sense. For our convenience, let us set $f_\lambda(t) = t^{-\alpha} + \lambda t$ for λ, $t > 0$.

Definition 7.2.1 *We say that $u \in H_0^1(\Omega)$ is a weak subsolution of problem (7.1) if $u > 0$ in Ω and*

$$\int_\Omega \nabla u \nabla \phi dx \leq \int_\Omega f_\lambda(u)\phi dx,$$

for all $\phi \in H_0^1(\Omega)$, $\phi \geq 0$.

Reversing the sign in the previous inequality, we obtain the definition of a supersolution in the weak sense. The main result of this section is the following.

Theorem 7.2.2 *Let $\underline{u}$ and $\overline{u}$ be a weak subsolution respectively a weak supersolution of problem (7.1) such that $\underline{u} \le \overline{u}$ in Ω. Then there exists a weak solution $u \in H_0^1(\Omega)$ of (7.1) such that $\underline{u} \le u \le \overline{u}$ in Ω.*

Because we are dealing with functions in the space $H_0^1(\Omega)$, in the statement of Theorem 7.2.2 and throughout this chapter, the relation $\underline{u} \le \overline{u}$ should be understood that it holds almost everywhere in Ω.

Proof The solution of (7.1) will be obtained by minimizing the functional J_λ over the set

$$M := \{u \in H_0^1(\Omega) : \underline{u} \le u \le \overline{u} \text{ in } \Omega\}.$$

We first observe that M is convex and closed with respect to the H^1 topology. Furthermore, for all $u \in H_0^1(\Omega)$ we have

$$J_\lambda(u) \ge \frac{1}{2}\|u\|^2_{H_0^1(\Omega)} - \frac{1}{1-\alpha}\int_\Omega |\overline{u}|^{-\alpha} dx - \frac{\lambda}{p+1}\int_\Omega |\overline{u}|^p dx,$$

which implies that J_λ is coercive.

To apply Theorem C.1.1 we need to show that J_λ is weakly lower semicontinuous on M. To this aim, let $(u_n)_{n\ge 1} \subset M$ be an arbitrary sequence that converges weakly to u in M. Then, $u_n \to u$ almost everywhere in Ω as $n \to \infty$. Because $\underline{u} \le u_n \le \overline{u}$ in Ω, by Lebesgue's theorem on dominated convergence we obtain

$$\lim_{n\to\infty} \int_\Omega u_n^{1-\alpha} dx = \int_\Omega u^{1-\alpha} dx$$

and

$$\lim_{n\to\infty} \int_\Omega u_n^{p+1} dx = \int_\Omega u^{p+1} dx.$$

It follows that

$$\liminf_{n\to\infty} J_\lambda(u_n) \le J_\lambda(u).$$

Thus, J_λ verifies the hypotheses of Theorem C.1.1. According to this one, there exists a relative minimizer $u_\lambda \in M$ of J_λ. We show in what follows that u_λ is a solution of (7.1).

Let $\phi \in H_0^1(\Omega)$, $\varepsilon > 0$, and consider

$$\overline{w}_\varepsilon := (u_\lambda + \varepsilon\phi - \overline{u})^+, \quad \underline{w}_\varepsilon := (u_\lambda + \varepsilon\phi - \underline{u})^-.$$

Set $v_\varepsilon := u_\lambda + \varepsilon\phi - \overline{w}_\varepsilon + \underline{w}_\varepsilon$. Then, $v_\varepsilon \in M$, which implies that $u_\lambda + t(v_\varepsilon - u_\lambda) \in M$, for all $0 < t < 1$. This yields

$$J_\lambda(u_\lambda + t(v_\varepsilon - u_\lambda)) - J_\lambda(u_\lambda) \ge 0 \quad \text{for all } 0 < t < 1.$$

Dividing by t in the previous inequality and passing to the limit, by the mean value theorem we have

$$\int_\Omega \nabla u_\lambda \nabla(v_\varepsilon - u_\lambda)dx - \lim_{t\to 0}\int_\Omega (u_\lambda + \theta t(v_\varepsilon - u_\lambda))^{-\alpha}(v_\varepsilon - u_\lambda)dx \\ - \lambda \int_\Omega u_\lambda^p (v_\varepsilon - u_\lambda)dx \geq 0, \tag{7.3}$$

for some $0 < \theta < 1$. Notice that

$$|(u_\lambda + \theta t(v_\varepsilon - u_\lambda))^{-\alpha}(v_\varepsilon - u_\lambda)| \leq \underline{u}^{-\alpha}|v_\varepsilon - u_\lambda| \quad \text{in } \Omega.$$

Because $\underline{u}$ is a subsolution of (7.1), we have $\underline{u}^{-\alpha}|v_\varepsilon - u_\lambda| \in L^1(\Omega)$. Hence, by Lebesgue's theorem we derive

$$\lim_{t\to 0}\int_\Omega (u_\lambda + \theta t(v_\varepsilon - u_\lambda))^{-\alpha}(v_\varepsilon - u_\lambda)dx = \int_\Omega u_\lambda^{-\alpha}(v_\varepsilon - u_\lambda)dx. \tag{7.4}$$

From (7.3) and (7.4) we obtain

$$\int_\Omega \nabla u_\lambda \nabla(v_\varepsilon - u_\lambda)dx - \int_\Omega f_\lambda(u_\lambda)(v_\varepsilon - u_\lambda)dx \geq 0.$$

That is,

$$\int_\Omega (\nabla u_\lambda \nabla \phi - f_\lambda(u_\lambda)\phi)dx \geq \frac{A_\varepsilon - B_\varepsilon}{\varepsilon}, \tag{7.5}$$

where

$$A_\varepsilon := \int_\Omega \Big(\nabla u_\lambda \nabla \overline{w}_\varepsilon - f_\lambda(u_\lambda)\overline{w}_\varepsilon\Big)dx,$$

$$B_\varepsilon := \int_\Omega \Big(\nabla u_\lambda \nabla \underline{w}_\varepsilon - f_\lambda(u_\lambda)\underline{w}_\varepsilon\Big)dx.$$

Let us first evaluate A_ε. Because $\overline{w}_\varepsilon = (u_\lambda + \varepsilon\phi - \overline{u})\chi_{\{u_\lambda + \varepsilon\phi \geq \overline{u}\}}$, we have

$$\begin{aligned} A_\varepsilon &= \int_\Omega \nabla(u_\lambda - \overline{u})\nabla \overline{w}_\varepsilon dx + \int_\Omega \Big(\nabla \overline{u}\nabla \overline{w}_\varepsilon - f_\lambda(u_\lambda)\overline{w}_\varepsilon\Big)dx \\ &\geq \int_{\{u_\lambda + \varepsilon\phi \geq \overline{u}\}} |\nabla(u_\lambda - \overline{u})|^2 dx + \varepsilon \int_{\{u_\lambda + \varepsilon\phi \geq \overline{u}\}} \nabla(u_\lambda - \overline{u})\nabla\phi dx \\ &\quad + \int_{\{u_\lambda + \varepsilon\phi \geq \overline{u}\}} (f_\lambda(\overline{u}) - f_\lambda(u_\lambda))\overline{w}_\varepsilon dx \\ &\geq \varepsilon \int_{\{u_\lambda + \varepsilon\phi \geq \overline{u}\}} \nabla(u_\lambda - \overline{u})\nabla\phi dx + \int_{\{u_\lambda + \varepsilon\phi \geq \overline{u}\}} (\overline{u}^{-\alpha} - u_\lambda^{-\alpha})\overline{w}_\varepsilon dx. \end{aligned}$$

The last integral in the right-hand side can be evaluated as

$$
\begin{aligned}
\int_{\{u_\lambda+\varepsilon\phi\geq\overline{u}\}} (\overline{u}^{-\alpha} - u_\lambda^{-\alpha})\overline{w}_\varepsilon dx = & \int_{\{u_\lambda+\varepsilon\phi\geq\overline{u}\}} (\overline{u}^{-\alpha} - u_\lambda^{-\alpha})(u_\lambda - \overline{u}) dx \\
& + \varepsilon \int_{\{u_\lambda+\varepsilon\phi\geq\overline{u}\}} (\overline{u}^{-\alpha} - u_\lambda^{-\alpha})\phi dx \\
\geq & \, \varepsilon \int_{\{u_\lambda+\varepsilon\phi\geq\overline{u}\}} (\overline{u}^{-\alpha} - u_\lambda^{-\alpha})\phi dx \\
\geq & - \varepsilon \|\phi\|_\infty \int_{\{u_\lambda+\varepsilon\phi\geq\overline{u}\}} (\underline{u}^{-\alpha} - \overline{u}^{-\alpha}) dx.
\end{aligned}
$$

Thus, we obtain

$$
\frac{A_\varepsilon}{\varepsilon} \geq \int_{\{u_\lambda+\varepsilon\phi\geq\overline{u}\}} \nabla(u_\lambda - \overline{u})\nabla\phi dx - \|\phi\|_\infty \int_{\{u_\lambda+\varepsilon\phi\geq\overline{u}\}} (\underline{u}^{-\alpha} - \overline{u}^{-\alpha}) dx.
$$

Because the Lebesgue measure of the set $\{u_\lambda + \varepsilon\phi \geq \overline{u}\}$ tends to zero as $\varepsilon \to 0$, the previous inequality leads us to $A_\varepsilon/\varepsilon \geq o(1)$ as $\varepsilon \to 0$. Similarly, we obtain $B_\varepsilon/\varepsilon \leq o(1)$ as $\varepsilon \to 0$. Therefore, from (7.5) we find

$$
\int_\Omega \Big(\nabla u_\lambda \nabla\phi - f_\lambda(u_\lambda)\phi\Big) dx \geq o(1) \qquad \text{as } \varepsilon \to 0.
$$

Replacing ϕ with $-\phi$ in the previous relations and letting $\varepsilon \to 0$, we deduce

$$
\int_\Omega \Big(\nabla u_\lambda \nabla\phi - f_\lambda(u_\lambda)\phi\Big) dx = 0 \qquad \text{for all } \phi \in H_0^1(\Omega).
$$

Hence, u_λ is a weak solution of (7.1). The proof is now complete. □

7.3 H^1 local minimizers

An interesting property of the weak solutions obtained by the method described in Section 7.2 is that these are H^1 local minimizers for the energy functional J_λ. To be more specific, a function $u \in H_0^1(\Omega)$ is called a *local minimizer* for J_λ in the H^1 topology if there exists $r > 0$ such that

$$
J_\lambda(u) \leq J_\lambda(v) \qquad \text{for all } v \in H_0^1(\Omega) \text{ with } \|v - u\|_{H_0^1(\Omega)} < r.
$$

The main result of this section is the following.

Theorem 7.3.1 *If u_λ is the solution of (7.1) obtained in Theorem 7.2.2, then u_λ is a local minimizer of J_λ in the H^1 topology.*

Proof Supposing to the contrary, there exists $(u_n)_{n\geq 1} \subset H_0^1(\Omega)$ such that $u_n \to u_\lambda$ in $H_0^1(\Omega)$ as $n \to \infty$ and $J_\lambda(u_n) < J_\lambda(u_\lambda)$. Define

$$
\overline{w}_n := (u_n - \overline{u})^+, \quad \underline{w}_n := (u_n - \underline{u})^-.
$$

Let also $E_n = \operatorname{supp} \overline{w}_n$ and $F_n = \operatorname{supp} \underline{w}_n$. We first need the following result.

Lemma 7.3.2 $\lim_{\varepsilon\to 0}|E_n| = \lim_{\varepsilon\to 0}|F_n| = 0$.

Proof Let $\varepsilon > 0$. Then there exists $\delta > 0$ such that $|\Omega \setminus \Omega_\delta| < \varepsilon/2$, where $\Omega_\delta = \{x \in \Omega : \mathrm{dist}(x, \partial\Omega) > \delta\}$. By Corollary 1.3.8 we have

$$\underline{u} \geq c\,\mathrm{dist}(x, \partial\Omega) \geq \frac{c\delta}{2} > 0 \quad \text{for all } x \in \Omega_{\delta/2}.$$

Then,

$$-\Delta(\overline{u} - u_\lambda) \geq -\alpha \underline{u}^{-1-\alpha}(\overline{u} - u_\lambda) \geq -\alpha\left(\frac{c\delta}{2}\right)^{-1-\alpha}(\overline{u} - u_\lambda) \quad \text{in } \Omega_{\delta/2}.$$

Again by Corollary 1.3.8 we obtain

$$\overline{u} - u_\lambda \geq c_1 \mathrm{dist}(x, \partial\Omega_{\delta/2}) \geq c_2 > 0 \quad \text{for all } x \in \Omega_\delta.$$

Hence,

$$\begin{aligned} |E_n| &\leq |\Omega \setminus \Omega_\delta| + |E_n \cap \Omega_\delta| \\ &< \frac{\varepsilon}{2} + \frac{1}{c_2^2}\int_{E_n\cap\Omega_\delta}(u_n - u_\lambda)^2 dx \\ &\leq \frac{\varepsilon}{2} + \frac{1}{c_2^2}\int_\Omega (u_n - u_\lambda)^2 dx. \end{aligned}$$

Because $u_n \to u_\lambda$ in $H_0^1(\Omega)$, for n sufficiently large, the previous estimate yields $|E_n| < \varepsilon$. Thus, $\lim_{\varepsilon\to 0}|E_n| = 0$, and similarly we obtain $\lim_{\varepsilon\to 0}|F_n| = 0$. This finishes the proof. □

Because $(\overline{w}_n)_{n\geq 1}$ and $(\underline{w}_n)_{n\geq 1}$ are bounded in $H_0^1(\Omega)$, by Lemma 7.3.2 we have

$$\overline{w}_n \to 0 \quad \text{and} \quad \underline{w}_n \to 0 \quad \text{in } H_0^1(\Omega) \quad \text{as } n \to \infty. \tag{7.6}$$

Set $v_n := \max\{\underline{u}, \min\{u_n, \overline{u}\}\}$. Then $v_n \in M$, $v_n = \overline{u}$ in E_n, and $v_n = \underline{u}$ in F_n. Furthermore,

$$J_\lambda(u_n) = J_\lambda(v_n) + A_n + B_n, \tag{7.7}$$

where

$$\begin{aligned} A_n =& \frac{1}{2}\int_{E_n}\left(|\nabla u_n|^2 - |\nabla \overline{u}|^2\right)dx - \frac{1}{1-\alpha}\int_{E_n}\left(|u_n|^{1-\alpha} - \overline{u}^{1-\alpha}\right)dx \\ &- \frac{\lambda}{p+1}\int_{E_n}\left(|u_n|^{p+1} - \overline{u}^{p+1}\right)dx, \end{aligned}$$

$$\begin{aligned} B_n =& \frac{1}{2}\int_{F_n}\left(|\nabla u_n|^2 - |\nabla \underline{u}|^2\right)dx - \frac{1}{1-\alpha}\int_{F_n}\left(|u_n|^{1-\alpha} - \underline{u}^{1-\alpha}\right)dx \\ &- \frac{\lambda}{p+1}\int_{F_n}\left(|u_n|^{p+1} - \underline{u}^{p+1}\right)dx. \end{aligned}$$

Because u_λ is a minimizer of J_λ restricted to M, it follows that

$$J_\lambda(u_n) \geq J_\lambda(u_\lambda) + A_n + B_n. \tag{7.8}$$

To evaluate A_n, we use the fact that $u_n = \overline{u} + \overline{w}_n$ in E_n. By the mean value theorem we have

$$\begin{aligned}
A_n =& \frac{1}{2}\int_{E_n}\Big(|\nabla(\overline{u}+\overline{w}_n)|^2 - |\nabla\overline{u}|^2\Big)dx - \frac{1}{1-\alpha}\int_{E_n}\Big(|\overline{u}+\overline{w}_n|^{1-\alpha} - \overline{u}^{1-\alpha}\Big)dx \\
&- \frac{\lambda}{p+1}\int_{E_n}\Big(|\overline{u}+\overline{w}_n|^{p+1} - \overline{u}^{p+1}\Big)dx \\
\geq& \frac{1}{2}\|\overline{w}_n\|^2_{H_0^1(\Omega)} + \int_{E_n}\nabla\overline{u}\nabla\overline{w}_n dx - \int_{E_n} f_\lambda(\overline{u}+\theta\overline{w}_n)\overline{w}_n dx,
\end{aligned}$$

for some $0 < \theta < 1$. Because $\overline{u}$ is a weak supersolution of (7.1), we deduce

$$\begin{aligned}
A_n \geq& \frac{1}{2}\|\overline{w}_n\|^2_{H_0^1(\Omega)} + \int_{E_n}(\overline{u}^{-\alpha} + \lambda\overline{u}^p)\overline{w}_n dx \\
&- \int_{E_n}\Big((\overline{u}+\theta\overline{w}_n)^{-\alpha} + \lambda(\overline{u}+\theta\overline{w}_n)^p\Big)\overline{w}_n dx \\
\geq& \frac{1}{2}\|\overline{w}_n\|^2_{H_0^1(\Omega)} - \lambda\int_{E_n}\Big((\overline{u}+\theta\overline{w}_n)^p - \overline{u}^p\Big)\overline{w}_n dx \\
\geq& \frac{1}{2}\|\overline{w}_n\|^2_{H_0^1(\Omega)} - c\lambda\int_{E_n}(\overline{u}^{p-1} + \overline{w}_n^{p-1})\overline{w}_n^2 dx.
\end{aligned}$$

By (7.6), Lemma 7.3.2, and Sobolev's inequality, for n large enough we derive

$$\begin{aligned}
A_n \geq& \frac{1}{2}\|\overline{w}_n\|^2_{H_0^1(\Omega)} - C\left(\int_{E_n}\overline{u}^{p+1}dx\right)^{(p-1)/(p+1)}\|\overline{w}_n\|^2_{H_0^1(\Omega)} - C\|\overline{w}_n\|^{p+1}_{H_0^1(\Omega)} \\
=& \frac{1}{2}\|\overline{w}_n\|^2_{H_0^1(\Omega)} - o(1)\|\overline{w}_n\|^2_{H_0^1(\Omega)} \geq 0,
\end{aligned}$$

for some positive constant $C > 0$. In the same way we obtain $B_n \geq 0$. Hence, by (7.7) it follows that $J_\lambda(u_n) \geq J_\lambda(u_\lambda)$, which contradicts our assumption. This finishes the proof. □

7.4 Existence of the first solution

By means of the weak sub- and supersolution method described in Section 7.2 we establish the existence of the first solution of (7.1).

The core of this section is the following result, which is the first step in proving Theorem 7.1.2.

Theorem 7.4.1 *There exists $0 < \lambda^* < \infty$ such that for all $0 < \lambda \leq \lambda^*$ problem (7.1) has at least one weak solution u_λ, and no solutions exist if $\lambda > \lambda^*$.*

Because the embedding $H_0^1(\Omega) \hookrightarrow L^p(\Omega)$ is compact for $1 < p < (N+2)/(N-2)$, we consider only the critical case $p = (N+2)/(N-2)$. As we did in previous chapters, let us define

$$A := \{\lambda > 0 : \text{ problem (7.1) has at least one weak solution}\}$$

and set $\lambda^* := \sup A$.

Proposition 7.4.2 *The set A is nonempty and λ^* is finite.*

Proof By Sobolev's and Hölder's inequality we derive

$$J_\lambda(u) \geq \frac{1}{2}\|u\|^2_{H_0^1(\Omega)} - C\|u\|^{1-\alpha}_{H_0^1(\Omega)} - C\lambda\|u\|^{p+1}_{H_0^1(\Omega)},$$

for all $u \in H_0^1(\Omega)$. Thus, we can choose $\lambda > 0$ small enough and $r, \delta > 0$ such that

$$\begin{aligned} \frac{1}{2}\|u\|^2_{H_0^1(\Omega)} - \frac{\lambda}{p+1}\|u\|^{p+1}_{p+1} &\geq 2\delta && \text{for all } u \in \partial B_r, \\ \frac{1}{2}\|u\|^2_{H_0^1(\Omega)} - \frac{\lambda}{p+1}\|u\|^{p+1}_{p+1} &\geq 0 && \text{for all } u \in B_r, \\ J_\lambda(u) &\geq \delta && \text{for all } u \in \partial B_r, \end{aligned} \tag{7.9}$$

where $B_r := \{u \in H_0^1(\Omega) : \|u\|_{H_0^1(\Omega)} < r\}$. Set

$$c := \inf_{u \in B_r} J_\lambda(u).$$

Because $0 < 1 - \alpha < 1$, for all $u \in B_r \setminus \{0\}$ there exists $t > 0$ small enough such that $J_\lambda(tu) < 0$. This yields $c < 0$. Let $(u_n)_{n\geq 1} \subset B_r$ be a minimizing sequence for c. Then there exists $u \in B_r$ such that, up to a subsequence, we have as $n \to \infty$

$$\begin{aligned} u_n &\rightharpoonup u && \text{weakly in } H_0^1(\Omega), \\ u_n &\to u && \text{strongly in } L^q(\Omega) \text{ for all } 2 \leq q < (N+2)/(N-2), \\ u_n &\to u && \text{almost everywhere in } \Omega. \end{aligned}$$

Because $J_\lambda(u_n) = J_\lambda(|u_n|)$, replacing if necessary u_n by u_n^+ we may assume that $u_n \geq 0$ in Ω, for all $n \geq 1$. By Hölder's inequality we have as $n \to \infty$

$$\begin{aligned} \int_\Omega u_n^{1-\alpha}dx &\leq \int_\Omega u^{1-\alpha}dx + \int_\Omega |u_n - u|^{1-\alpha}dx \\ &\leq \int_\Omega u^{1-\alpha}dx + C\|u_n - u\|_2^{1-\alpha} \\ &= \int_\Omega u^{1-\alpha}dx + o(1). \end{aligned}$$

In the same manner as shown earlier we obtain

$$\int_\Omega u^{1-\alpha}dx \le \int_\Omega u_n^{1-\alpha}dx + \int_\Omega |u_n - u|^{1-\alpha}dx = \int_\Omega u_n^{1-\alpha}dx + o(1).$$

Thus,

$$\int_\Omega u^{1-\alpha}dx = \int_\Omega u_n^{1-\alpha}dx + o(1) \quad \text{as } n \to \infty. \tag{7.10}$$

A similar relation holds in the L^p norm, even in a more general setting.

Lemma 7.4.3 (Brezis–Lieb) *Let $1 \le p < \infty$ and $(u_n)_{n\ge1} \subset L^p(\Omega)$ be a bounded sequence such that $u_n \to u$ almost everywhere in Ω as $n \to \infty$. Then $u \in L^p(\Omega)$ and*

$$\|u\|_p^p = \lim_{n\to\infty} (\|u_n\|_p^p - \|u_n - u\|_p^p).$$

Proof Let $\varepsilon > 0$ and $M := \sup_{n\ge1} \|u_n\|_p$. Let us first notice that there exists $C_\varepsilon > 0$ such that

$$\Big||t+1|^p - |t|^p - 1\Big| \le \varepsilon|t|^p + C_\varepsilon \quad \text{for all } t \in \mathbb{R}. \tag{7.11}$$

This comes from the fact that

$$\lim_{|t|\to\infty} \frac{|t+1|^p - |t|^p - 1}{|t|^p} = 0.$$

From (7.11) we derive

$$\Big||a+b|^p - |a|^p - |b|^p\Big| \le \varepsilon|a|^p + C_\varepsilon|b|^p \quad \text{for all } a, b \in \mathbb{R}. \tag{7.12}$$

Let

$$\begin{aligned} v_n &:= \Big||u_n|^p - |u_n - u|^p - |u|^p\Big|, \\ w_n &:= (v_n - \varepsilon|u_n - u|^p)^+. \end{aligned}$$

By (7.12) we have $w_n \to 0$ almost everywhere in Ω as $n \to \infty$ and $0 \le w_n \le C_\varepsilon|u|^p$ in Ω. Thus, by Lebesgue's theorem we obtain $\|w_n\|_1 \to 0$ as $n \to \infty$. On the other hand,

$$0 \le v_n \le \varepsilon|u_n - u|^p + w_n \quad \text{almost everywhere in } \Omega,$$

which yields

$$\|v_n\|_1 \le \varepsilon\|u_n - u\|_p^p + \|w_n\|_1 \le \varepsilon 2^p M^p + \|w_n\|_1.$$

Hence, $\|v_n\|_1 \to 0$ as $n \to \infty$, which concludes the proof. □

By Lemma 7.4.3 we have

$$\|u_n\|_{p+1}^{p+1} = \|u\|_{p+1}^{p+1} + \|u_n - u\|_{p+1}^{p+1} + o(1). \tag{7.13}$$

Note also that

$$\|u_n\|_{H_0^1(\Omega)}^2 = \|u\|_{H_0^1(\Omega)}^2 + \|u_n - u\|_{H_0^1(\Omega)}^2 + o(1). \tag{7.14}$$

Because $J_\lambda \geq \delta$ on ∂B_r and $c < 0$, there exist $\varepsilon_0 > 0$ small enough and $n_0 \geq 1$ such that

$$\|u_n\|_{H_0^1(\Omega)} \leq r - \varepsilon_0 \quad \text{for all } n \geq n_0.$$

Now from (7.14) we find $u_n - u \in B_r$ for all $n \geq n_0$. Therefore, by (7.9) we have

$$\frac{1}{2}\|u_n - u\|_{H_0^1(\Omega)}^2 - \frac{\lambda}{p+1}\|u_n - u\|_{p+1}^{p+1} \geq 0. \tag{7.15}$$

From (7.10) and (7.13) through (7.15) we obtain

$$\begin{aligned} c &= J_\lambda(u_n) + o(1) \\ &= J_\lambda(u) + \frac{1}{2}\|u_n - u\|_{H_0^1(\Omega)}^2 - \frac{1}{p+1}\|u_n - u\|_{p+1}^{p+1} + o(1) \\ &\geq J_\lambda(u) + o(1) \geq c + o(1), \end{aligned}$$

for all $n \geq n_0$. Therefore $J_\lambda(u) = c$—that is, u is a local minimizer of J_λ in the H^1 topology.

Let $\phi \in H_0^1(\Omega)$, $\phi \geq 0$. Because $J_\lambda(u) < 0$ and $J_\lambda > 0$ on ∂B_r, it follows that $\|u\|_{H_0^1(\Omega)} < r$. Hence, for $t > 0$ small enough we still have $u + t\phi \in B_r$. This implies

$$J_\lambda(u + t\phi) - J_\lambda(u) \geq 0, \tag{7.16}$$

which in turn implies

$$\frac{1}{2}\int_\Omega |\nabla(u + t\phi)|^2 dx - \frac{1}{2}\int_\Omega |\nabla u|^2 dx \geq 0.$$

Dividing by $t > 0$ in the last inequality and then passing to the limit we find

$$\int_\Omega \nabla u \nabla \phi dx \geq 0 \quad \text{for all } \phi \in H_0^1(\Omega), \phi \geq 0.$$

This means that $-\Delta u \geq 0$ weakly in Ω. Because $u \geq 0$ in $\overline{\Omega}$, by Corollary 1.3.8 we obtain $u > 0$ in Ω.

We claim that u is a weak solution of (7.1). This assertion will be proved in three steps.

Step 1: For all $\phi \in H_0^1(\Omega)$, $\phi \geq 0$ we have

$$\int_\Omega (\nabla u \nabla \phi - f_\lambda(u)\phi)dx \geq 0. \tag{7.17}$$

From (7.16) we obtain

$$\frac{1}{1-\alpha}\int_\Omega \Big((u+t\phi)^{1-\alpha} - u^{1-\alpha}\Big)dx \leq \frac{1}{2}\int_\Omega \Big(|\nabla(u+t\phi)|^2 - |\nabla u|^2\Big)dx \\ - \frac{\lambda}{1+p}\int_\Omega \Big((u+t\phi)^{p+1} - u^{p+1}\Big)dx.$$

Dividing by $t > 0$ and then passing to the limit we have

$$\frac{1}{1-\alpha}\liminf_{t\searrow 0}\int_\Omega \frac{(u+t\phi)^{1-\alpha} - u^{1-\alpha}}{t}dx \leq \int_\Omega (\nabla u \nabla \phi - \lambda u^p \phi)dx. \tag{7.18}$$

Notice that

$$\frac{1}{1-\alpha}\int_\Omega \frac{(u+t\phi)^{1-\alpha} - u^{1-\alpha}}{t}dx = \int_\Omega (u+\theta t\phi)^{-\alpha}\phi dx,$$

for some $0 < \theta < t$. Because $(u+\theta t\phi)^{-\alpha} \to u^{-\alpha}$ almost everywhere in Ω, by virtue of Fatou's lemma and (7.18) we retrieve (7.17).

Step 2: We have

$$\int_\Omega |\nabla u|^2 dx - \int_\Omega u^{1-\alpha}dx - \lambda\int_\Omega u^{p+1}dx = 0. \tag{7.19}$$

Let $h : \mathbb{R} \to \mathbb{R}$, $h(t) = J_\lambda(u+tu)$. Then $t = 0$ is a local minimum of h, which yields $h'(0) = 0$. This leads us to (7.19).

Step 3: u verifies (7.2). Let $\varepsilon > 0$ and $\phi \in H_0^1(\Omega)$ be arbitrary. Replacing ϕ with $(u+\varepsilon\phi)^+$ in (7.17), in view of (7.19) we obtain

$$\begin{aligned} 0 &\leq \int_{\{u+\varepsilon\phi\geq 0\}} \Big(\nabla u \nabla(u+\varepsilon\phi) - f_\lambda(u)(u+\varepsilon\phi)\Big)dx \\ &= \varepsilon\int_\Omega \Big(\nabla u\nabla\phi - f_\lambda(u)\phi\Big)dx - \int_{\{u+\varepsilon\phi<0\}} \Big(\nabla u\nabla(u+\varepsilon\phi) - f_\lambda(u)(u+\varepsilon\phi)\Big)dx \\ &\leq \varepsilon\int_\Omega \Big(\nabla u\nabla\phi - f_\lambda(u)\phi\Big)dx - \varepsilon\int_{\{u+\varepsilon\phi<0\}} \nabla u\nabla\phi dx. \end{aligned}$$

Next, we divide by ε in the previous estimates and then let $\varepsilon \to 0$. Taking into account the fact that $|\{u+\varepsilon\phi < 0\}| \to 0$ as $\varepsilon \to 0$, we deduce

$$\int_\Omega (\nabla u\nabla\phi - f_\lambda(u)\phi)dx \geq 0.$$

Replacing ϕ with $-\phi$ in the last estimate we finally derive (7.2).

To end the proof, it remains only to show that λ^* is finite. Because $p > 1$, there exists $\Lambda > 0$ such that $f_\lambda(t) > \lambda_1 t$, for all $\lambda > \Lambda$ and $t > 0$. We claim that (7.1) has no weak solutions for $\lambda \geq \Lambda$. Indeed, if u would be such a solution, multiplying by φ_1 in (7.1) we derive

$$\lambda_1 \int_\Omega u\varphi_1 dx = \int_\Omega f_\lambda(u)\varphi_1 dx > \lambda_1 \int_\Omega u\varphi_1 dx,$$

which is a contradiction. This means that $\lambda^* \leq \Lambda < \infty$, and the proof of Proposition 7.4.2 is now complete. □

The following result will end the proof of Theorem 7.4.1.

Proposition 7.4.4 *For all* $0 < \lambda \leq \lambda^*$, *problem (7.1) has at least one weak solution.*

Proof Assume first $0 < \lambda < \lambda^*$. By the definition of A, there exist $\mu > \lambda$ and $u_\mu \in H_0^1(\Omega)$, a weak solution of

$$\begin{cases} -\Delta u = f_\mu(u) & \text{in } \Omega, \\ u > 0 & \text{in } \Omega, \\ u = 0 & \text{on } \partial\Omega. \end{cases}$$

Clearly, $\overline{u} := u_\mu$ is a weak supersolution of (7.1). On the other hand, by Theorem 4.3.2 there exists $v \in C^2(\Omega) \cap C^1(\overline{\Omega})$ such that

$$\begin{cases} -\Delta v = v^{-\alpha} & \text{in } \Omega, \\ v > 0 & \text{in } \Omega, \\ v = 0 & \text{on } \partial\Omega. \end{cases} \tag{7.20}$$

Obviously, $\underline{u} := v$ is a subsolution of (7.1). Moreover, because $0 < \alpha < 1$, multiplying by $\underline{u}$ in (7.20) we infer that $J_\lambda(\underline{u}) < 0$.

To prove that $\underline{u} \leq \overline{u}$ in Ω, let $\theta \in C^1(\mathbb{R})$ be a nondecreasing function such that $\theta(t) \geq 1$ for all $t \geq 1$ and $\theta(t) = 0$ for all $t \leq 0$. For $\varepsilon > 0$ we set $\theta_\varepsilon(t) = \theta(t/\varepsilon)$. Using $\theta_\varepsilon(\underline{u} - \overline{u})$ as a test function in (7.1) and (7.20), by subtraction we obtain

$$\begin{aligned} 0 &\geq -\int_\Omega |\nabla(\overline{u} - \underline{u})|^2 \theta_\varepsilon'(\underline{u} - \overline{u})dx \\ &\geq \int_\Omega (f_\lambda(\overline{u}) - \underline{u}^{-\alpha})\theta_\varepsilon(\underline{u} - \overline{u})dx \\ &\geq \int_\Omega (\overline{u}^{-\alpha} - \underline{u}^{-\alpha})\theta_\varepsilon(\underline{u} - \overline{u})dx. \end{aligned}$$

Letting $\varepsilon \to 0$ it follows that

$$\int_{\{\underline{u} - \overline{u} > 0\}} (\overline{u}^{-\alpha} - \underline{u}^{-\alpha})dx \leq 0,$$

which means that the set $\{\underline{u} - \overline{u} > 0\}$ has the Lebesgue measure zero. Hence, $\underline{u} \leq \overline{u}$ in Ω, and by Theorem 7.2.2 we obtain that problem (7.1) has at least one solution.

For $\lambda = \lambda^*$, let $(\lambda_n)_{n\geq 1} \subset A$ be an increasing sequence such that $\lambda_n \nearrow \lambda^*$ as $n \to \infty$ and let u_n be a solution of (7.1) for $\lambda = \lambda_n$. Then

$$J_\lambda(u_n) = \frac{1}{2}\int_\Omega |\nabla u_n|^2 dx - \frac{1}{1-\alpha}\int_\Omega u_n^{1-\alpha} dx - \frac{\lambda_n}{p+1}\int_\Omega u_n^{p+1} dx < 0$$

and

$$\|u_n\|^2_{H_0^1(\Omega)} - \int_\Omega u_n^{1-\alpha} dx - \lambda_n \int_\Omega u_n^{p+1} dx = 0.$$

From the previous relations it is easy to see that $(u_n)_{n\geq 1}$ is bounded in $H_0^1(\Omega)$. Thus, there exists $u^* \in H_0^1(\Omega)$ such that, passing to a subsequence, we have as $n \to \infty$

$$u_n \rightharpoonup u^* \quad \text{weakly in } H_0^1(\Omega),$$
$$u_n \to u^* \quad \text{almost everywhere in } \Omega.$$

As before, we derive that $u_n \geq v$ in Ω. Because u_n is a weak solution of (7.1) with $\lambda = \lambda_n$, we have

$$\int_\Omega \nabla u_n \nabla \phi dx = \int_\Omega f_\lambda(u_n)\phi dx \qquad \text{for all } \phi \in H_0^1(\Omega).$$

Passing to the limit in the previous equality and using Lebesgue's theorem, we deduce that u^* is a weak solution of (7.1) for $\lambda = \lambda^*$. This ends the proof. □

7.5 Existence of the second solution

We have seen so far that there exists $\lambda^* > 0$ such that problem (7.1) has at least one solution u_λ if $0 < \lambda \leq \lambda^*$, and no solutions exist if $\lambda > \lambda^*$. As a result of the superlinear character of (7.1), we are able to show that there exists another solution v_λ of (7.1) provided $0 < \lambda < \lambda^*$. Moreover, we have $0 < u_\lambda < v_\lambda$ in Ω. We use Ekeland's variational principle, as stated in Theorem C.2.1.

Define

$$\mathcal{A} := \{u \in H_0^1(\Omega) : u \geq u_\lambda \text{ in } \Omega\}.$$

Because u_λ is an H^1 local minimizer of J_λ, there exists $0 < r_0 < \|u_\lambda\|_{H_0^1(\Omega)}$ such that

$$J_\lambda(u) \geq J_\lambda(u_\lambda) \qquad \text{for all } u \in H_0^1(\Omega) \ \text{ with } \|u - u_\lambda\|_{H_0^1(\Omega)} \leq r_0.$$

In our further analysis, the following two cases may occur:

Case 1: $\inf\{J_\lambda(u) : u \in \mathcal{A}, \|u - u_\lambda\|_{H_0^1(\Omega)} = r\} = J_\lambda(u_\lambda)$, for all $0 < r < r_0$.

Case 2: There exists $0 < r_1 < r_0$ such that

$$\inf\{J_\lambda(u) : u \in \mathcal{A}, \|u - u_\lambda\|_{H_0^1(\Omega)} = r_1\} > J_\lambda(u_\lambda).$$

We shall discuss these two situations separately.

7.5.1 *First case*

Proposition 7.5.1 *Let $0 < \lambda < \lambda^*$. Then, for all $0 < r < r_0$ there exists a weak solution v_λ of (7.1) such that $0 < u_\lambda < v_\lambda$ in Ω and $\|u_\lambda - v_\lambda\|_{H_0^1(\Omega)} = r$.*

As a consequence, in this case, problem (7.1) has infinitely many solutions.

Proof Fix $0 < r < r_0$ and let $\rho > 0$ be such that $0 < r - \rho < r + \rho < r_0$. Define

$$\mathcal{B} := \{u \in \mathcal{A} : 0 < r - \rho \leq \|u - u_\lambda\|_{H_0^1(\Omega)} \leq r + \rho\}.$$

Clearly, $\mathcal{B}$ is closed in $H_0^1(\Omega)$ and by our assumption, $\inf_{\mathcal{B}} J_\lambda = J_\lambda(u_\lambda)$. Let $(u_n)_{n\geq 1}$ be a minimizing sequence of u_λ. Then, by Ekeland's variational principle (see Theorem C.2.1), for all $n \geq 1$ there exists $v_n \in \mathcal{B}$ such that

$$J_\lambda(v_n) \leq J_\lambda(u_n) \leq J_\lambda(u_\lambda) + \frac{1}{n}, \tag{7.21a}$$

$$\|u_n - v_n\|_{H_0^1(\Omega)} \leq \frac{1}{n}, \tag{7.21b}$$

$$J_\lambda(v_n) \leq J_\lambda(v) + \frac{1}{n}\|v - v_n\|_{H_0^1(\Omega)} \quad \text{for all } v \in \mathcal{B}. \tag{7.21c}$$

Because $(v_n)_{n\geq 1}$ is bounded in $H_0^1(\Omega)$, there exists $v_\lambda \in H_0^1(\Omega)$ such that, up to a subsequence, we have $v_n \rightharpoonup v_\lambda$ in $H_0^1(\Omega)$ and $v_n \to v_\lambda$ almost everywhere in Ω as $n \to \infty$. The definition of the set $\mathcal{A}$ also implies $v_\lambda \geq u_\lambda$ in Ω.

Lemma 7.5.2 *v_λ is a weak solution of (7.1).*

Proof We first claim that for all $w \in \mathcal{B}$ we have

$$-\frac{1}{n}\|w - v_n\|_{H_0^1(\Omega)} \leq \int_\Omega \nabla v_n \nabla(w - v_n)dx - \int_\Omega f_\lambda(v_n)(w - v_n)dx. \tag{7.22}$$

Let $w \in \mathcal{B}$. Then, for $\varepsilon > 0$ small enough we have $v_n + \varepsilon(w - v_n) \in \mathcal{B}$. Furthermore, by (7.21c) we derive

$$-\frac{1}{n}\|w - v_n\|_{H_0^1(\Omega)} \leq \frac{J_\lambda(v_n + \varepsilon(w - v_n)) - J_\lambda(v_n)}{\varepsilon} \quad \text{for all } n \geq 1.$$

Passing to the limit with $\varepsilon \to 0$ in the previous relation we obtain

$$-\frac{1}{n}\|w-v_n\|_{H_0^1(\Omega)} \leq \int_\Omega \nabla v_n \nabla(w-v_n)dx$$
$$-\frac{1}{1-\alpha}\liminf_{\varepsilon\to 0}\int_\Omega \frac{(v_n+\varepsilon(w-v_n))^{1-\alpha}-v_n^{1-\alpha}}{\varepsilon}dx$$
$$-\lambda\int_\Omega v_n^p(w-v_n)dx$$
$$=\int_\Omega \nabla v_n \nabla(w-v_n)dx$$
$$-\liminf_{\varepsilon\to 0}\int_\Omega (v_n+\varepsilon\theta(w-v_n))^{-\alpha}(w-v_n)dx, \quad (0<\theta<1)$$
$$-\lambda\int_\Omega v_n^p(w-v_n)dx.$$

As we have argued for (7.4), we deduce

$$\liminf_{\varepsilon\to 0}\int_\Omega (v_n+\varepsilon\theta(w-v_n))^{-\alpha}(w-v_n)dx = \int_\Omega v_n^{-\alpha}(w-v_n)dx.$$

This proves the claim. Let $\phi \in H_0^1(\Omega)$ and $\varepsilon > 0$. Set

$$w_{n,\varepsilon} := (v_n+\varepsilon\phi-u_\lambda)^-, \quad \text{and} \quad w_\varepsilon := (v_\lambda+\varepsilon\phi-u_\lambda)^-.$$

Therefore, as $n\to\infty$ we have

$$w_{n,\varepsilon} \rightharpoonup w_\varepsilon \quad \text{weakly in } H_0^1(\Omega),$$
$$w_{n,\varepsilon} \to w_\varepsilon \quad \text{almost everywhere in } \Omega.$$

Obviously, $v_n+\varepsilon\phi+w_{n,\varepsilon}\in\mathcal{B}$. Then, replacing w with $v_n+\varepsilon\phi+w_{n,\varepsilon}$ in (7.22) we obtain

$$-\frac{1}{n}\|\varepsilon\phi+w_{n,\varepsilon}\|_{H_0^1(\Omega)} \leq \int_\Omega \nabla v_n\nabla(\varepsilon\phi+w_{n,\varepsilon})dx - \int_\Omega f_\lambda(v_n)(\varepsilon\phi+w_{n,\varepsilon})dx. \quad (7.23)$$

By Lebesgue's theorem on dominated convergence we obtain

$$\int_\Omega f_\lambda(v_n)(\varepsilon\phi+w_{n,\varepsilon})dx \to \int_\Omega f_\lambda(v_\lambda)(\varepsilon\phi+w_\varepsilon)dx \quad \text{as } n\to\infty. \quad (7.24)$$

Because the measure of the set $\{v_n+\varepsilon\phi<u_\lambda\leq v_\lambda+\varepsilon\phi\}$ tends to zero as $n\to\infty$, we also have

$$\begin{aligned}\int_\Omega \nabla v_n\nabla w_{n,\varepsilon}dx &= \int_{\{u_\lambda>v_n+\varepsilon\phi\}} \nabla v_n\nabla(u_\lambda-v_n-\varepsilon\phi)dx\\ &\leq \int_{\{u_\lambda>v_\lambda+\varepsilon\phi\}} \nabla v_n\nabla(u_\lambda-v_\lambda-\varepsilon\phi)dx\\ &\quad+\int_{\{u_\lambda>v_n+\varepsilon\phi\}} \nabla v_n\nabla(v_\lambda-v_n)dx+o(1)\\ &=\int_\Omega \nabla v\nabla w_\varepsilon dx+o(1).\end{aligned} \quad (7.25)$$

Using now (7.24) and (7.25) in (7.23) and then letting $n \to \infty$ we obtain

$$\begin{aligned}\int_\Omega \Big(\nabla v_\lambda \nabla \phi - f_\lambda(v_\lambda)\phi\Big)dx &\geq -\frac{1}{\varepsilon}\int_\Omega \Big(\nabla v_\lambda \nabla w_\varepsilon - f_\lambda(v_\lambda)w_\varepsilon\Big)dx \\ &= \frac{1}{\varepsilon}\int_\Omega \nabla(u_\lambda - v_\lambda)\nabla w_\varepsilon dx + \frac{1}{\varepsilon}\int_\Omega (f_\lambda(v_\lambda) - f_\lambda(u_\lambda))w_\varepsilon dx \\ &\geq \frac{1}{\varepsilon}\int_{\{u_\lambda > v_\lambda + \varepsilon\phi\}} \nabla(u_\lambda - v_\lambda)\nabla(u_\lambda - v_\lambda - \varepsilon\phi)dx \\ &\quad + \frac{1}{\varepsilon}\int_{\{u_\lambda > v_\lambda + \varepsilon\phi\}} (v_\lambda^{-\alpha} - u_\lambda^{-\alpha})(u_\lambda - v_\lambda - \varepsilon\phi)dx \\ &\geq \int_{\{u_\lambda > v_\lambda + \varepsilon\phi\}} \nabla(v_\lambda - u_\lambda)\nabla\phi dx + \int_{\{u_\lambda > v_\lambda + \varepsilon\phi\}} (v_\lambda^{-\alpha} - u_\lambda^{-\alpha})\phi dx.\end{aligned}$$

Using again the fact that the Lebesgue's measure of the set $\{u_\lambda > v_\lambda + \varepsilon\phi\}$ tends to zero as $\varepsilon \to 0$, the last estimate yields

$$\int_\Omega \Big(\nabla v_\lambda \nabla \phi - f_\lambda(v_\lambda)\Big)dx \geq o(1) \quad \text{as } \varepsilon \to 0.$$

Letting $\varepsilon \to 0$ and then reversing the sign of ϕ we derive that v_λ is a weak solution of (7.1). □

Lemma 7.5.3 *The sequence $(v_n)_{n\geq 1}$ converges strongly to v_λ in $H_0^1(\Omega)$.*

Proof Arguing as in the proof of Proposition 7.4.2, as $n \to \infty$ we have

$$\|v_n\|^2_{H_0^1(\Omega)} = \|v_n - v_\lambda\|^2_{H_0^1(\Omega)} + \|v_\lambda\|^2_{H_0^1(\Omega)} + o(1), \tag{7.26}$$

$$\|v_n\|^{p+1}_{p+1} = \|v_n - v_\lambda\|^{p+1}_{p+1} + \|v_\lambda\|^{p+1}_{p+1} + o(1), \tag{7.27}$$

$$\int_\Omega v_n^{1-\alpha}dx = \int_\Omega v_\lambda^{1-\alpha}dx + o(1), \tag{7.28}$$

and

$$\int_\Omega |v_n - v_\lambda|^{1-\alpha}dx = o(1). \tag{7.29}$$

Let us take $w = v_\lambda$ in (7.22). We have

$$\begin{aligned}\int_\Omega |\nabla(v_n - v_\lambda)|^2 dx + \int_\Omega v_n^{-\alpha}v_\lambda dx &\leq \int_\Omega v_\lambda^{1-\alpha}dx + \lambda\int_\Omega v_n^p(v_n - v_\lambda)dx + o(1) \\ &\leq \int_\Omega v_\lambda^{1-\alpha}dx + \lambda\|v_n - v_\lambda\|^{p+1}_{p+1} + o(1).\end{aligned}$$

By Lebesgue's theorem, the previous estimate produces

$$\|v_n - v_\lambda\|^2_{H_0^1(\Omega)} \leq \lambda\|v_n - v_\lambda\|^{p+1}_{p+1} + o(1) \quad \text{as } n \to \infty. \tag{7.30}$$

Letting now $w = 2v_n$ in (7.22) we have

$$\|v_n\|^2_{H^1_0(\Omega)} - \lambda\|v_n\|^{p+1}_{p+1} - \int_\Omega v_n^{1-\alpha}dx \geq o(1) \quad \text{as } n \to \infty. \tag{7.31}$$

Because v_λ is a solution of (7.1) we also have

$$\|v_\lambda\|^2_{H^1_0(\Omega)} - \lambda\|v_\lambda\|^{p+1}_{p+1} - \int_\Omega v_\lambda^{1-\alpha}dx = 0. \tag{7.32}$$

From (7.26) through (7.28) and (7.31), (7.32) we find

$$\|v_n - v_\lambda\|^2_{H^1_0(\Omega)} \geq \lambda\|v_n - v_\lambda\|^{p+1}_{p+1} + o(1) \quad \text{as } n \to \infty. \tag{7.33}$$

Combining (7.30) and (7.33) it follows that

$$\|v_n - v_\lambda\|^2_{H^1_0(\Omega)} = \lambda\|v_n - v_\lambda\|^{p+1}_{p+1} + o(1) \quad \text{as } n \to \infty. \tag{7.34}$$

The last equality together with (7.26) through (7.29) provides

$$J_\lambda(v_n - v_\lambda) = J_\lambda(v_n) - J_\lambda(v_\lambda) + o(1) \quad \text{as } n \to \infty. \tag{7.35}$$

Recall that $v_n \rightharpoonup v_\lambda$ in $H^1_0(\Omega)$ as $n \to \infty$. Hence, by the definition of $\mathcal{B}$ we deduce

$$\|v_\lambda - u_\lambda\|_{H^1_0(\Omega)} \leq \liminf_{n\to\infty} \|v_n - u_\lambda\|_{H^1_0(\Omega)} \leq r + \rho < r_0.$$

Therefore, according to the hypotheses in Case 1 we have $J_\lambda(u_\lambda) \leq J_\lambda(v_\lambda)$. Furthermore, by (7.21a) it follows that

$$J_\lambda(v_n) - J_\lambda(v_\lambda) \leq J_\lambda(u_\lambda) - J_\lambda(v_\lambda) + \frac{1}{n} \leq o(1) \quad \text{as } n \to \infty. \tag{7.36}$$

From (7.29), (7.35), and (7.36) we finally obtain

$$\frac{1}{2}\|v_n - v_\lambda\|^2_{H^1_0(\Omega)} - \frac{\lambda}{p+1}\|v_n - v_\lambda\|^{p+1}_{p+1} \leq o(1) \quad \text{as } n \to \infty. \tag{7.37}$$

Now, by (7.33) and (7.37) it follows that $v_n \to v_\lambda$ in $H^1_0(\Omega)$ as $n \to \infty$. □

Let us come back to the proof of Proposition 7.5.1.

Because $\|v_n - u_\lambda\|_{H^1_0(\Omega)} = r$ and $v_n \to v_\lambda$ in $H^1_0(\Omega)$ as $n \to \infty$, it follows that $\|v_\lambda - u_\lambda\|_{H^1_0(\Omega)} = r$. This yields $v_\lambda \not\equiv u_\lambda$. Notice that $u_\lambda \leq v_\lambda$ in Ω. Furthermore, applying Corollary 1.3.8 as we did in the proof of Lemma 7.3.2 we derive that $v_\lambda - u_\lambda > 0$ in any subdomain $\omega \subset\subset \Omega$. Hence, $u_\lambda < v_\lambda$ in Ω. This completes the proof. □

7.5.2 *Second case*

Similar to Proposition 7.5.1 we have the following.

Proposition 7.5.4 *Let* $0 < \lambda < \lambda^*$. *Then there exists a solution* v_λ *of problem (7.1) such that* $0 < u_\lambda < v_\lambda$ *in* Ω.

Proof Consider the complete metric space

$$\mathcal{P} = \left\{ \eta \in C([0,1], \mathcal{A}) : \begin{array}{r} \eta(0) = u_\lambda, J_\lambda(\eta(1)) < J_\lambda(u_\lambda) \\ \|\eta(1) - u_\lambda\|_{H_0^1(\Omega)} > r_1 \end{array} \right\}$$

endowed with the distance

$$d(\eta, \eta') := \max_{t \in [0,1]} \|\eta'(t) - \eta(t)\| \quad \text{for all } \eta, \eta' \in \mathcal{P}.$$

Set

$$c_{\mathcal{P}} := \inf_{\eta \in \mathcal{P}} \max_{t \in [0,1]} J_\lambda(\eta(t)).$$

We first prove that $\mathcal{P}$ is nonempty. When this claim is established, we may proceed to determine the second solution v_λ of (7.1). This will be done by making use of Ekeland's variational principle, but in a quite different manner to that in the previous case.

To show that $\mathcal{P}$ is nonempty, let S be the best Sobolev constant for the embedding $H_0^1(\Omega) \hookrightarrow L^{p+1}(\Omega)$—that is,

$$S = \inf\{\|\nabla u\|_2^2 : u \in H_0^1(\Omega),\ \|u\|_{p+1} = 1\}. \tag{7.38}$$

Then, S is independent of Ω and depends only on N. Furthermore, the infimum in (7.38) is never achieved in bounded domains. If $\Omega = \mathbb{R}^N$ then the infimum is achieved by

$$U(x) = \frac{C_N}{(1 + |x|^2)^{(N-2)/2}},$$

for some normalization constant $C_N > 0$. Also we have

$$S = \frac{B}{A^{2/(p+1)}}, \tag{7.39}$$

where

$$A = \int_{\mathbb{R}^N} U^{p+1} dx, \quad B = \int_{\mathbb{R}^N} |\nabla U|^2 dx. \tag{7.40}$$

Let us fix $y \in \Omega$ and $\phi \in C_0^\infty(\Omega)$ such that $\phi = 1$ in a neighborhood of y, and define

$$U_\varepsilon(x) = \frac{C_N \varepsilon^{(N-2)/2}}{(\varepsilon^2 + |x - y|^2)^{(N-2)/2}} \phi(x).$$

Lemma 7.5.5 *We have*

(i) $\|\nabla U_\varepsilon\|_2^2 = B + O(\varepsilon^{N-2})$ *and* $\|U_\varepsilon\|_{p+1}^{p+1} = A + O(\varepsilon^N)$;

(ii) *for all* $v \in H_0^1(\Omega)$ *and* $\rho > 0$, *the following estimate holds:*

$$\begin{aligned}\|v+\rho U_\varepsilon\|_{p+1}^{p+1} &= \|v\|_{p+1}^{p+1} + \rho^{p+1}\|U_\varepsilon\|_{p+1}^{p+1} + p\rho\int_\Omega U_\varepsilon v^p dx \\ &\quad + p\rho^p\int_\Omega U_\varepsilon^p v dx + o(\varepsilon^{(N-2)/2}).\end{aligned} \tag{7.41}$$

Proof We give here a complete proof of (i). Remark first that

$$\nabla U_\varepsilon = C_N\varepsilon^{(N-2)/2}\left\{\frac{\nabla\phi(x)}{(\varepsilon^2+|x-y|^2)^{(N-2)/2}} - \frac{(N-2)\phi(x)(x-y)}{(\varepsilon^2+|x-y|^2)^{N/2}}\right\}.$$

Because $\phi = 1$ in a neighborhood of y we obtain

$$\begin{aligned}\|\nabla U_\varepsilon\|_2^2 &= C_N^2\varepsilon^{N-2}(N-2)^2\int_{\mathbb{R}^N}\frac{|x-y|^2}{(\varepsilon^2+|x-y|^2)^N}dx + O(\varepsilon^{N-2}) \\ &= C_N^2(N-2)^2\int_{\mathbb{R}^N}\frac{|z|^2}{(1+|z|^2)^N}dx + O(\varepsilon^{N-2}) \\ &= B + O(\varepsilon^{N-2}).\end{aligned}$$

We also have

$$\begin{aligned}\|U_\varepsilon\|_{p+1}^{p+1} &= C_N^{2N/(N-2)}\varepsilon^N\int_{\mathbb{R}^N}\frac{\phi^{2N/(N-2)}(x)}{(\varepsilon^2+|x-y|^2)^N}dx \\ &= C_N^{2N/(N-2)}\varepsilon^N\left\{\int_{\mathbb{R}^N}\frac{1}{(\varepsilon^2+|x-y|^2)^N}dx + \int_{\mathbb{R}^N}\frac{\phi^{2N/(N-2)}-1}{(\varepsilon^2+|x-y|^2)^N}dx\right\} \\ &= C_N^{2N/(N-2)}\varepsilon^N\int_{\mathbb{R}^N}\frac{1}{(\varepsilon^2+|x-y|^2)^N}dx + O(\varepsilon^N) \\ &= C_N^{2N/(N-2)}\int_{\mathbb{R}^N}\frac{1}{(1+|z|^2)^N}dz + O(\varepsilon^N) \\ &= A + O(\varepsilon^N).\end{aligned}$$

Estimate (7.41) follows with similar arguments. □

Lemma 7.5.6 *There exist* $\varepsilon_0 > 0$ *and* $\rho_0 \geq 1$ *such that for all* $0 < \varepsilon < \varepsilon_0$ *we have*

$$\begin{aligned}J_\lambda(u_\lambda + \rho U_\varepsilon) &< J_\lambda(u_\lambda) && \textit{for all } \rho \geq \rho_0, \\ J_\lambda(u_\lambda + \rho U_\varepsilon) &< J_\lambda(u_\lambda) + \frac{S^{N/2}}{N\lambda^{(N-2)/2}} && \textit{for all } 0 < \rho < \rho_0.\end{aligned}$$

Proof A straightforward computation yields

$$\begin{aligned} J_\lambda(u_\lambda+\rho U_\varepsilon) =&\frac{1}{2}\int_\Omega |\nabla u_\lambda|^2dx+\rho\int_\Omega \nabla u_\lambda \nabla U_\varepsilon dx+\frac{\rho^2}{2}\int_\Omega |\nabla U_\varepsilon|^2dx \\ &-\frac{1}{1-\alpha}\int_\Omega (u_\lambda+\rho U_\varepsilon)^{1-\alpha}dx-\frac{\lambda}{p+1}\|u_\lambda+\rho U_\varepsilon\|_{p+1}^{p+1}. \end{aligned} \tag{7.42}$$

Because u_λ is a solution of (7.1) we have

$$\int_\Omega \nabla u_\lambda \nabla U_\varepsilon dx=\int_\Omega u_\lambda^{-\alpha}U_\varepsilon dx+\lambda\int_\Omega u_\lambda^p U_\varepsilon dx. \tag{7.43}$$

Next, from (7.41) through (7.43) we obtain

$$\begin{aligned} J_\lambda(u_\lambda+\rho U_\varepsilon) =&J_\lambda(u_\lambda)+\frac{\rho^2 B}{2}-\frac{\lambda}{p+1}\rho^{p+1}A-\lambda\rho^p\int_\Omega U_\varepsilon^p u_\lambda dx \\ &+D_\varepsilon+\rho^\beta o(\varepsilon^{(N-2)/2}), \end{aligned} \tag{7.44}$$

for some $0<\beta\le p+1$, and

$$D_\varepsilon:=\int_\Omega\left(u_\lambda^{-\alpha}\rho U_\varepsilon+\frac{1}{1-\alpha}u_\lambda^{1-\alpha}-\frac{1}{1-\alpha}(u_\lambda+\rho U_\varepsilon)^{1-\alpha}\right)dx.$$

Let us estimate D_ε. In this sense we fix $0<\tau<1/4$ and we decompose

$$\begin{aligned} D_\varepsilon=&\int_{|x-y|\le\varepsilon^\tau}\left(u_\lambda^{-\alpha}\rho U_\varepsilon+\frac{1}{1-\alpha}u_\lambda^{1-\alpha}-\frac{1}{1-\alpha}(u_\lambda+\rho U_\varepsilon)^{1-\alpha}\right)dx \\ &+\int_{|x-y|>\varepsilon^\tau}\left(u_\lambda^{-\alpha}\rho U_\varepsilon+\frac{1}{1-\alpha}u_\lambda^{1-\alpha}-\frac{1}{1-\alpha}(u_\lambda+\rho U_\varepsilon)^{1-\alpha}\right)dx. \end{aligned}$$

Denote by I and II the two integrals in the previous equality. Taking $\varepsilon>0$ small enough, I can be evaluated as

$$\begin{aligned} I\le&c_1\rho\int_{|x-y|\le\varepsilon^\tau}U_\varepsilon dx=c_1\rho\int_{|x-y|\le\varepsilon^\tau}\frac{C_N\varepsilon^{(N-2)/2}}{(\varepsilon^2+|x-y|^2)^{(N-2)/2}}dx \\ \le&c_2\rho\varepsilon^{(N-2)/2}\int_0^{\varepsilon^\tau}rdr\le c_6\rho\varepsilon^{(N-2)/2+2\tau}. \end{aligned} \tag{7.45}$$

To evaluate the second integral we use the mean value theorem. We have

$$\begin{aligned} II\le&\int_{|x-y|>\varepsilon^\tau}\left(u_\lambda^{-\alpha}-(u_\lambda+\theta_1\rho U_\varepsilon)^{-\alpha}\right)\rho U_\varepsilon dx \\ \le&\alpha\int_{\operatorname{supp}\phi\cap\{|x-y|>\varepsilon^\tau\}}(u_\lambda+\theta_2\rho U_\varepsilon)^{-1-\alpha}(\rho U_\varepsilon)^2dx, \end{aligned}$$

for some $0<\theta_1,\theta_2<1$. Using the fact that

$$(u_\lambda + \theta_2\rho\theta U_\varepsilon)^{-1-\alpha} \le u_\lambda^{-1-\alpha} \le M \quad \text{in supp}\,\phi,$$

we deduce

$$\begin{aligned} II \le & c_3\rho^2 \int_{|x-y|>\varepsilon^\tau} U_\varepsilon^2 dx \le c_4\rho^2 \int_{|x-y|>\varepsilon^\tau} \frac{\varepsilon^{(N-2)/2}}{(\varepsilon^2+|x-y|^2)^{(N-2)/2}} dx \\ \le & c_5\rho^2\varepsilon^{N-2-2\tau(N-2)}. \end{aligned} \tag{7.46}$$

Therefore, from (7.45) and (7.46) we finally obtain

$$D_\varepsilon \le c(\rho+\rho^2)o(\varepsilon^{(N-2)/2}) \quad \text{as } \varepsilon \to 0. \tag{7.47}$$

Let us now evaluate the integral in the right-hand side of (7.44). For this purpose, we extend u_λ by zero outside Ω. Then $u_\lambda \in C(\mathbb{R}^N)$ and

$$\begin{aligned} \int_\Omega U_\varepsilon^p u_\lambda dx &= \varepsilon^{(N+2)/2} \int_{\mathbb{R}^N} \frac{C_N^p u_\lambda(x)\phi^p(x)}{(\varepsilon^2+|x-y|^2)^{(N+2)/2}} dx \\ &= \varepsilon^{-(N+2)/2} \int_{\mathbb{R}^N} u_\lambda(x)\phi^p(x)\psi\Big(\frac{x-y}{\varepsilon}\Big) dx, \end{aligned}$$

where $\psi(x) = C_N^p(1+|x|^2)^{-(N+2)/2} \in L^1(\mathbb{R}^N)$. Notice that

$$\frac{1}{\varepsilon^N}\int_{\mathbb{R}^N} u_\lambda(x)\phi^p(x)\psi\Big(\frac{x-y}{\varepsilon}\Big)dx \to u_\lambda(y)\int_{\mathbb{R}^N}\psi(x)dx \quad \text{as } \varepsilon \to 0.$$

Hence, we have obtained

$$\lambda\rho^p \int_\Omega U_\varepsilon^p u_\lambda dx = \rho^p T\varepsilon^{(N-2)/2} + \rho^p o(\varepsilon^{(N-2)/2}), \tag{7.48}$$

where $T = \lambda u_\lambda(y)\int_{\mathbb{R}^N}\psi(x)dx$.

Now, from (7.44), (7.47), and (7.48) we find

$$J_\lambda(u_\lambda+\rho U_\varepsilon) \le c_{\mathcal{P}} + \frac{\rho^2 B}{2} - \frac{\lambda\rho^{p+1}A}{p+1} - \rho^p T\varepsilon^{(N-2)/2} + \rho^\gamma o(\varepsilon^{(N-2)/2}), \tag{7.49}$$

for some $0 < \gamma \le p+1$. Thus, we can take $\rho_0 > 1$ large enough such that $J_\lambda(u_\lambda+\rho U_\varepsilon) < J_\lambda(u_\lambda)$ for all $\rho \ge \rho_0$. This establishes the first claim in the statement of Lemma 7.5.6. To prove the second one, let us consider

$$q:(0,\infty)\to(0,\infty), \quad q(t) = \frac{t^2 B}{2} - \frac{\lambda t^{p+1}A}{p+1} - t^p T\varepsilon^{(N-2)/2}.$$

Then q achieves its maximum at a point $t_\varepsilon > 0$. This obviously implies that

$$t_\varepsilon B - \lambda t_\varepsilon^p A = pT\varepsilon^{(N-2)/2}t_\varepsilon^{p-1}. \tag{7.50}$$

Define

$$S_0 := \left(\frac{B}{\lambda A}\right)^{1/(p-1)}.$$

Then, (7.50) yields $0 < t_\varepsilon < S_0$ as $\varepsilon \to 0$. Let us set $t_\varepsilon := S_0(1-\delta_\varepsilon)$. We need to find the rate at which $\delta_\varepsilon \to 0$ as $\varepsilon \to 0$. From (7.50) we derive

$$\left(\frac{B^p}{\lambda A}\right)^{1/(p-1)} \Big((1-\delta_\varepsilon) - (1-\delta_\varepsilon)^p\Big) = p\frac{TB}{\lambda A}(1-\delta_\varepsilon)^{p-1}\varepsilon^{(N-2)/2}.$$

Expanding for δ_ε we obtain

$$(p-1)\left(\frac{B^p}{\lambda A}\right)^{1/(p-1)} \delta_\varepsilon = p\frac{TB}{\lambda A}\varepsilon^{(N-2)/2} + o(\varepsilon^{(N-2)/2}). \tag{7.51}$$

Now, from (7.49) and (7.51) we deduce

$$\begin{aligned} J_\lambda(u_\lambda + \rho U_\varepsilon) &\le J_\lambda(u_\lambda) + \frac{t_\varepsilon^2 B}{2} - \frac{\lambda t_\varepsilon^{p+1} A}{p+1} - t_\varepsilon^p T \varepsilon^{(N-2)/2} + o(\varepsilon^{(N-2)/2}) \\ &= J_\lambda(u_\lambda) + \frac{S_0^2 B}{2} - \frac{\lambda S_0^{p+1} A}{p+1} \\ &+ (S_0^2 B - \lambda S_0^{p+1} A)\delta_\varepsilon - S_0^p T \varepsilon^{(N-2)/2} + o(\varepsilon^{(N-2)/2}). \end{aligned} \tag{7.52}$$

Remark that $S_0^2 B = \lambda S_0^{p+1} A$ and

$$\frac{S_0^2 B}{2} - \frac{\lambda S_0^{p+1} A}{p+1} = \frac{S^{N/2}}{N\lambda^{(N-2)/2}}.$$

Furthermore, taking $\varepsilon > 0$ small enough, estimate (7.52) produces

$$J_\lambda(u_\lambda + \rho U_\varepsilon) < J_\lambda(u_\lambda) + \frac{S^{N/2}}{N\lambda^{(N-2)/2}} \quad \text{for all } 0 < \rho < \rho_0.$$

This completes the proof. □

From Lemma 7.5.6 it follows that $\eta_0(t) := u_\lambda + t\rho_0 U_\varepsilon$, $t \in [0,1]$ is an element of $\mathcal{P}$. Thus, $\mathcal{P}$ is nonempty and

$$c_{\mathcal{P}} \le \max_{t\in[0,1]} J_\lambda(\eta_0(t)) < J_\lambda(u_\lambda) + \frac{S^{N/2}}{N\lambda^{(N-2)/2}}.$$

Notice that by the definition of $\mathcal{P}$, for all $\eta \in \mathcal{P}$ there exists $t_0 \in [0,1]$ such that $\|\eta(t_0) - u_\lambda\|_{H_0^1(\Omega)} = r_1$. By virtue of our assumption in Case 2 we have

$$\max_{t\in[0,1]} J_\lambda(\eta(t)) \ge J_\lambda(\eta(t_0)) > J_\lambda(u_\lambda).$$

This yields $c_{\mathcal{P}} > J_\lambda(u_\lambda)$. Hence, we have obtained

$$J_\lambda(u_\lambda) < c_{\mathcal{P}} < J_\lambda(u_\lambda) + \frac{S^{N/2}}{N\lambda^{(N-2)/2}}. \tag{7.53}$$

In view of Ekeland's variational principle applied to the functional

$$\mathcal{P} \ni \eta \longmapsto \max_{t\in[0,1]} J_\lambda(\eta(t)) \in \mathbb{R},$$

there exists a sequence $(\eta_k)_{k\geq 1} \subset \mathcal{P}$ such that

$$\max_{t\in[0,1]} J_\lambda(\eta_k(t)) \leq c_{\mathcal{P}} + \frac{1}{k} \quad \text{for all } k \geq 1, \tag{7.54}$$

and for all $\eta \in \mathcal{P}$ there holds

$$\max_{t\in[0,1]} J_\lambda(\eta_k(t)) \leq \max_{t\in[0,1]} J_\lambda(\eta(t)) + \frac{1}{k} \max_{t\in[0,1]} \|\eta(t) - \eta_k(t)\|_{H_0^1(\Omega)}. \tag{7.55}$$

Set

$$M_k := \{t \in (0,1) : J_\lambda(\eta_k(t)) = \max_{s\in[0,1]} J_\lambda(\eta_k(s))\}.$$

The key result in this case is the following lemma.

Lemma 7.5.7 *For all $k \geq 1$ there exists $t_k \in M_k$ such that $v_k := \eta_k(t_k) \in \mathcal{A}$ satisfies*

$$\int_\Omega \Big(\nabla v_k \nabla(w - v_k) - f_\lambda(v_k)(w - v_k)\Big)dx \geq -\frac{\max\{1, \|w - v_k\|_{H_0^1(\Omega)}\}}{k}, \tag{7.56}$$

for all $w \in \mathcal{A}$.

Proof Suppose to the contrary that the conclusion in Lemma 7.5.7 does not hold. Then there exists $k \geq 1$ such that for all $t \in M_k$ we can find $\xi_t \in \mathcal{A}$ with the property

$$\int_\Omega \Big(\nabla\eta_k(t)\nabla(\xi_t - \eta_k(t)) - f_\lambda(\eta_k(t))(\xi_t - \eta_k(t))\Big)dx < -\frac{a_{t,k}}{k}, \tag{7.57}$$

where $a_{t,k} := \max\{1, \|\xi_t - \eta_k(t)\|_{H_0^1(\Omega)}\}$. We claim that ξ_t may be assumed locally constant on M_k. Indeed, by continuity arguments, for all $t \in M_k$ there exists a neighborhood V_t of t such that for all $s \in V_t \cap M_k$ we have

$$\int_\Omega \Big(\nabla\eta_k(s)\nabla(\xi_t - \eta_k(s)) - f_\lambda(\eta_k(s))(\xi_t - \eta_k(s))\Big)dx < -\frac{b_{t,k}(s)}{k}, \tag{7.58}$$

where $b_{t,k}(s) := \max\{1, \|\xi_t - \eta_k(s)\|_{H_0^1(\Omega)}\}$. We choose $t_1, t_2, \ldots, t_q \in U_k$ such that $M_k \subseteq \cup_{i=1}^q V_{t_i}$ and set $\xi_i := \xi_{t_i}$, $1 \leq i \leq q$. Let also $\{\chi_1, \chi_2, \ldots, \chi_q\}$ be a partition of unity subordinated to the covering $\{V_{t_1}, V_{t_2}, \ldots, V_{t_q}\}$. Define now

$$\xi(s) := \sum_{i=1}^{q} \chi_i(s)\xi_i \quad 0 \le s \le 1.$$

Then, ξ is continuous and $\xi(s) \in \mathcal{A}$, for all $0 \le s \le 1$. From (7.58) we deduce

$$\int_\Omega \Big(\nabla\eta_k(s)\nabla(\xi(s) - \eta_k(s)) - f_\lambda(\eta_k(s))(\xi(s) - \eta_k(s))\Big)dx < -\frac{c_k(s)}{k}, \tag{7.59}$$

where $c_k(s) := \max\{1, \|\xi(s) - \eta_k(s)\|_{H_0^1(\Omega)}\}$.

Now let $z : [0,1] \to [0,1]$ be a continuous function such that $z(t) = 1$ in a neighborhood of M_k and $z(0) = z(1) = 0$. For $0 < \varepsilon < 1$ define

$$\eta_\varepsilon(t) := \eta_k(t) + \varepsilon z(t)\frac{\xi(t) - \eta_k(t)}{c_k(t)} \qquad \text{for all } 0 \le t \le 1.$$

Then $\eta_\varepsilon(t) \in \mathcal{A}$ for all $0 \le t \le 1$. By (7.55) we obtain

$$\max_{t\in[0,1]} J_\lambda(\eta_k(t)) \le \max_{t\in[0,1]} J_\lambda(\eta_\varepsilon(t)) + \frac{\varepsilon}{k}\max_{t\in[0,1]} z(t)\frac{\|\xi(t) - \eta_k(t)\|_{H_0^1(\Omega)}}{c_k(t)}. \tag{7.60}$$

Let $t_{k,\varepsilon} \in [0,1]$ be such that

$$J_\lambda(\eta_\varepsilon(t_{k,\varepsilon})) = \max_{t\in[0,1]} J_\lambda(\eta_\varepsilon(t)).$$

Passing to a subsequence if necessary, we may assume that $t_{k,\varepsilon} \to t_k$ as $\varepsilon \to 0$. Obviously $t_k \in M_k$, which yields $z(t_{k,\varepsilon}) = 1$ for small $\varepsilon > 0$. Set

$$v_k := \eta_k(t_k), \quad v_{k,\varepsilon} := \eta_k(t_{k,\varepsilon}).$$

Then, for small $\varepsilon > 0$, in view of (7.60) we have

$$J_\lambda(v_{k,\varepsilon}) \le J_\lambda(v_k) \le J_\lambda(\eta_\varepsilon(t_{k,\varepsilon})) + \frac{\varepsilon}{k} \qquad \text{for all } k \ge 1.$$

This yields

$$\frac{J_\lambda(\eta_\varepsilon(t_{k,\varepsilon})) - J_\lambda(v_{k,\varepsilon})}{\varepsilon} \ge -\frac{1}{k} \qquad \text{for all } k \ge 1. \tag{7.61}$$

On the other hand,

$$\begin{aligned}\eta_\varepsilon(t_{k,\varepsilon}) &= \eta_k(t_{k,\varepsilon}) + \varepsilon\frac{\xi(t_{k,\varepsilon}) - \eta_k(t_{k,\varepsilon})}{c_k(t_{k,\varepsilon})}\\ &= v_{k,\varepsilon} + \varepsilon\frac{\xi(t_k) - \eta_k(t_k)}{c_k(t_k)}\\ &\quad + \varepsilon\Big(\frac{\xi(t_{k,\varepsilon}) - \eta_k(t_{k,\varepsilon})}{c_k(t_{k,\varepsilon})} - \frac{\xi(t_k) - \eta_k(t_k)}{c_k(t_k)}\Big).\end{aligned} \tag{7.62}$$

Notice that $\xi(t_{k,\varepsilon}) \to \xi(t_k)$ and $c(t_{k,\varepsilon}) \to c(t_k)$ as $\varepsilon \to 0$. Thus, from (7.62) we obtain

$$\eta_\varepsilon(t_{k,\varepsilon}) = v_{k,\varepsilon} + \varepsilon\frac{\xi(t_k) - \eta_k(t_k)}{c_k(t_k)} + o(\varepsilon) \quad \text{as } \varepsilon \to 0.$$

Using this relation in (7.61) and passing to the limit with $\varepsilon \to 0$, we deduce

$$\int_\Omega \Big(\nabla v_k \nabla(\xi(t_k) - v_k) - f_\lambda(v_k)(\xi(t_k) - v_k)\Big)dx \geq -\frac{\max\{1, \|\xi(t_k) - v_k\|_{H_0^1(\Omega)}\}}{k}.$$

The previous estimate contradicts (7.59). This concludes the proof. □

By Lemma 7.5.7 and (7.54) we also have

$$J_\lambda(v_k) \to c_{\mathcal{P}} \quad \text{as } k \to \infty. \tag{7.63}$$

Furthermore, letting $w = 2u_\lambda$ in (7.56) we deduce

$$\begin{aligned} c_{\mathcal{P}} + o(1) &= \frac{1}{2}\|v_k\|_{H_0^1(\Omega)} - \frac{1}{1-\alpha}\int_\Omega |v_k|^{1-\alpha}dx - \frac{\lambda}{p+1}\int_\Omega |v_k|^{p+1}dx \\ &\geq \Big(\frac{1}{2} - \frac{1}{p+1}\Big)\|v_k\|^2_{H_0^1(\Omega)} - \Big(\frac{1}{1-\alpha} - \frac{1}{p+1}\Big)\int_\Omega |v_k|^{1-\alpha}dx \\ &\quad - \frac{1}{k(p+1)}(1 + \|v_k\|_{H_0^1(\Omega)}). \end{aligned}$$

Therefore, $(v_k)_{k\geq 1}$ is bounded in $H_0^1(\Omega)$. Hence, there exists $v_\lambda \in H_0^1(\Omega)$ such that, up to a subsequence, we have $v_k \rightharpoonup v_\lambda$ in $H_0^1(\Omega)$ and $v_k \to v_\lambda$ almost everywhere in Ω as $k \to \infty$.

At this point, we can proceed as in the proof of Proposition 7.5.1 to derive that $(v_k)_{k\geq 1}$ converges weakly to v_λ in $H_0^1(\Omega)$ and v_λ is a weak solution of (7.1). Let us set $\phi_k := v_k - v_\lambda$, $k \geq 1$. Then, as $k \to \infty$ we have

$$\|\phi_k\|^2_{H_0^1(\Omega)} = \lambda\|\phi_k\|^{p+1}_{p+1} + o(1), \tag{7.64}$$

$$\frac{1}{2}\|\phi_k\|^2_{H_0^1(\Omega)} - \frac{\lambda}{p+1}\|\phi_k\|^{p+1}_{p+1} \leq c_{\mathcal{P}} - J_\lambda(v_\lambda) + o(1). \tag{7.65}$$

By (7.53) and for a suitable ε_0 we obtain

$$\frac{1}{2}\|\phi_k\|^2_{H_0^1(\Omega)} - \frac{\lambda}{p+1}\|\phi_k\|^{p+1}_{p+1} \leq \frac{S^{N/2}}{N\lambda^{(N-2)/2}} - \varepsilon_0 + o(1). \tag{7.66}$$

We claim that $\phi_k \to 0$ in $H_0^1(\Omega)$ as $k \to \infty$. Supposing, to the contrary, it follows that, up to a subsequence, $(\|\phi_k\|_{H_0^1(\Omega)})_{k\geq 1}$ is bounded away from zero. Furthermore, from (7.38) we have

$$\|\phi_k\|^2_{H_0^1(\Omega)} \geq S\|\phi_k\|^2_{p+1}.$$

Therefore, by (7.64) we deduce

$$\|\phi_k\|_{p+1} \geq \left(\frac{S}{\lambda}\right)^{1/(p-1)} + o(1) \quad \text{as } k \to \infty. \tag{7.67}$$

Combining (7.64), (7.66), and (7.67), it follows that

$$\begin{aligned}\frac{S^{N/2}}{N\lambda^{(N-2)/2}} - \varepsilon_0 &\geq \left(\frac{1}{2} - \frac{1}{p+1}\right)\lambda\|\phi_k\|_{p+1}^2 + o(1)\\ &\geq \frac{S^{N/2}}{N\lambda^{(N-2)/2}} + o(1) \quad \text{as } k \to \infty,\end{aligned}$$

which is a contradiction. Therefore, $v_k \to v_\lambda$ strongly in $H_0^1(\Omega)$ as $k \to \infty$. By (7.63) this yields

$$J_\lambda(v_\lambda) = c_{\mathcal{P}} > J_\lambda(u_\lambda).$$

Hence, $v_\lambda \neq u_\lambda$ in Ω. Moreover, by Corollary 1.3.8 we have $u_\lambda < v_\lambda$ in Ω. This completes the proof of Proposition 7.5.4 and Theorem 7.1.2. □

7.6 C^1 regularity of solution

The C^1 regularity of solution is based upon standard procedure for elliptic equations related to this issue. However, a restriction on the exponent of the singular term $u^{-\alpha}$ is required.

Before we start, let us point out that the C^1 regularity is a particular feature of positive weak solutions to (7.1) in the sense of Definition 7.1.1. More precisely, if a weak solution $u \in H_0^1(\Omega)$ is nonnegative and vanishes at $x_0 \in \Omega$, then u is not differentiable at x_0. Indeed, supposing to the contrary it follows that $\nabla u(x_0) = 0$ and for each $\varepsilon > 0$ we can find $r > 0$ such that $B_r(x_0) \subset\subset \Omega$ and

$$u(x) = u(x) - u(x_0) \leq \varepsilon|x - x_0| \quad \text{for all } x \in B_r(x_0).$$

Let $\phi \in C_0^2(B_r(x_0))$ be such that $0 \leq \phi \leq 1$ in $B_r(x_0)$, $\phi = 1$ on $B_{r/2}(x_0)$, and $\Delta\phi \leq cr^{-1-\alpha}$ in $B_r(x_0)$ for some constant $c > 0$.

Then, by (7.2) we have

$$\int_\Omega u\Delta\phi dx = \int_\Omega f_\lambda(u)\phi dx. \tag{7.68}$$

Note that

$$\begin{aligned}\int_{B_r(x_0)} u\Delta\phi dx &\geq \int_{B_{r/2}(x_0)} f_\lambda(u)\phi dx\\ &\geq \int_{B_{r/2}(x_0)} u^{-\alpha}\phi dx\\ &\geq \varepsilon^{-\alpha}\int_{B_{r/2}(x_0)} |x - x_0|^{-\alpha} dx\\ &\geq \varepsilon^{-\alpha}c_1 r^{N-\alpha},\end{aligned} \tag{7.69}$$

where $c_1 > 0$ is independent of r. On the other hand,

$$\int_{B_r(x_0)} u\Delta\phi dx \le c\varepsilon r^{-1-\alpha}\int_{B_r(x_0)}|x-x_0|dx \le \varepsilon c_2 r^{N-\alpha}. \tag{7.70}$$

Thus, from (7.68) through (7.70) we obtain $\varepsilon^{1+\alpha} \ge c_1/c_2$, which is a contradiction if ε is small enough. Hence, u is not differentiable in x_0.

The main result in this section is the following.

Theorem 7.6.1 *Assume* $0 < \alpha < 1/N$ *and* $1 < p < (N-2)/(N+2)$. *Then* u *is a weak solution of (7.1) if and only if* u *is a classical solution of (7.1).*

Moreover, there exist a positive integer $m > 0$, $0 < \gamma < 1$, *and* $C > 0$ *that are independent of* p *such that any solution* u *of (7.1) satisfies* $u \in C^2(\Omega)\cap C^{1,\gamma}(\overline{\Omega})$ *and*

$$\|u\|_{C^{1,\gamma}(\overline{\Omega})} \le C\Big(1+\|u\|^{p^m}_{H_0^1(\Omega)}\Big). \tag{7.71}$$

Proof Assume first that $u \in C^2(\Omega)\cap C(\overline{\Omega})$ is a classical solution of (7.1). To obtain $u \in H_0^1(\Omega)$, it suffices to prove that for all $\phi \in H_0^1(\Omega)$ there holds

$$\int_\Omega u^{-\alpha}\phi dx < \infty.$$

Let $v \in C^1(\overline{\Omega})$ be the unique solution of problem (7.20). With similar arguments to those in Proposition 7.4.4 and by virtue of Corollary 1.3.8 we have

$$u(x) \ge v(x) \ge c\,\mathrm{dist}(x,\partial\Omega) \quad \text{in } \Omega, \tag{7.72}$$

for some positive constant $c > 0$ that does not depend on p. Because $0 < \alpha < 1/N$, by Hölder's inequality we obtain

$$\begin{aligned}\int_\Omega u^{-\alpha}\phi dx &\le \left(\int_\Omega u^{-2N\alpha/(N+2)}dx\right)^{(N+2)/(2N)}\|\phi\|_{2^*}\\ &\le \left(\int_\Omega \mathrm{dist}(x,\partial\Omega)^{-2N\alpha/(N+2)}dx\right)^{N+2)/(2N)}\|\phi\|_{2^*} < \infty,\end{aligned}$$

where $2^* = 2N/(N-2)$ denotes the critical Sobolev exponent.

Conversely, let $u \in H_0^1(\Omega)$ be a weak solution of (7.1). To achieve the regularity in the statement of Theorem 7.6.1, we use Moser's iteration technique. Without losing the generality we may assume that $2^*\alpha < p$; if not, we replace p with another value close to $(N+2)/(N-2)$. Let us first choose $q_0 > N$ such that $\alpha q_0 < 1$. In view of (7.72) this yields $u^{-\alpha} \in L^{q_0}(\Omega)$.

By standard elliptic estimates, there exists $c_1 > 0$ such that

$$\begin{aligned}\|u\|_{W^{2,2^*/p}(\Omega)} &\le c_1(\|u^{-\alpha}\|_{2^*/p} + \lambda\|u^p\|_{2^*/p})\\ &\le c_1(\|v^{-\alpha}\|_{2^*/p} + \lambda\|u^p\|_{2^*/p})\\ &\le c_2(1+\|u\|^p_{2^*}).\end{aligned} \tag{7.73}$$

Therefore, by Sobolev's inequality we find

$$\|u\|_{q_1} \le c_3(1 + \|u\|_{2^*}^p), \tag{7.74}$$

for some $c_3 > 0$, where

$$q_1 = \frac{N \cdot 2^*/p}{N - 2 \cdot 2^*/p} = 2^*\delta_0, \ \ \delta_0 = \left(p - 2 \cdot \frac{2^*}{N}\right)^{-1} > 1.$$

If $q_1\alpha < p$, we replace 2^* with q_1 in (7.73) and (7.74). We obtain

$$\|u\|_{q_2} \le c_4(1 + \|u\|_{q_1}^p) \le c_4(1 + \|u\|_{2^*}^{p^2}),$$

where

$$q_2 = \frac{N \cdot q_1/p}{N - 2 \cdot q_1/p} > 2^*\delta_0^2.$$

After a finite number of such iterations we find a positive integer m, which is independent of p, and $q_m > 1$ such that $q_m\alpha \ge p$ and

$$\|u\|_{q_m} \le c(1 + \|u\|_{2^*}^{p^m}). \tag{7.75}$$

Notice that $q_m\alpha \ge p$ produces $q_m/p > N$. Using the fact that the Sobolev space $W^{2,\min\{q_0,q_m/p\}}(\Omega)$ is continuously embedded in $C^{1,\gamma}(\overline{\Omega})$ for some $0 < \gamma < 1$, by (7.75) it follows that

$$\begin{aligned}
\|u\|_{C^{1,\gamma}(\overline{\Omega})} &\le C_1\|u\|_{W^{2,\min\{q_0,q_m/p\}}(\Omega)} \\
&\le C_2(\|u^{-\alpha}\|_{q_0} + \lambda\|u^p\|_{q_m/p}) \\
&\le C_2(\|v^{-\alpha}\|_{q_0} + \lambda\|u^p\|_{q_m/p}) \\
&\le C_3(1 + \|u\|_{q_m}^p) \\
&\le C_4(1 + \|u\|_{H_0^1(\Omega)}^{p^{m+1}}),
\end{aligned}$$

where the constants $C_1, \ldots, C_4$ are independent of p. Now, by standard regularity theory we derive $u \in C^2(\Omega)$. This completes the proof. □

Remark 7.6.2 *Let* $0 < \alpha < 2/N$ *and* $1 < p < (N-2)/(N+2)$. *With the same method as that used earlier we can prove that any solution* u *of (7.1) satisfies* $u \in C^2(\Omega) \cap C^\gamma(\overline{\Omega})$ *for some* $0 < \gamma < 1$.

7.7 Asymptotic behavior of solutions

In this section we study the asymptotic behavior of solutions to (7.1) as $p \searrow 1$. According to Theorem 7.1.2, for all $1 < p \le (N-2)/(N+2)$ there exists $0 < \lambda_p^* < \infty$ such that (7.1) has at least one solution if $0 < \lambda \le \lambda_p^*$, and no solution exists for $\lambda > \lambda_p^*$. The asymptotic behavior of λ_p^* is described in the following result.

Theorem 7.7.1 *We have*

$$\lim_{p\searrow 1} \lambda_p^* = \lambda_1(-\Delta).$$

Proof Let $0 < \lambda < \lambda_1 := \lambda_1(-\Delta)$ and fix $\delta > 0$ such that $\lambda(1+\delta) < \lambda_1$. By virtue of Theorem 5.6.1, there exists $w \in C^2(\Omega) \cap C^{1,1-\alpha}(\overline{\Omega})$ such that

$$\begin{cases} -\Delta w = w^{-\alpha} + \lambda(1+\delta)w & \text{in } \Omega, \\ w > 0 & \text{in } \Omega, \\ w = 0 & \text{on } \partial\Omega. \end{cases}$$

Hence,

$$\begin{aligned} -\Delta w &\geq w^{-\alpha} + \lambda w^p + \lambda\Big((1+\delta) - \|w\|_\infty^{p-1}\Big)w \\ &\geq w^{-\alpha} + \lambda w^p \quad \text{in } \Omega \text{ as } p \searrow 1. \end{aligned}$$

Therefore, w is a supersolution of (7.1). Notice that v defined in (7.20) is a subsolution of (7.1). As in the proof of Proposition 7.4.4, we find $v \leq w$ in Ω. Thus, by Theorem 7.2.2, problem (7.1) has at least one solution. Therefore, $\liminf_{p\searrow 1} \lambda_p^* \geq \lambda$. Because $\lambda < \lambda_1$ was arbitrary, we obtain $\liminf_{p\searrow 1} \lambda_p^* \geq \lambda_1$.

Now let $\lambda > \lambda_1$. Then we can find $p_0 > 1$ such that

$$t^{-\alpha} + \lambda t^p > \lambda_1 t \quad \text{for all } t > 0 \text{ and } 1 < p < p_0. \tag{7.76}$$

Indeed, it suffices to take, for instance, $p_0 > 1$ such that $\lambda > \lambda_1^{1+p_0/(1+\alpha)}$. We claim that problem (7.1) has no solutions for $1 < p < p_0$. Indeed, if u would be a solution of (7.1), multiplying in (7.1) by the first eigenvalue φ_1 of $(-\Delta)$ we have

$$\int_\Omega \Big(f_\lambda(u) - \lambda_1 u\Big)\varphi_1 dx = 0.$$

Using (7.76) and the fact that $\varphi_1 > 0$ in Ω, the previous equality leads us to a contradiction. Hence, $\lambda_p^* \leq \lambda$ for all $1 < p < p_0$, which yields $\limsup_{p\searrow 1} \lambda_p^* \leq \lambda$. Because $\lambda > \lambda_1$ was arbitrary, we deduce $\limsup_{p\searrow 1} \lambda_p^* \leq \lambda_1$. This completes the proof. □

Let $0 < \lambda < \lambda_1$ and $0 < \alpha < 1/N$ be fixed. By Theorem 7.7.1, for all $1 < p < (N+2)/(N-2)$, problem (7.1) has at least two solutions. Let u_p be the solution of (7.1) obtained by Theorem 7.4.4. Recall that $J_\lambda(u_p) \leq 0$, which means that

$$\frac{1}{2}\|u_p\|_{H_0^1(\Omega)}^2 - \frac{1}{1-\alpha}\int_\Omega u_p^{1-\alpha} dx - \frac{\lambda}{p+1}\|u_p\|_{p+1}^{p+1} \leq 0$$

and

$$\|u_p\|_{H_0^1(\Omega)}^2 = \int_\Omega u_p^{1-\alpha} dx + \lambda\|u_p\|_{p+1}^{p+1}. \tag{7.77}$$

Combining these relations, by Hölder's inequality we deduce

$$\left(\frac{1}{2}-\frac{1}{p+1}\right)\|u_p\|^2_{H_0^1(\Omega)} \le \left(\frac{1}{1-\alpha}-\frac{1}{p+1}\right)\int_\Omega |u_p|^{1-\alpha}dx$$
$$\le c_1\|u_p\|_2^{1-\alpha} \le c_2\|u_p\|^{1-\alpha}_{H_0^1(\Omega)},$$

where $c_1, c_2 > 0$ do not depend on p. Thus, we may find $c_3 > 0$ such that

$$\|u_p\|_{H_0^1(\Omega)} \le c_3(p-1)^{-1/(1+\alpha)}. \tag{7.78}$$

Now, according to estimate (7.71), as $p \searrow 1$ we derive

$$\|u_p\|_{C^{1,\gamma}(\overline{\Omega})} \le c_4\Big(1+(p-1)^{-p^m/(1+\alpha)}\Big) \le c_5(p-1)^{-1}, \tag{7.79}$$

where $c_4, c_5 > 0$ are independent of p.

Furthermore, we have the following theorem.

Theorem 7.7.2 *Assume that* $0<\lambda<\lambda_1$ *and* $0<\alpha<1/N$. *Then, the sequence* $(u_p)_{p>1}$ *is bounded in* $H_0^1(\Omega)\cap C^{1,\gamma}(\overline{\Omega})$ *as* $p\searrow 1$.

Proof We argue by contradiction. Assume that there exists a sequence $(p_k)_{k\ge 1}$ such that $p_k \searrow 1$ and $\|u_{p_k}\|_{H_0^1(\Omega)} \to \infty$ as $k\to\infty$. Set $M_k := \|u_{p_k}\|_{H_0^1(\Omega)}$ and let $w_k := u_{p_k}/M_k$, $k \ge 1$. Because $(w_k)_{k\ge1}$ is bounded in $H_0^1(\Omega)$, there exists $0 \le w_0 \in H_0^1(\Omega)$ such that, up to a subsequence, $w_k \rightharpoonup w_0$ weakly in $H_0^1(\Omega)$ and $w_k \to w_0$ strongly in $L^2(\Omega)$ as $k\to\infty$.

We first claim that $w_0 \ne 0$. From (7.77) we have

$$M_k^2 = \int_\Omega u_{p_k}^{1-\alpha}dx + \lambda\int_\Omega u_{p_k}^{p_k+1}dx.$$

Dividing by M_k^2 in the previous equality and using the fact that $M_k \to \infty$ as $k\to\infty$, we have

$$\begin{aligned} 1 &= o(1) + \lambda\int_\Omega \frac{u_{p_k}^{p_k+1}}{M_k^2}dx \\ &= \lambda\|w_0\|_2^2 + \lambda\int_\Omega \frac{u_{p_k}^{p_k+1}-u_{p_k}^2}{M_k^2}dx + o(1) \quad \text{as } k\to\infty. \end{aligned} \tag{7.80}$$

Let us evaluate the last integral in (7.80). By the mean value theorem and estimate (7.79), we have

$$\begin{aligned} \left|\int_\Omega \frac{u_{p_k}^{p_k+1}-u_{p_k}^2}{M_k^2}dx\right| &\le \frac{(p_k-1)}{M_k^2}\int_\Omega u_{p_k}^{1+\theta_k}\ln u_{p_k}dx \quad (1<\theta_k<p_k) \\ &\le \frac{(p_k-1)\|u_{p_k}\|_\infty^{\theta_k-1}\|u_{p_k}\|_2^2}{M_k^2}\int_\Omega \ln u_{p_k}dx \\ &\le C(p_k-1)^{2-\theta_k}\ln\Big(C(p_k-1)^{-1}\Big) \quad \text{as } k\to\infty, \end{aligned} \tag{7.81}$$

where $C > 0$ does not depend on k. Therefore,

$$\int_\Omega \frac{u_{p_k}^{p_k+1} - u_{p_k}^2}{M_k^2} dx \to 0 \quad \text{as } k \to \infty.$$

Passing to the limit in (7.80), by (7.81) we obtain $\lambda \|w_0\|_2^2 = 1$. This proves the claim.

Now let $\phi \in H_0^1(\Omega)$. Because u_{p_k} is the solution of (7.1), we have

$$\begin{aligned} \int_\Omega \nabla w_k \nabla \phi dx &= \frac{1}{M_k} \int_\Omega \nabla u_{p_k} \nabla \phi dx \\ &= \frac{1}{M_k} \int_\Omega f_\lambda(u_{p_k}) \phi dx \\ &= \lambda \int_\Omega w_k \phi dx + \int_\Omega \frac{u_{p_k}^{-\alpha}}{M_k} \phi dx + \lambda \int_\Omega \frac{u_{p_k}^{p_k} - u_{p_k}}{M_k} \phi dx. \end{aligned} \tag{7.82}$$

Let v be the unique solution of problem (7.20). Because $v \le u_{p_k}$ in Ω it follows that

$$\left| \int_\Omega \frac{u_{p_k}^{-\alpha}}{M_k} \phi dx \right| \le \int_\Omega \frac{v^{-\alpha}}{M_k} |\phi| dx \to 0 \quad \text{as } k \to \infty. \tag{7.83}$$

On the other hand, with the same method as in (7.81) we obtain

$$\int_\Omega \frac{u_{p_k}^{p_k} - u_{p_k}}{M_k} \phi dx \to 0 \quad \text{as } k \to \infty. \tag{7.84}$$

Passing to the limit in (7.82) with $k \to \infty$, by (7.83) and (7.84) we deduce

$$\int_\Omega \nabla w_0 \nabla \phi dx = \lambda \int_\Omega w_0 \phi dx.$$

That is, $-\Delta w_0 = \lambda w_0$ in Ω. But this is a contradiction because $\lambda < \lambda_1$. Hence $(u_p)_{p>1}$ is bounded in $H_0^1(\Omega) \cap C^{1,\gamma}(\overline{\Omega})$ as $p \searrow 1$. □

Next, we shall be concerned with the asymptotic behavior of solutions to (7.1) as $p \searrow 1$. Let us consider the problem

$$\begin{cases} -\Delta w = w^{-\alpha} + \lambda w & \text{in } \Omega, \\ w > 0 & \text{in } \Omega, \\ w = 0 & \text{on } \partial\Omega. \end{cases} \tag{7.85}$$

Recall that by Theorem 5.6.1, the previous problem has a solution if and only if $\lambda < \lambda_1$. Moreover, for all $\lambda < \lambda_1$, problem (7.85) has a unique solution $w \in C^2(\Omega) \cap C^{1,1-\alpha}(\Omega)$.

The asymptotic behavior of solutions to (7.1) is described in the following result.

Theorem 7.7.3 *Assume $0 < \lambda < \lambda_1$, $0 < \alpha < 1/N$, and $1 < p < (N+2)/(N-2)$. Then the following hold:*

(i) *The solution u_p obtained in Theorem 7.4.1 is the minimal solution of (7.1).*

(ii) *$u_p \to w$ in $C^1(\overline{\Omega})$ as $p \searrow 1$, where w is the unique solution of problem (7.85).*

(iii) *For any other solution v_p of (7.1) we have $\|v_p\|_\infty \to \infty$ as $p \searrow 1$.*

Proof (i) We first show that (7.1) has a minimal solution. To this aim, let v be the unique solution of (7.20) and consider the sequence $(u_n)_{n\geq 0}$ defined by $u_0 \equiv v$, and for all $n \geq 1$, u_n satisfies

$$\begin{cases} -\Delta u_n = u_n^{-\alpha} + \lambda u_{n-1}^p & \text{in } \Omega, \\ u_n > 0 & \text{in } \Omega, \\ u_n = 0 & \text{on } \partial\Omega. \end{cases} \tag{7.86}$$

By Theorem 4.3.2 it follows that $u_n \in C^2(\Omega) \cap C^{1,1-\alpha}(\overline{\Omega})$ for all $n \geq 0$. If u is an arbitrary solution of (7.1), it is easy to check that $u_{n-1} \leq u_n \leq u$ in Ω. As in Theorem 7.7.2 we prove that $(u_n)_{n\geq 0}$ is bounded in $H_0^1(\Omega) \cap C^{1,\gamma}(\overline{\Omega})$. Hence, $(u_n)_{n\geq 0}$ converges to the minimal solution of (7.1). To show that u_p is the minimal solution of (7.1), it is enough to prove (ii) and (iii).

(ii) Let $(u_{p_k})_{k\geq 1}$ be a subsequence of $(u_p)_{p>1}$ with $p_k \searrow 1$ as $k \to \infty$. According to Theorem 7.7.2 we have that $(u_{p_k})_{k\geq 1}$ is bounded in $H_0^1(\Omega) \cap C^{1,\gamma}(\overline{\Omega})$. By the Arzelà–Ascoli theorem, there exists $u_0 \in H_0^1(\Omega) \cap C^1(\overline{\Omega})$ such that, up to a sequence, $u_{p_k} \to u_0$ in $C^1(\overline{\Omega})$ as $k \to \infty$.

Let $\phi \in H_0^1(\Omega)$ be arbitrary. By (7.1) we have

$$\int_\Omega \nabla u_{p_k} \nabla \phi dx = \int_\Omega (u_{p_k}^{-\alpha} + \lambda u_{p_k}^{p_k})\phi dx.$$

Passing to the limit in the previous equality we deduce

$$\int_\Omega \nabla u_0 \nabla \phi dx = \int_\Omega (u_0^{-\alpha} + \lambda u_0)\phi dx.$$

This means that u_0 is the unique weak solution of (7.85). By standard regularity theory, we easily derive that u_0 is a classical solution of (7.85), so $u_0 \equiv w$. Therefore, we have proved that any subsequence of $(u_p)_{p>1}$ has a subsequence that converges in $C^1(\overline{\Omega})$ to the unique solution w of problem (7.85). It follows that the whole sequence $(u_p)_{p>1}$ converges to w in $C^1(\overline{\Omega})$ as $p \searrow 1$.

(iii) Let v_p be an arbitrary solution of (7.1) such that $v_p \not\equiv u_p$, for all $1 < p < (N+2)/(N-2)$. Assume by contradiction that there exists a subsequence of $(v_p)_{p>1}$ still denoted by $(v_p)_{p>1}$ such that $\|v_p\|_\infty < M < \infty$ as $p \searrow 1$. Note that by Corollary 1.3.8 there exists $c_p > 0$ such that

$$u_p, v_p \geq c_p \operatorname{dist}(x, \partial\Omega) \quad \text{in } \Omega.$$

Because $u_p, v_p \in C^2(\Omega) \cap C^{1,\gamma}(\overline{\Omega})$, it follows that

$$\frac{u_p^2 - v_p^2}{u_p}, \frac{u_p^2 - v_p^2}{v_p} \in C^2(\Omega) \cap C^1(\overline{\Omega}).$$

Then,

$$\begin{aligned}
0 &\leq \int_\Omega \left(\left| \nabla u_p - \frac{u_p}{v_p} \nabla v_p \right|^2 + \left| \nabla v_p - \frac{v_p}{u_p} \nabla u_p \right|^2 \right) dx \\
&\leq \int_\Omega \left(-\frac{\Delta u_p}{u_p} + \frac{\Delta v_p}{v_p} \right) (u_p^2 - v_p^2) dx \\
&= \int_\Omega \Big((u_p^{-1-\alpha} + \lambda u_p^{p-1}) - (v_p^{-1-\alpha} + \lambda v_p^{p-1}) \Big) (u_p^2 - v_p^2) dx \qquad (7.87) \\
&= \int_\Omega \Big(-(1+\alpha)\theta_p^{-2-\alpha} + \lambda(p-1)\theta_p^{p-2} \Big) (u_p - v_p)^2 (u_p + v_p) dx \\
&= \int_\Omega \Big(-(1+\alpha) + \lambda(p-1)\theta_p^{p+\alpha} \Big) \theta_p^{-2-\alpha} (u_p - v_p)^2 (u_p + v_p) dx,
\end{aligned}$$

for some θ_p between u_p and v_p. Notice that by our assumption and Theorem 7.7.2 we have

$$\|u_p\|_\infty, \ \|v_p\|_\infty < C,$$

for some $C > 0$ independent of $p > 1$. Hence,

$$-(1+\alpha) + \lambda(p-1)\theta_p^{p+\alpha} \leq -(1+\alpha) + \lambda(p-1)C^{p+\alpha} < 0 \qquad \text{as } p \searrow 1.$$

This means that the right-hand side in (7.87) is negative, which is a contradiction. Therefore $\|v_p\|_\infty \to \infty$ as $p \searrow 1$ for any solution v_p of problem (7.1) with $v_p \not\equiv u_p$. As we have already argued in the first part of the proof, this also implies that u_p is the minimal solution of (7.1). The proof is now complete. □

Remark 7.7.4 *From estimate (7.71) in Theorem 7.6.1 and Theorem 7.7.3* (iii), *it follows that any nonminimal solution v_p of problem (7.1) satisfies*

$$\|v_p\|_{H_0^1(\Omega)} \to \infty \qquad \textit{as } p \searrow 1.$$

7.8 Comments and historical notes

Variational methods offer a deep insight into the nature of elliptic equations. The interested reader may find a valuable introduction in this sense in the book by Struwe [184]. These methods are a very useful tool when comparison principles fail to hold. This is the case, for instance, of elliptic problems involving smooth nonlinearities that have a superlinear growth at infinity. As we have seen so far in this book, this behavior is associated with the issue of multiplicity. In our case, the use of variational arguments provided us with the second solution, which was obtained by Ekeland's variational principle.

The existence of weak solutions for singular elliptic equations was first investigated by Lair and Shaker [125] for the problem

$$\begin{cases} -\Delta u = p(x)f(u) & \text{in } \Omega, \\ u > 0 & \text{in } \Omega, \\ u = 0 & \text{on } \partial\Omega \end{cases} \tag{7.88}$$

in a smooth bounded domain $\Omega \subset \mathbb{R}^N$ $(N \geq 1)$, $p \in L^2(\Omega)$ is nonnegative, and f is a positive decreasing function such that $\int_0^1 f(t)dt < \infty$. It was shown that (7.88) has a unique weak solution $u \in H_0^1(\Omega)$.

The approach given here follows an idea of Haitao [101], but the sharp estimates in proving the existence of the second solution originate in the works of Brezis and Nirenberg [33], Tarantello [187], and Badiale and Tarantello [10]. For a detailed proof of the estimate (7.41), the interested reader may consult Brezis and Nirenberg [33], [34].

This chapter completes the study of bifurcation and asymptotic analysis for the model problem

$$\begin{cases} -\Delta u = u^{-\alpha} + \lambda u^p & \text{in } \Omega, \\ u > 0 & \text{in } \Omega, \\ u = 0 & \text{on } \partial\Omega, \end{cases} \tag{7.89}$$

where $0 < p \leq (N+2)/(N-2)$, $0 < \alpha < 1$, $\lambda > 0$. The results obtained so far concerning (7.89) can be summarized as follows:

- If $0 < p < 1$, then problem (7.89) has a unique solution $u_\lambda \in C^2(\Omega) \cap C^{1,1-\alpha}(\overline{\Omega})$ (see Theorem 4.3.2).
- If $p = 1$, then problem (7.89) has solutions if and only if $\lambda < \lambda_1$. Moreover, for all $0 < \lambda < \lambda_1$ there exists a unique solution $u_\lambda \in C^2(\Omega) \cap C^{1,1-\alpha}(\overline{\Omega})$ of (7.89) that satisfies $\lim_{\lambda \nearrow \lambda_1} u_\lambda = \infty$ uniformly on compact subsets of Ω (see Theorem 5.6.1).
- If $1 < p \leq (N+2)/(N-2)$ and $N \geq 3$, then there exists $\lambda^* > 0$ such that
  - (i) for all $0 < \lambda < \lambda^*$, problem (7.89) has at least two weak solutions u_λ and v_λ such that $u_\lambda < v_\lambda$ in Ω;
  - (ii) for $\lambda = \lambda^*$, problem (7.89) has at least one weak solution u_λ;
  - (iii) for all $\lambda > \lambda^*$, problem (7.89) has no weak solution (see Theorem 7.1.2).

  Furthermore, if $0 < \alpha < 1/N$ and $1 < p < (N+2)/(N-2)$, then any weak solution of (7.89) is actually a $C^2(\Omega) \cap C^{1,\gamma}(\overline{\Omega})$ $(0 < \gamma < 1)$ solution and
  - (iv) $\lim_{p \searrow 1} \lambda^* = \lambda_1(-\Delta)$ (see Theorem 7.7.1);
  - (v) problem (7.89) has a minimal solution;
  - (vi) any other solution of (7.89) blows up in the L^∞ norm as $p \searrow 1$ (see Theorem 7.7.3).

  As remarked by Meadows [138], the $C^{1,\gamma}$ regularity is a particular feature of the fact that the weak solutions of (7.89) are positive inside the domain.

8

STABILITY OF THE SOLUTION OF A SINGULAR PROBLEM

> All the mathematical sciences are founded on relations between physical laws and laws of numbers, so that the aim of exact science is to reduce the problems of nature to the determination of quantities by operations with numbers.
>
> James C. Maxwell (1831–1879)

The issue of the stability of a solution was raised in Chapters 5 and 6 for elliptic equations involving smooth nonlinearities. In this chapter we provide more results related to this matter, being mainly interested in the case of singular nonlinearities. The analysis we develop here is carried out in a general framework that includes the classical singular elliptic problems already considered so far and for which we have discussed the existence, bifurcation, and regularity. This fact will be illustrated by some examples presented in the last part of this chapter.

8.1 Stability of the solution in a general singular setting

We are mainly interested in the stability of solutions corresponding to the problem

$$\begin{cases} -\Delta u = f(x,u) & \text{in } \Omega, \\ u > 0 & \text{in } \Omega, \\ u = 0 & \text{on } \partial\Omega, \end{cases} \tag{8.1}$$

where $\Omega \subset \mathbb{R}^N$ ($N \geq 1$) is a smooth bounded domain of class $C^{3,\gamma}$, $0 < \gamma < 1$. We assume that the nonlinearity $f : \Omega \times (0,\infty) \to \mathbb{R}$ satisfies the following:

($f1$) There exists an integer $m > 0$ such that

$$f,\ \frac{\partial^j f}{\partial t^j},\ \frac{\partial^j f}{\partial t^{j-1}\partial x_k} \in C(\Omega \times (0,\infty)),$$

for all $1 \leq j \leq m+1$ and $1 \leq k \leq N$.

($f2$) There exists $-1 < \alpha < 1$ such that for all $u : \Omega \to \mathbb{R}$ satisfying $0 < c_1 d(x) < u(x) < c_2 d(x)$ in Ω, we have

$$|f(x, u(x))| < C_0(1 + d(x)^\alpha) \quad \text{in } \Omega$$

and

$$\left|\frac{\partial^j f(x, u(x))}{\partial t^j}\right| + \sum_{k=1}^{N} \left|\frac{\partial^j f(x, u(x))}{\partial t^{j-1} \partial x_k}\right| < C_j d^{\alpha - j}(x) \quad \text{in } \Omega,$$

for all $1 \le j \le m+1$ and $1 \le k \le N$, where the constants $C_0, C_1, \ldots, C_{m+1}$ do not depend on u.

Typical examples of nonlinearities that fulfill ($f1$) and ($f2$) are those already encountered in the previous chapters. For instance, we may consider any combination of pure powers like

$$f(x, u) = a(x)u^\alpha + u^p \quad (x, u) \in \Omega \times (0, \infty),$$

with $-1 < \alpha < 1$, $0 < p < 1$, and $a \in C^1(\Omega)$ is such that $d(x)^{2-\alpha}|\nabla a(x)|$ is bounded in Ω.

Throughout this chapter, by $C_0^{k,\alpha}(\overline{\Omega})$ we mean the set of all functions $u \in C^{k,\alpha}(\overline{\Omega})$ such that $u = 0$ on $\partial\Omega$.

We assume that problem (8.1) has solutions in $C^2(\Omega) \cap C^1(\overline{\Omega})$, and we study the stability of these solutions in the sense given by Definition 5.1.1. This means that we have to deal with the linearized problem

$$\begin{cases} -\Delta v - f_t(x, u(x))v = \lambda v & \text{in } \Omega, \\ v = 0 & \text{on } \partial\Omega. \end{cases} \tag{8.2}$$

The analysis of (8.2) will be carried out in a more general setting by considering the following linear eigenvalue problem

$$\begin{cases} -\Delta v - a(x)v = \lambda b(x)v & \text{in } \Omega, \\ v = 0 & \text{on } \partial\Omega. \end{cases} \tag{8.3}$$

Taking into account the hypotheses ($f1$) and ($f2$) on f, a natural assumption on a is

(H) $\quad a \in C^1(\Omega)$ and $d(x)^{2-\alpha}|\nabla a(x)|$ is bounded in Ω.

As a consequence, if a satisfies (H) then the mapping $x \longmapsto d(x)^2 a(x)$ belongs to $C^{0,\beta}(\overline{\Omega})$ for all $0 < \beta < \min\{1 + \alpha, \gamma\}$, and $d(x)^{1-\alpha}a(x)$ is bounded in Ω.

Note that $a(x)$ may change sign near the boundary. To control the sign of a near $\partial\Omega$, we state the following result.

Lemma 8.1.1 *There exist two functions $\varphi^{\pm} \in C^{2,\beta}(\overline{\Omega})$ with $\varphi^{\pm} > 0$ in Ω such that if $v \in C^2(\Omega) \cap C^{1,\beta}(\overline{\Omega})$ is a solution of (8.3) then the functions $v^{\pm} := \varphi^{\pm} v$ belong to $C^2(\Omega) \cap C^{1,\beta}(\overline{\Omega})$ and verify*

$$\begin{cases} \mathcal{L}^{\pm} v^{\pm} = a^{\pm}(x) v^{\pm} + \lambda b(x) v^{\pm} & \text{in } \Omega, \\ v^{\pm} = 0 & \text{on } \partial\Omega, \end{cases} \tag{8.4}$$

where $\mathcal{L}^{\pm}$ is defined as

$$\mathcal{L}^{\pm} w = -\Delta w \pm \Phi(x) \cdot \nabla w \quad \text{for all } w \in C^{1,\beta}(\Omega), \tag{8.5}$$

and $\Phi : \overline{\Omega} \to \mathbb{R}^N$ is a $C^{1,\beta}$ function. Moreover, $a^{\pm}$ satisfies the assumption (H) and $\pm a^{\pm} > 0$ in a neighborhood of $\partial\Omega$.

Proof As a result of the regularity of the domain, we can find $\delta > 0$ such that $d(x) \in C^{2,\gamma}(\overline{\Omega}_\delta)$, where $\Omega_\delta := \{x \in \Omega : d(x) < \delta\}$.

Let us fix $-1 < p < \alpha$ and consider $\psi \in C^3[0, \infty)$ such that

$$\begin{aligned} p\psi(t) &\geq 0 && \text{for all } t \geq 0, \\ p\psi(t) &= t^{p+1} && \text{for all } 0 \leq t \leq \varepsilon < \rho, \\ \psi(t) &= 0 && \text{for all } t \geq \delta, \end{aligned}$$

where $\varepsilon > 0$ will be defined later in this proof. Consider now

$$\varphi^{\pm}(x) := e^{\mp\psi(d(x))} \quad x \in \overline{\Omega}$$

and

$$v^{\pm}(x) := \varphi^{\pm}(x) v(x) \quad x \in \overline{\Omega}.$$

Then $v^{\pm}$ verifies

$$\begin{aligned} v^{\pm} &= 0 \quad \text{on } \partial\Omega, \\ \frac{\partial v^{\pm}}{\partial n} &= \frac{\partial v}{\partial n} \quad \text{on } \partial\Omega, \end{aligned}$$

and

$$-\Delta v^{\pm} \mp 2\nabla v^{\pm} \cdot \nabla(\psi \circ d)(x) = a^{\pm}(x) v^{\pm} + \lambda b(x) v^{\pm} \quad \text{in } \Omega,$$

where

$$a^{\pm}(x) := a(x) + \Big((\psi'(d(x))^2 \pm \psi''(d(x))\Big)|\nabla d(x)|^2 \pm \psi'(d(x))\Delta d(x) \quad \text{in } \Omega.$$

Setting $\Phi(x) := -2\nabla(\psi \circ d)(x)$, $x \in \overline{\Omega}$, we obtain (8.4). Now, to have $\pm a^{\pm} > 0$ in a neighborhood of $\partial\Omega$, it suffices to take $\varepsilon > 0$ such that for all $x \in \Omega$ with $d(x) < \varepsilon$ there holds

$$\pm a(x) d(x)^{1-p} + (p+1)\Big(1 - \frac{p+1}{p^2} d(x)^{p+1}\Big)|\nabla d(x)|^2 + \frac{p+1}{p} d(x)\Delta d(x) > 0.$$

Because $|\nabla d(x)|$ is bounded away from zero in the neighborhood of $\partial\Omega$ and the mapping $x \longmapsto d(x)^{1-\alpha}|a(x)|$ is bounded in Ω according to assumption (H), we easily can find $\varepsilon > 0$ with the previous property. This concludes the proof. □

Remark 8.1.2 *Lemma 8.1.1 implies that the spectrum of (8.3) coincides with the spectrum of (8.4), provided that the eigenfunctions are required to be in* $C^2(\Omega) \cap C_0^1(\overline{\Omega})$. *Moreover, we have:*

(i) $v > 0$ *in* Ω *if and only if* $v^{\pm} > 0$ *in* Ω.

(ii) $v = 0$ *in* Ω *if and only if* $v^{\pm} = 0$ *in* Ω.

(iii) $\frac{\partial v}{\partial n} < 0$ *on* $\partial\Omega$ *if and only if* $\frac{\partial v^{\pm}}{\partial n} < 0$ *on* $\partial\Omega$.

(iv) $\frac{\partial v}{\partial n} = 0$ *on* $\partial\Omega$ *if and only if* $\frac{\partial v^{\pm}}{\partial n} = 0$ *on* $\partial\Omega$.

The following result states that the strong maximum principle holds for the operators of type $\mathcal{L}^{\pm} + a(x)$.

Proposition 8.1.3 *Let* $a \geq 0$ *satisfy (H) and let* $\mathcal{L}$ *be a differential operator defined as*

$$\mathcal{L}v := -\Delta v + \Phi(x) \cdot \nabla v \qquad \text{for all } v \in C^2(\Omega),$$

where $\Phi \in C^1(\overline{\Omega})$ *(in particular,* $\mathcal{L}$ *may be any of two operators* $\mathcal{L}^{\pm}$ *defined in (8.5)). Assume that* $v \in C^2(\Omega) \cap C^{1,\beta}(\overline{\Omega})$ *satisfies*

$$\begin{cases} \mathcal{L}v + a(x)v \geq 0 & \text{in } \Omega, \\ v \geq 0 & \text{in } \Omega, \end{cases}$$

and let $x_0 \in \overline{\Omega}$ *be such that* $v(x_0) = 0$.

(i) *If* $x_0 \in \Omega$ *then* $v \equiv 0$.

(ii) *If* $x_0 \in \partial\Omega$ *and* $v > 0$ *in* Ω *then* $\frac{\partial v(x_0)}{\partial n} < 0$.

Proof Because the coefficients of the linear operator $\mathcal{L}+a(x)$ are locally bounded in Ω, the proof of (i) follows directly by the strong maximum principle as stated in Theorem 1.3.5.

(ii) Assume by contradiction that $v > 0$ in Ω and $v(x_0) = \partial v(x_0)/\partial n = 0$. Because $v \in C^{1,\beta}(\overline{\Omega})$, there exists $c > 0$ such that

$$v(x) = |v(x) - v(x_0)| \leq c|x - x_0|^{1+\beta} \qquad \text{for all } x \in \Omega. \tag{8.6}$$

Note also that Ω satisfies the interior sphere condition at x_0. Hence, there exists an open ball $B \subset \Omega$ with a center at y_0 and a radius $r > 0$ such that $B \cap \partial\Omega = \{x_0\}$. Define $w(x) := (r - |x - y_0|)^{1+\beta/2}$ for all $x \in B$. Then,

$$\begin{aligned} \mathcal{L}w + a(x)w = &-\Delta w + \Phi(x) \cdot \nabla w + a(x)w \\ = &\frac{2+\beta}{2}\frac{N-1}{|x-y_0|}(r - |x - y_0|)^{\beta/2} \\ &- \frac{\beta(2+\beta)}{2}(r - |x - y_0|)^{-1+\beta/2} \\ &\frac{2+\beta}{2}\frac{(r - |x - y_0|)}{|x - y_0|}^{\beta/2}\Phi(x) \cdot (x - y_0) + a(x)w \qquad \text{in } B. \end{aligned} \tag{8.7}$$

Because $d(x)^{1-\beta}a(x)$ is bounded, it is readily seen that there exists r_0 close to r such that $\mathcal{L}w + a(x)w < 0$ in the annulus

$$A := \{x \in \Omega : r_0 < |x - x_0| < r\}.$$

Let $\varepsilon > 0$ and set $v_\varepsilon := v - \varepsilon w$. Then $\mathcal{L}v_\varepsilon + a(x)v_\varepsilon > 0$ in A. By the standard maximum principle it follows that v_ε attains its minimum at a point $x_1 \in \partial A$. Thus, if $|x_1 - y_0| = r$, then $v_\varepsilon(x_1) = v(x_1) \geq 0$, and if $|x_1 - y_0| = r_0$, then $v_\varepsilon(x_1) = v(x_1) - \varepsilon w(x_1)$. We can choose $\varepsilon > 0$ small enough such that $v_\varepsilon(x_1) > 0$. Hence, in both cases we have $v_\varepsilon \geq 0$ in A. This property holds true in particular on the rectilinear segment $S \subset \overline{A}$ that joins y_0 with x_0 and for which we have $r - |x - y_0| = |x - x_0|$. Therefore,

$$v(x) \geq \varepsilon |x - x_0|^{1+\beta/2} \quad \text{for all } x \in S \cap A.$$

Because $\varepsilon > 0$, the previous relation contradicts (8.6). This concludes the proof. □

For $w \in C_0^{0,1}(\overline{\Omega})$, consider the linear problem

$$\begin{cases} -\Delta v = a(x)w & \text{in } \Omega, \\ v = 0 & \text{on } \partial\Omega. \end{cases} \tag{8.8}$$

The following result provides a continuous dependence on data for solutions of problem (8.8) even when a is a singular potential. We omit the proof, which is lengthy and beyond the purposes of this chapter.

Proposition 8.1.4 *Let a satisfy (H). Then, for all $w \in C_0^{0,1}(\overline{\Omega})$, there exists a unique solution $v \in C^2(\Omega) \cap C^{1,\beta}(\overline{\Omega})$, $0 < \beta < \min\{1 + \alpha, \gamma\}$ of problem (8.8). Moreover, there exists a positive constant $C > 0$ that is independent of w such that*

$$\|v\|_{C^{1,\beta}(\overline{\Omega})} \leq C \|w\|_{C^{0,1}(\overline{\Omega})}. \tag{8.9}$$

Let $\mathcal{G}_a$ be Green's operator of (8.8)—that is, for all $w \in C_0^{0,1}(\overline{\Omega})$, $\mathcal{G}_a(w)$ is the unique solution $v \in C^2(\Omega) \cap C(\overline{\Omega})$ of the problem (8.8). Thus, Proposition 8.1.4 states that the Green operator $\mathcal{G}_a : C_0^{0,1}(\overline{\Omega}) \to C^{1,\beta}(\overline{\Omega})$ is continuous.

Proposition 8.1.5 *Assume that a and b satisfy the assumption (H) and $b > 0$ in Ω. Then, there exists $k_0 \in \mathbb{R}$ such that for all $k > k_0$ and all $w \in C_0^{0,1}(\overline{\Omega})$, the problem*

$$\begin{cases} -\Delta v - a(x)v + kb(x)v = b(x)w & \text{in } \Omega, \\ v = 0 & \text{on } \partial\Omega \end{cases} \tag{8.10}$$

has a unique solution $v \in C^2(\Omega) \cap C^{1,\beta}(\overline{\Omega})$ that fulfills

$$\|v\|_{C^{1,\beta}(\overline{\Omega})} \leq C \|w\|_{C^{0,1}(\overline{\Omega})} \tag{8.11}$$

for some positive constant $C > 0$ that does not depend on w.

Moreover, if $w \geq 0$ in Ω and w is not identically zero, then $\partial v/\partial n < 0$ on $\partial\Omega$.

Proof We first choose k_0 which ensures the uniqueness. To this aim we rewrite problem (8.10) in the form

$$\begin{cases} \mathcal{L}^- v^- - a^-(x)v^- + kb(x)v^- = b(x)\varphi^- w & \text{in } \Omega, \\ v^- = 0 & \text{on } \partial\Omega, \end{cases} \tag{8.12}$$

where $\mathcal{L}^-, a^-, v^-$, and φ^- are defined in Lemma 8.1.1 and $v^- = \varphi^- v$. Because $a^- > 0$ in a neighborhood Ω_δ of $\partial\Omega$ and $b > 0$ in Ω, we can define

$$k_0 := \sup_{x \in \Omega \setminus \Omega_\delta} \frac{a^-(x)}{b(x)}.$$

This yields $kb(x) - a^-(x) > 0$ in Ω for all $k > k_0$. Furthermore, by Proposition 8.1.3 we derive the uniqueness of the solution to problem (8.12) and hence the uniqueness of (8.10). Moreover, $\partial v^- / \partial n < 0$ on $\partial\Omega$ provided that $w \geq 0$ is not identically zero. Using Remark 8.1.2, this yields $\partial v / \partial n < 0$ on $\partial\Omega$.

Let us prove now the existence for all $k > k_0$. For this purpose let

$$\mathcal{G}_a, \mathcal{G}_b : C_0^{0,1}(\overline{\Omega}) \to C_0^{1,\beta}(\overline{\Omega})$$

be the Green's operators associated with problems (8.8) and

$$\begin{cases} -\Delta v = b(x)w & \text{in } \Omega, \\ v = 0 & \text{on } \partial\Omega, \end{cases}$$

respectively. Then problem (8.10) may be formulated as

$$\mathcal{H}(v) = \mathcal{G}_b(w), \tag{8.13}$$

where

$$\mathcal{H}(v) := v - \mathcal{G}_a(v) + k\mathcal{G}_b(v) \quad v \in C^{1,\beta}(\overline{\Omega}).$$

Because the embedding

$$i : C_0^{1,\beta}(\overline{\Omega}) \to C_0^{0,1}(\overline{\Omega}) \tag{8.14}$$

is compact, we derive that $\widehat{\mathcal{H}} := \mathcal{H} \circ i$ is a compact perturbation of the identity in $C_0^{1,\beta}(\overline{\Omega})$. Furthermore, $\widehat{\mathcal{H}}$ is injective for all $k > k_0$. By standard Riesz theory on compact operators, we note that $\widehat{\mathcal{H}}$ is a linear homeomorphism. Thus, if $w \in C_0^{0,1}(\overline{\Omega})$, then problem (8.13) has a unique solution $v \in C^{1,\beta}(\overline{\Omega})$. Moreover, according to Proposition 8.1.4 we also have

$$\|v\|_{C^{1,\beta}(\overline{\Omega})} \leq C_1 \|\mathcal{G}_b(w)\|_{C^{1,\beta}(\overline{\Omega})} \leq C_2 \|w\|_{C^{0,1}(\overline{\Omega})},$$

where $C_1, C_2 > 0$ are independent of w. This finishes the proof. □

Theorem 8.1.6 *Assume that a and b satisfy the assumption (H) and $b > 0$ in Ω. Then the spectrum of the linear eigenvalue problem (8.3) with $v \in C^2(\Omega) \cap C_0^{1,\beta}(\overline{\Omega})$ has the following properties:*

(i) *It consists of at most a countable set of eigenvalues which are isolated.*
(ii) *The first eigenvalue is simple and the associated eigenfunction v_1 satisfies $v_1 > 0$ in Ω and $\partial v_1/\partial n < 0$ on $\partial\Omega$.*
(iii) *It does not change when the eigenfunctions are required to be in $C^2(\Omega) \cap H_0^1(\Omega)$.*

Proof (i) Let k_0 be as in Proposition 8.1.5. Then, for all $k > k_0$, problem (8.10) defines a Green's operator $\mathcal{G} : C_0^{0,1}(\overline{\Omega}) \to C_0^{1,\beta}(\overline{\Omega})$, which, in view of (8.11), is bounded. Furthermore, if i is the compact embedding defined in (8.14), then $\widehat{\mathcal{G}} = \mathcal{G} \circ i : C_0^{1,\beta}(\overline{\Omega}) \to C_0^{1,\beta}(\overline{\Omega})$ is compact. Thus, (i) follows by the spectral characterization of the compact operators and the fact that the eigenvalues of (8.3) and the eigenvalues μ of $\widehat{\mathcal{G}}$ are related by

$$\mu(\lambda + k) = 1. \tag{8.15}$$

(ii) By Proposition 8.1.5, the compact operator $\widehat{\mathcal{G}}$ maps the positive cone of $C_0^{1,\beta}(\overline{\Omega})$ into its interior—that is, into the set of those functions $u \in C_0^{1,\beta}(\overline{\Omega})$ that fulfill $u > 0$ in Ω and $\partial u/\partial n < 0$ on $\partial\Omega$. Hence, by the Krein–Rutman theorem (see Theorem A.2.3 in Appendix A), it follows that the first eigenvalue μ_1 of $\widehat{\mathcal{G}}$ is simple and positive. Taking into account the relation (8.15), we derive that $\lambda_1 = 1/\mu_1 - k$ is also simple. Moreover, the first eigenfunction v_1 is positive in Ω and satisfies $\partial v_1/\partial n < 0$ on $\partial\Omega$.

(iii) Let X be the completion of $C_0^1(\overline{\Omega})$ with respect to the norm

$$\|v\|_b^2 := \int_\Omega b(x) v^2 dx.$$

It turns out that X is a Hilbert space with respect to the inner product

$$\langle v, w\rangle = \int_\Omega b(x) v w dx \quad v, w \in X.$$

Furthermore, we claim that the embedding $j : H_0^1(\Omega) \to X$ is compact. Taking into account the standard compact embedding theorems into L^p spaces with weights, we need only to prove that j is continuous. Indeed, by the assumption (H) it follows that $d(x)^2 b(x)$ is bounded in Ω. Hence, there exists $c > 0$ such that

$$b(x) \le \frac{c}{d^2(x)} \quad \text{in } \Omega.$$

By the Hardy–Sobolev inequality there exists $c_0 > 0$, which is independent of v such that

$$\int_\Omega \frac{v(x)}{d(x)^2} dx \le c_0 \int_\Omega |\nabla v|^2 dx. \tag{8.16}$$

Hence,

$$\|v\|_b^2 \le c \int_\Omega \frac{v(x)}{d(x)^2} dx \le C \int_\Omega |\nabla v|^2 dx, \tag{8.17}$$

and the claim follows.

Now let $w \in C_0^{0,1}(\overline{\Omega})$ and $v \in C^2(\Omega) \cap C_0^1(\overline{\Omega})$ be a solution of (8.10). Multiplying by v in (8.10) and then integrating by parts, we obtain

$$\int_\Omega |\nabla v|^2 dx - \int_\Omega a(x)v^2 dx + k\int_\Omega b(x)v^2 dx = \int_\Omega b(x)vw dx. \tag{8.18}$$

On the other hand, by virtue of the assumption (H) there exists $\omega \subset\subset \Omega$ and $c_1 > 0$ such that

$$d(x)^2|a(x)| \le \frac{1}{2c_0} \quad \text{in } \Omega \setminus \omega \tag{8.19}$$

and

$$|a(x)| \le c_1 b(x) \quad \text{in } \omega, \tag{8.20}$$

where c_0 is the constant from (8.16). This yields

$$\begin{aligned} \left|\int_\Omega a(x)v^2 dx\right| &\le \frac{1}{2c_0}\int_\Omega \frac{v(x)}{d(x)^2}dx + c_1\int_\Omega b(x)v^2 dx \\ &\le \frac{1}{2}\int_\Omega |\nabla v|^2 dx + c_1\int_\Omega b(x)v^2 dx. \end{aligned} \tag{8.21}$$

Choosing $k > |c_1|$, from (8.21) and (8.18) we obtain

$$\int_\Omega |\nabla v|^2 dx \le c_2\int_\Omega b(x)vw dx,$$

for some positive constant $c_2 > 0$. Next, by Hölder's inequality and (8.17) we derive

$$\|v\|_{H_0^1} \le c\|w\|_b,$$

for some constant $c > 0$ independent of v and w. This means that Green's operator $\mathcal{G} : X \to H_0^1(\Omega)$ of problem (8.10) is bounded. Then $\widetilde{\mathcal{G}} := \mathcal{G} \circ j : H_0^1(\Omega) \to H_0^1(\Omega)$ is compact. From now on, we proceed as in the proof of (i) and use the fact that the eigenvalues of (8.3) and those of $\widetilde{\mathcal{G}}$ are related by (8.15). This finishes the proof of Theorem 8.1.6. □

Theorem 8.1.6 does not give a precise answer on the sign of the first eigenvalue of (8.2). In other words, we cannot decide whether the solution u of (8.1) is stable. In this sense, the following result provides some natural and simple conditions to derive that the first eigenvalue of (8.2) is strictly positive. It is a very useful result when applying the implicit function theorem as we did in Theorem 5.2.1.

Proposition 8.1.7 *Suppose that (8.1) has a solution $u \in C^2(\Omega) \cap C^1(\overline{\Omega})$ such that*

(i) $\frac{\partial u}{\partial n} < 0$ *on* $\partial\Omega$;
(ii) $f(x,u) - uf_t(x,u) > 0$ *in* Ω.

Then the first eigenvalue λ_1 of (8.2) is positive—that is, u is a stable solution of problem (8.1).

8.2 A min–max characterization of the first eigenvalue for the linearized problem

We start this section with an equivalent condition for the sign of the first eigenvalue of problem (8.3).

Lemma 8.2.1 *The first eigenvalue λ_1 of the spectral problem (8.3) is positive if and only if the operator $-\Delta - a(x)$ satisfies the strong maximum principle—that is, for all $v \in C^2(\Omega) \cap C^1(\overline{\Omega})$ not identically zero such that*

$$\begin{cases} -\Delta v - a(x)v \geq 0 & \text{in } \Omega, \\ v \geq 0 & \text{on } \partial\Omega, \end{cases} \tag{8.22}$$

then $v > 0$ in Ω and $\partial v(x)/\partial n > 0$ for all $x \in \partial\Omega$ with $v(x) = 0$.

Proof If $-\Delta - a(x)$ satisfies the strong maximum principle, then we clearly have $\lambda_1 > 0$. To prove the converse, we transform (8.22) according to Lemma 8.1.1 in

$$\begin{cases} \mathcal{L}^- v^- - a^-(x)v^- \geq 0 & \text{in } \Omega, \\ v^- \geq 0 & \text{on } \partial\Omega, \end{cases} \tag{8.23}$$

and there exists $\omega \subset\subset \Omega$ such that $a^-(x) < 0$ in $\Omega \setminus \omega$. Let v_1 be the first eigenvalue of (8.3) corresponding to λ_1. Notice that $v_1 > 0$ in Ω. Set

$$k := \sup_{x \in \Omega \setminus \omega} \frac{2|a^-(x)|}{\lambda_1 b(x) v_1(x)} > 0, \tag{8.24}$$

and for $\varepsilon > 0$ define

$$w := v^- + \varepsilon + \varepsilon k v_1^- .$$

Thus, by (8.24) we have

$$\begin{aligned} \mathcal{L}^- w - a^-(x)w &= (\mathcal{L}^- v^- - a^-(x)v^-) + \varepsilon k(\mathcal{L}^- v_1^- - a^-(x)v_1^-) - \varepsilon a^-(x) \\ &\geq \varepsilon\Big(k\lambda_1 b(x) v_1^- - a^-(x)\Big) > 0 \quad \text{in } \Omega. \end{aligned}$$

Using the continuity of w, there exists $\delta = \delta(\varepsilon) > 0$ such that $w > 0$ in $\overline{\Omega}_\delta$. Moreover, by the strong maximum principle (see Theorem 1.3.5) applied to w in $\Omega \setminus \Omega_\delta$ we also have $w > 0$ in $\Omega \setminus \Omega_\delta$. Therefore, $w > 0$ in Ω and letting $\varepsilon \to 0$, we obtain $v^- \geq 0$ in Ω. Thus, we have obtained

$$\begin{cases} \mathcal{L}^- v^- - a^-(x)v^- \geq 0 & \text{in } \Omega, \\ v^- \geq 0 & \text{in } \Omega. \end{cases}$$

Because v^- is not identically zero, by Proposition 8.1.3 we find $v^- > 0$ in Ω and $\partial v^-(x)/\partial n > 0$ for all $x \in \partial\Omega$ with $v^-(x) = 0$. By virtue of Remark 8.1.2, the last conclusion also holds for v. This completes the proof. □

A min–max characterization of the first eigenvalue to (8.3) is stated in the following proposition.

Proposition 8.2.2 *Assume that a, b satisfy (H) and $b > 0$ in Ω. Then, the first eigenvalue λ_1 of (8.3) is stated by*

$$\lambda_1 = \sup_{\substack{v\in C^2(\Omega)\cap C_0^1(\overline{\Omega})\\ v>0 \text{ in } \Omega}} \inf_{x\in\Omega} \frac{-\Delta v - a(x)v}{b(x)v}. \tag{8.25}$$

Proof Denote by m the quantity in the right-hand side of (8.25) and let v_1 be the corresponding eigenfunction of the first eigenvalue λ_1 to (8.3). Then $v_1 \in C^2(\Omega) \cap C_0^1(\overline{\Omega})$, $v > 0$ in Ω, and

$$\lambda_1 = \inf_{x\in\Omega} \frac{-\Delta v_1 - a(x)v_1}{b(x)v_1}.$$

This means that $\lambda_1 \leq m$. To prove that $m \leq \lambda_1$, we argue by contradiction and we assume that there exists $\varepsilon > 0$ and $w \in C^2(\Omega) \cap C_0^1(\overline{\Omega})$, $w > 0$ in Ω, such that

$$\inf_{x\in\Omega} \frac{-\Delta w - a(x)w}{b(x)w} > \lambda_1 + \varepsilon. \tag{8.26}$$

Let us consider the spectral problem

$$\begin{cases} -\Delta v - a(x)v - (\lambda_1 - \varepsilon)b(x)v = \lambda b(x)v & \text{in } \Omega, \\ v = 0 & \text{on } \partial\Omega. \end{cases} \tag{8.27}$$

Clearly, $\varepsilon > 0$ is the first eigenvalue of (8.27). Furthermore, by Lemma 8.2.1 the operator $-\Delta v - a(x) - (\lambda_1 - \varepsilon)b(x)$ satisfies the strong maximum principle. To raise a contradiction, let $\tau := \min\{t_1, t_2\}$, where

$$t_1 := \sup\{t \geq 0 : w - tv_1 \geq 0 \text{ in } \Omega\},$$

$$t_2 := \sup\left\{t \geq 0 : \frac{\partial(w - tv_1)}{\partial n} \leq 0 \text{ on } \partial\Omega\right\}.$$

Define now $W := w - \tau v_1$. Clearly, $W \geq 0$ in Ω. Moreover, if $\tau = t_1$ then $W = 0$ at some point in Ω and if $\tau = t_2$ then $W = \partial W/\partial n = 0$ at some point on $\partial\Omega$. On the other hand, by (8.26) we have

$$-\Delta W - a(x)W - (\lambda_1 - \varepsilon)b(x)W > \varepsilon b(x)(w + W) > 0 \quad \text{in } \Omega.$$

By Lemma 8.2.1 we find that $W > 0$ in Ω and $\partial W/\partial n < 0$ on $\partial\Omega$, which is clearly a contradiction. Hence, $\lambda_1 = m$ and the proof is now complete. □

8.3 Differentiability of some singular nonlinear problems

In this section we are concerned with the differentiability of the semilinear elliptic problem (8.1) around positive solutions $u \in C^2(\Omega) \cap C^1(\overline{\Omega})$ that satisfy $\partial u/\partial n < 0$ on $\partial\Omega$. To this aim, let $\mathcal{G} : C_0^1(\overline{\Omega}) \to C_0^1(\overline{\Omega})$ be Green's operator of

$$\begin{cases} -\Delta v = w & \text{in } \Omega, \\ v = 0 & \text{on } \partial\Omega, \end{cases}$$

defined as $v = \mathcal{G}(w)$. By virtue of Proposition 8.1.4, $\mathcal{G}$ is bounded and it can be extended as a bounded operator to $C_0^{1,\beta}(\overline{\Omega})$. Furthermore, $u \in C^2(\Omega) \cap C^1(\overline{\Omega})$ is a solution of (8.1) if and only if $0 < u \in C_0^1(\overline{\Omega})$ satisfies

$$\mathcal{T}(u) = 0, \tag{8.28}$$

where

$$\mathcal{T}(u) := u - \mathcal{G}(f(\cdot, u)).$$

Let $\mathcal{C}$ be the positive cone of $C_0^1(\overline{\Omega})$—that is, the set of all functions $u \in C_0^1(\overline{\Omega})$ that fulfill $u > 0$ in Ω and $\partial u/\partial n < 0$ on $\partial\Omega$. We are concerned here with the Fréchet differentiability of the operator $\mathcal{T} : C_0^1(\overline{\Omega}) \to C_0^1(\overline{\Omega})$.

Theorem 8.3.1 *Under the assumptions* $(f1)$ *and* $(f2)$, *the operator* $\mathcal{T}$ *is of class* C^m *on the positive cone* $\mathcal{C}$, *and for all* $v \in \mathcal{C}$ *the linear operator*

$$\mathcal{T}'(v) : C_0^1(\Omega) \to C_0^1(\Omega)$$

is given by

$$\mathcal{T}'(v)(w) = w - \mathcal{G}(f_t(\cdot, v)w). \tag{8.29}$$

If $m > 1$ *and* $1 < j \leq m$ *then the* j *linear operator*

$$\mathcal{T}^{(j)}(v) : [C_0^1(\overline{\Omega})]^j \to C_0^1(\overline{\Omega})$$

is given by

$$\mathcal{T}^{(j)}(v)(w_1, w_2, \ldots, w_j) = -\mathcal{G}\Big(\frac{\partial^j f(\cdot, v)}{\partial t^j}(w_1, w_2, \ldots, w_j)\Big). \tag{8.30}$$

Proof Remark first that $\mathcal{T} = I - \widetilde{\mathcal{T}}$, where $I : C_0^1(\overline{\Omega}) \to C_0^1(\overline{\Omega})$ is the identity operator and $\widetilde{\mathcal{T}}(v) = \mathcal{G}(f(\cdot, v))$. Notice that I is linear and bounded; hence, I is of class C^∞. Furthermore, its first derivative is I and its higher order derivatives are zero. Thus, it suffices to show that for all $1 \leq j \leq m$ the jth derivative of $\widetilde{\mathcal{T}}$ is given by

$$\widetilde{\mathcal{T}}^{(j)}(v)(w_1, w_2, \ldots, w_j) = \mathcal{G}\Big(\frac{\partial^j f(\cdot, v)}{\partial t^j}(w_1, w_2, \ldots, w_j)\Big) \tag{8.31}$$

and the mth derivative of $\widetilde{\mathcal{T}}$ is continuous.

We prove (8.31) by induction on j. Consider first the case $j = 1$ and let $a \in C^2(\Omega)$ be such that $a \geq 1$ in Ω and $a(x) = d(x)^{\alpha-1}$ in a neighborhood of $\partial\Omega$. Then,

$$\frac{d(x)}{a(x)}^{\alpha-1} \quad \text{and } d(x)^{2-\alpha}|\nabla a(x)| \text{ are uniformly bounded in } \Omega. \tag{8.32}$$

Let $v \in \mathcal{C}$ and $w \in C_0^1(\overline{\Omega})$ with $\|w\|_{C_0^1(\Omega)}$ sufficiently small. Then there exist two positive constants c_1, c_2 that are independent of w such that

$$0 < c_1 d(x) < v + \theta w < c_2 d(x) \quad \text{in } \overline{\Omega}, \tag{8.33}$$

for all $0 \leq \theta \leq 1$. Also define

$$W(x) := \frac{1}{a(x)}\Big(f(x, v + w) - f(x, v) - f_t(x, v)w\Big).$$

By the mean value theorem, relation (8.33), and the hypotheses $(f1)$ and $(f2)$ we have

$$\begin{aligned} |W(x)| &= \frac{|f_t(x, v + \theta_1(x)w) - f_t(x, v)||w|}{a(x)} \quad (0 < \theta_1(x) < 1) \\ &\leq \frac{|f_{tt}(x, v + \theta_2(x)w)|w^2}{a(x)} \leq C_1\|w\|^2_{C_0^1(\overline{\Omega})} \quad (0 < \theta_2(x) < 1) \end{aligned} \tag{8.34}$$

and

$$\begin{aligned} |W_{x_k}(x)| \leq &\frac{|f_{tx_k}(x, v + \theta(x)w) - f_{tx_k}(x, v)||w|}{a(x)} \quad (0 < \theta(x) < 1) \\ &+ \frac{|f_{tt}(x, v + \theta(x)w) - f_{tt}(x, v)|v_{x_k} w|}{a(x)} \\ &+ \frac{|f_{tt}(x, v + \theta(x)w)||ww_{x_k}|}{a(x)} + \frac{|W(x)a_{x_k}(x)|}{a(x)^2} \\ \leq &C_2\|w\|_{C_0^1(\overline{\Omega})}\tau(\|w\|_{C_0^1(\overline{\Omega})}), \end{aligned} \tag{8.35}$$

for all $1 \leq k \leq N$. Here, $\tau : \mathbb{R} \to \mathbb{R}$ is such that $\tau(s) \to 0$ as $s \to 0$ and the positive constants C_1, C_2 are independent of w. It follows that

$$\|W\|_{C_0^1(\overline{\Omega})} \leq C_3\|w\|_{C_0^1(\overline{\Omega})}\tau(\|w\|_{C_0^1(\overline{\Omega})}).$$

Taking into account the previous estimate and Proposition 8.1.4 we have

$$\begin{aligned} \|\widetilde{T}(v + w) - \widetilde{T}(v) - \mathcal{G}(f_t(\cdot, v)w)\|_{C_0^1(\overline{\Omega})} &= \|\mathcal{G}(aW)\|_{C_0^1(\overline{\Omega})} \\ &\leq C_4\|w\|_{C_0^1(\overline{\Omega})}\tau(\|w\|_{C_0^1(\overline{\Omega})}), \end{aligned} \tag{8.36}$$

for all $w \in C_0^1(\overline{\Omega})$ such that $\|w\|_{C_0^1(\overline{\Omega})}$ is small and $\|w\|_{C_0^1(\overline{\Omega})} \leq \|v\|_{C_0^1(\overline{\Omega})}$. Hence, $\widetilde{T}'(v)$ exists and

$$\widetilde{\mathcal{T}}'(v)(w) = \mathcal{G}(f_t(\cdot, v)w) \qquad \text{for all } v \in \mathcal{C}, w \in C_0^1(\overline{\Omega}).$$

Next we assume that (8.31) holds for jth derivative of $\widetilde{\mathcal{T}}$ and let us prove that it also holds for its $(j+1)$th derivative. Let $w_1, w_2, \ldots, w_{j+1} \in C_0^1(\overline{\Omega})$ be such that $\|w_{j+1}\|_{C_0^1(\overline{\Omega})} \leq \|v\|_{C_0^1(\overline{\Omega})}$. Remark that (8.33) holds for $w = w_{j+1}$ and we retake the previous arguments for

$$W_1(x) := \frac{1}{a(x)}\Big(\frac{\partial^j f(x, v + w_{j+1})}{\partial t^j} - \frac{\partial^j f(x,v)}{\partial t^j} - \frac{\partial^{j+1} f(x,v)}{\partial t^{j+1}} w_{j+1}\Big) w_1 \cdots w_j.$$

We obtain

$$\|W_1\|_{C_0^1(\overline{\Omega})} \leq C_5 \|w_1\|_{C_0^1(\overline{\Omega})} \|w_2\|_{C_0^1(\overline{\Omega})} \cdots \|w_{j+1}\|_{C_0^1(\overline{\Omega})} \tau(\|w_{j+1}\|_{C_0^1(\overline{\Omega})}),$$

where $C_5 > 0$ is independent of $w_1, w_2, \ldots, w_{j+1}$. As in (8.36) we have

$$\begin{aligned}
&\Big\|\Big[\widetilde{\mathcal{T}}^{(j)}(v + w_{j+1}) - \widetilde{\mathcal{T}}^{(j)}(v)\Big](w_1 \ldots, w_j) - \mathcal{G}\Big(\frac{\partial^{j+1} f(\cdot, v)}{\partial t^{j+1}}\Big) w_1 \cdots w_j\Big\|_{C_0^1(\overline{\Omega})} \\
&\quad = \|\mathcal{G}(aW_1)\|_{C_0^1(\overline{\Omega})} \\
&\quad \leq C_5 \|w_1\|_{C_0^1(\overline{\Omega})} \|w_2\|_{C_0^1(\overline{\Omega})} \cdots \|w_{j+1}\|_{C_0^1(\overline{\Omega})} \tau(\|w_{j+1}\|_{C_0^1(\overline{\Omega})}),
\end{aligned}$$

with $C_6 > 0$ independent of $w_1, w_2, \ldots, w_{j+1}$.

This estimate implies that $\widetilde{\mathcal{T}}^{(j+1)}(v)$ exists and is given by (8.31). This completes the induction argument. Finally, the same reasoning that led us to the previous estimate yields

$$\begin{aligned}
&\Big\|\Big[\widetilde{\mathcal{T}}^{(m)}(v + w_{j+1}) - \widetilde{\mathcal{T}}^{(m)}(v)\Big](w_1, w_2, \ldots, w_j)\Big\|_{C_0^1(\overline{\Omega})} \\
&\qquad \leq C \|w_1\|_{C_0^1(\overline{\Omega})} \cdots \|w_{j+1}\|_{C_0^1(\overline{\Omega})} \tau(\|w_{j+1}\|_{C_0^1(\overline{\Omega})}), \qquad (8.37)
\end{aligned}$$

which implies that $\widetilde{\mathcal{T}}^{(m)}$ is continuous. This finishes the proof. □

Assume next that the nonlinearity f depends on a parameter $\lambda \in \Lambda$, where $\Lambda \subset \mathbb{R}$ is an interval, and the derivatives $\partial^p f / \partial \lambda^p$, $1 \leq p \leq r$, satisfy the assumption ($f2$). With the same arguments as used earlier we have the following corollary.

Corollary 8.3.2 *Using the previous assumptions, the operator*

$$\mathcal{T} : \Omega \times C_0^1(\overline{\Omega}) \times \Lambda \to C_0^1(\overline{\Omega}),$$

defined as

$$\mathcal{T}(u, \lambda) := u - \mathcal{G}(f(\cdot, u, \lambda)),$$

satisfies

(i) *for all* $1 \leq j \leq m$ *and* $1 \leq p \leq r$, *the derivative* $\partial^{j+p}\mathcal{T}/\partial t^j \partial \lambda^p$ *exists and is continuous at each point* $(v, \lambda) \in \mathcal{C} \times \Lambda$;

(ii) *the linear operator*

$$\frac{\partial^{j+p}\mathcal{T}(v,\lambda)}{\partial t^j \partial \lambda^p} : [C_0^1(\overline{\Omega})]^j \to C_0^1(\overline{\Omega})$$

is given by

$$\frac{\partial^{j+p}\mathcal{T}(v,\lambda)}{\partial t^j \partial \lambda^p}(w_1, w_2, \ldots, w_j) = -\mathcal{G}\Big(\frac{\partial^{j+p} f(\cdot, v, \lambda)}{\partial t^j \partial \lambda^p}(w_1, w_2, \ldots, w_j)\Big),$$

for all $w_1, w_2, \ldots, w_j \in C_0^1(\overline{\Omega})$.

8.4 Examples

The results obtained in this chapter concerning the stability and the differentiability in singular elliptic problems are illustrated by the following two examples.

Example 8.1 Let us first consider the problem

$$\begin{cases} -\Delta u = \lambda a(x) u^\alpha & \text{in } \Omega, \\ u > 0 & \text{in } \Omega, \\ u = 0 & \text{on } \partial\Omega \end{cases} \tag{8.38}$$

in a smooth bounded domain Ω with $-1 < \alpha < 1$, $\lambda > 0$ and

(i) $a \in C^1(\Omega)$ is positive and there exists $\beta \in (-1-\alpha, 1-\alpha)$ such that $a(x) < cd(x)^\beta$ in Ω for some $c > 0$;
(ii) the mapping $f(x,t) = a(x)t^\alpha$ satisfies $(f2)$.

If $0 < \alpha < 1$ then (8.38) arises in population dynamics when dealing with stationary solutions of the usual logistic equation with nonlinear diffusion. For $-1 < \alpha < 0$ and $a \in C(\overline{\Omega})$, $a > 0$ in $\overline{\Omega}$, problem (8.38) was considered in Chapter 4, where we proved the existence and uniqueness of a solution $u_\lambda \in C^2(\Omega) \cap C^{1,\gamma}(\overline{\Omega})$, $(0 < \gamma < 1)$ for all $\lambda > 0$. Similar results hold in the case $0 < \alpha < 1$. Concerning the stability and the differentiability of (8.38) we have the following.

Proposition 8.4.1 *For all* $\lambda > 0$, *problem (8.38) has a unique solution* $u_\lambda \in C^2(\Omega) \cap C^{1,\gamma}(\overline{\Omega})$, $0 < \gamma < \min\{1, 1+\alpha+\beta\}$, *such that*

(i) $\frac{\partial u_\lambda}{\partial n} < 0$ *on* $\partial\Omega$;
(ii) u_λ *is stable;*
(iii) *the mapping* $(0,\infty) \ni \lambda \longmapsto u_\lambda \in C^{1,\gamma}(\overline{\Omega})$ *is of class* C^∞.

The existence follows in a similar way as in the proof of Theorem 4.3.2. The conclusion in statements (ii) and (iii) follows by Proposition 8.1.7 and Theorem 8.3.1, respectively.

If $a(x)$ changes sign in Ω, the analysis of (8.38) becomes more delicate. A detailed study in this sense was carried out by Bandle, Pozio, and Tesei [17].

Because $0 < \alpha < 1$, the nonlinearity f is not locally Lipschitz near the origin, which may give rise to dead core solutions. Existence of this kind of solutions was obtained in [17] by the sub- and supersolution method. Also, the uniqueness is provided under some additional hypotheses.

Example 8.2 Our second example concerns the problem

$$\begin{cases} -\Delta u = a(x)u^{-\alpha} + \lambda u^p & \text{in } \Omega, \\ u = 0 & \text{on } \partial\Omega, \end{cases} \tag{8.39}$$

with $0 < \alpha < 1$, $0 < p \leq 1$, $\lambda > 0$, and

(i) $a \in C^1(\Omega)$ is positive and there exists $\beta \in (\alpha - 1, \alpha + 1)$ such that $a(x) < cd(x)^\beta$ in Ω for some $c > 0$;

(ii) the mapping $f(x,t) = a(x)t^{-\alpha} + \lambda t^p$ satisfies $(f2)$.

If $0 < p < 1$, in view of Theorem 4.3.2 we have the same conclusion as in Proposition 8.4.1. If $p = 1$, taking into account the bifurcation results in Theorem 5.6.1 we obtain the following.

Proposition 8.4.2 *Assume $p = 1$. Then, for all $0 < \lambda < \lambda_1(-\Delta)$, problem (4.71) has a unique solution $u_\lambda \in C^2(\Omega) \cap C^{1,\gamma}(\overline{\Omega})$, $0 < \gamma < \min\{1, 1 + \alpha + \beta\}$ such that*

(i) $\frac{\partial u_\lambda}{\partial n} < 0$ *on* $\partial\Omega$;

(ii) u_λ *is stable;*

(iii) *the mapping* $(0, \lambda_1) \ni \lambda \longmapsto u_\lambda \in C^{1,\gamma}(\overline{\Omega})$ *is of class* C^∞ *and* $u_\lambda \to \infty$ *uniformly on compact subsets of* Ω.

8.5 Comments and historical notes

In this chapter we pointed out new features of singular elliptic problems, being concerned with stability and differentiability of such types of problems.

First we studied the spectrum of some linearized singular elliptic problems in a general setting involving singular weights. In this sense we considered the general problem (8.3), where the (possible singular) potential b is positive in Ω. If the exponent α in the assumptions $(f2)$ and (H) satisfies $\alpha > -1/N$, then the standard regularity theory for elliptic equations applies to obtain eigenfunctions in the class $C_0^1(\overline{\Omega})$. However, we preferred to state our problems in an integral form by means of Green's operator. Thus, $C_0^1(\overline{\Omega})$ turns out to be a suitable space in our analysis for all exponents $-1 < \alpha < 1$.

The min–max characterization of the first eigenvalue in Theorem 8.2.2 was first introduced in Donsker and Varadhan [68] and then extended in Berestycki, Nirenberg, and Varadhan [24].

Next we have studied the Fréchet differentiability of the associated integral problem with respect to u and parameters.

The analysis presented in this chapter follows the general line in Hernández and Mancebo [102] or Hernández, Mancebo, and Vega [103], [104] for general elliptic operators. We also mention here the work of Bertsch and Rostamian [25], where similar singular eigenvalue problems in divergence form have been studied. By means of Hardy–Sobolev–type inequalities, it is established in [25] that the eigenfunctions belong to the class $C^2(\Omega) \cap H_0^1(\Omega)$.

9

THE INFLUENCE OF A NONLINEAR CONVECTION TERM IN SINGULAR ELLIPTIC PROBLEMS

To doubt everything or to believe everything are two equally convenient solutions; both dispense with the necessity of reflection.

Henri Poincaré (1854–1912),
Science et Hypothèse, 1902

9.1 Introduction

In the previous chapters we discussed the existence, the bifurcation, and the stability of classical solutions for elliptic problems involving smooth or singular nonlinearities. Our aim in this chapter is to study the following class of singular elliptic problems

$$\begin{cases} -\Delta u = p(d(x))g(u) + \lambda|\nabla u|^a + \mu f(x,u) & \text{in } \Omega, \\ u > 0 & \text{in } \Omega, \\ u = 0 & \text{on } \partial\Omega, \end{cases} \tag{9.1}$$

in a smooth bounded domain $\Omega \subset \mathbb{R}^N$ ($N \geq 1$). Here, $d(x) = \operatorname{dist}(x, \partial\Omega)$, $\lambda \in \mathbb{R}$, $\mu > 0$, and $0 < a \leq 2$.

Throughout this chapter we assume that $g \in C^1(0,\infty)$ verifies

($g1$) g is a positive decreasing function such that $\lim_{t\searrow 0} g(t) = +\infty$.

This hypothesis shows that g has a singular behavior at the origin. The standard example of such a nonlinearity is $g(t) = t^{-\alpha}$, with $\alpha > 0$.

We also assume that $f : \overline{\Omega} \times [0,\infty) \to [0,\infty)$ is a Hölder continuous function that is nondecreasing with respect to the second variable and such that f is positive in $\Omega \times (0,\infty)$. The analysis we develop in the sequel concerns the case in which f is either linear or sublinear with respect to its second variable.

The main feature of this chapter is the presence of the convection term $|\nabla u|^a$ combined with the possible singular weight $p : (0,\infty) \to \mathbb{R}$, which is supposed to be Hölder continuous.

At this stage, two questions arise:

1. Does the gradient term $|\nabla u|^a$ play a major role in the bifurcation results with respect to the parameters λ and μ?

2. Is the presence of the possible singular weight $p(d(x))$ significant in the existence of classical solutions to (9.1)?

To investigate these two questions, we first provide a general nonexistence result in terms of the decay rate of p near the origin. The existence part relies essentially on the sub- and supersolution method as described in Theorem 1.2.3. This method has the advantage of providing boundary estimates of solutions in case of some special nonlinearities. This matter will be emphasized in Section 9.6.

As a result of the presence of the gradient term and the lack of an adequate comparison result, we are not able to decide whether the solution of (9.1) (if exists) is unique. However, if $f(x,t)$ depends only on x variable, a standard argument based on the maximum principle and the fact that the gradient vanishes in all the extremum points that lie inside the domain allow us to derive the uniqueness of the solution once the existence has been proved.

The results obtained in this chapter allow us to extend the study of problem (9.1) in case $\Omega = \mathbb{R}^N$, as we started in Section 4.7.

9.2 A general nonexistence result

In this section we state a general nonexistence result related to problem (9.1). The following property strongly requires a quite strong decay rate of the variable potential at infinity.

Theorem 9.2.1 *Assume that g verifies $(g1)$ and $p:(0,\infty)\to(0,\infty)$ is a Hölder continuous function that is nonincreasing and*

$$\int_0^1 tp(t)dt = \infty. \tag{9.2}$$

Then the problem

$$\begin{cases} -\Delta u + \lambda|\nabla u|^2 \geq p(d(x))g(u) & \text{in } \Omega, \\ u > 0 & \text{in } \Omega, \\ u = 0 & \text{on } \partial\Omega \end{cases} \tag{9.3}$$

has no classical solutions.

Proof Without losing the generality, it suffices to consider $\lambda > 0$. We argue by contradiction and assume that there exists $u_\lambda \in C^2(\Omega)\cap C(\overline{\Omega})$ a solution of (9.3). Then, from $(g1)$, we can find $c>0$ small enough such that $\underline{u} := c\varphi_1$ verifies

$$-\Delta\underline{u} + \lambda|\nabla\underline{u}|^2 \leq p(d(x))g(\underline{u}) \quad \text{in } \Omega.$$

Indeed, it suffices to consider $c>0$ such that

$$\lambda_1\varphi_1 + \lambda c^2|\nabla\varphi_1|^2 \leq p(d(x))g(c\|\varphi_1\|_\infty) \quad \text{in } \Omega.$$

This is obviously possible if we take into account the hypotheses on g and $p(d(x))$. Furthermore, we claim that

$$u_\lambda \geq \underline{u} \quad \text{in } \Omega. \tag{9.4}$$

Assuming the contrary, it follows that $\min_{\overline{\Omega}}(u_\lambda - \overline{u})$ is achieved in Ω. At this point, say x_0, we have $\overline{u}(x_0) > u_\lambda(x_0)$, $\nabla(u_\lambda - \overline{u})(x_0) = 0$, and $\Delta(u_\lambda - \overline{u})(x_0) \geq 0$. Because g is decreasing, we obtain

$$\begin{aligned}
0 \geq & -\Delta(u_\lambda - \overline{u})(x_0) \\
= & \Big(-\Delta u_\lambda + \lambda |\nabla u_\lambda|^2 \Big)(x_0) - \Big(-\Delta \overline{u} + \lambda |\nabla \overline{u}|^2 \Big)(x_0) \\
\geq & p(d(x_0)) \Big(g(u_\lambda(x_0)) - g(\overline{u}(x_0)) \Big) \\
> & 0,
\end{aligned}$$

which is a contradiction. Hence, $u_\lambda \geq \underline{u}$ in Ω.

Let us perform the change of variable $v_\lambda = 1 - e^{-\lambda u_\lambda}$ in (9.3). Then, $0 \leq v_\lambda \leq 1$ in Ω and

$$-\Delta v_\lambda = \lambda(1 - v_\lambda) \left(\lambda |\nabla u_\lambda|^2 - \Delta u_\lambda \right) \quad \text{in } \Omega.$$

Furthermore, from (9.3) we have

$$\begin{cases}
-\Delta v_\lambda \geq \lambda(1 - v_\lambda) p(d(x)) g\left(-\dfrac{\ln(1 - v_\lambda)}{\lambda} \right) & \text{in } \Omega, \\
v_\lambda > 0 & \text{in } \Omega, \\
v_\lambda = 0 & \text{on } \partial\Omega.
\end{cases} \tag{9.5}$$

To avoid the singularities in (9.5), let us consider the approximated problem

$$\begin{cases}
-\Delta v = \lambda(1 - v) p(d(x)) g\left(\varepsilon - \dfrac{\ln(1 - v)}{\lambda} \right) & \text{in } \Omega, \\
v > 0 & \text{in } \Omega, \\
v = 0 & \text{on } \partial\Omega,
\end{cases} \tag{9.6}$$

with $0 < \varepsilon < 1$. Clearly, v_λ is a supersolution of (9.6). Moreover, by (9.4) and the fact that $\lim_{t \searrow 0}(1 - e^{-\lambda t})/t = \lambda > 0$, there exists $\tilde{c} > 0$ such that $v_\lambda \geq \tilde{c}\varphi_1$ in Ω. On the other hand, there exists $0 < m < \tilde{c}$ small enough such that $m\varphi_1$ is a subsolution of (9.6) and obviously $m\varphi_1 \leq v_\lambda$ in Ω. Then, problem (9.6) has a solution $v_\varepsilon \in C^2(\overline{\Omega})$ such that

$$m\varphi_1 \leq v_\varepsilon \leq v_\lambda \quad \text{in } \Omega. \tag{9.7}$$

Multiplying by φ_1 in (9.6) and then integrating over Ω we find

$$\lambda_1 \int_\Omega \varphi_1 v_\varepsilon dx = \lambda \int_\Omega (1 - v_\varepsilon) \varphi_1 p(d(x)) g\left(\varepsilon - \frac{\ln(1 - v_\varepsilon)}{\lambda} \right) dx.$$

Using (9.7) we obtain

$$\begin{aligned}\lambda_1 \int_\Omega \varphi_1 v_\lambda dx &\geq \lambda \int_\Omega (1-v_\lambda)\varphi_1 p(d(x)) g\left(1 - \frac{\ln(1-v_\lambda)}{\lambda}\right) dx \\ &\geq C \int_{\Omega_\delta} \varphi_1 p(d(x)) dx,\end{aligned} \tag{9.8}$$

where $C > 0$ and $\Omega_\delta = \{x \in \Omega : \ d(x) < \delta\}$, for some $\delta > 0$ sufficiently small. Because $\varphi_1(x)$ behaves like $d(x)$ in Ω_δ and $\int_0^1 tp(t)dt = \infty$, by (9.8) we find a contradiction. It follows that problem (9.3) has no classical solutions and the proof is now complete. □

A direct consequence of Theorem 9.2.1 is the following corollary.

Corollary 9.2.2 *Assume that $\int_0^1 tp(t)dt = \infty$ and conditions (g1), $0 < a \leq 2$ are fulfilled. Then for all $\mu > 0$ and $\lambda \in \mathbb{R}$, problem (9.1) has no classical solutions.*

9.3 A singular elliptic problem in one dimension

We are concerned in this section with the following singular elliptic problem in one dimension

$$\begin{cases} -y''(t) = q(t)g(y(t)) & 0 < t < 1, \\ y(t) > 0 & 0 < t < 1, \\ y(0) = y(1) = 0, \end{cases} \tag{9.9}$$

where g verifies (g1) and $q : (0,1) \to (0, \infty)$ is a Hölder continuous function that may be singular at 0 or 1. In the particular case $q \equiv 1$ and $g(t) = t^{-\alpha}$, $\alpha > 1$, we obtained in Section 4.1 some qualitative properties of the solution to (9.9). The main novelty here is the presence of the (possible singular) potential q. The existence results for problem (9.9) are of particular interest in the study of (9.1), because any solution of (9.9) may provide a supersolution in higher dimensions for problem (9.1). The main ingredient here is Theorem 1.1.2. To this aim we assume that q verifies

$$\int_0^1 t(1-t)q(t)dt < \infty. \tag{9.10}$$

Theorem 9.3.1 *Assume that (g1) and (9.10) hold. Then problem (9.9) has a unique solution $y \in C^2(0,1) \cap C[0,1]$.*

Proof Let us first remark that the mapping

$$\Phi : [0, \infty) \to [0, \infty), \quad \Phi(t) = \int_0^t \frac{ds}{g(s)} \tag{9.11}$$

is bijective. Thus, by (9.10) we can choose $M > 0$ and $\varepsilon > 0$ small enough such that

$$\int_{\varepsilon}^{M} \frac{ds}{g(s)} > 2\int_0^1 t(1-t)q(t)dt. \tag{9.12}$$

We fix an integer $k_0 \geq 1$ with the property $k_0\varepsilon > 1$, and for all $k \geq k_0$ consider the problem

$$\begin{cases} -y''(t) = q(t)G_k(y(t)) & 0 < t < 1, \\ y(t) > 0 & 0 < t < 1, \\ y(0) = y(1) = \dfrac{1}{k}, \end{cases} \tag{9.13}$$

where

$$G_k(t) = \begin{cases} g(t) & t \geq \dfrac{1}{k}, \\ g\Big(\dfrac{1}{k}\Big) & 0 < t < \dfrac{1}{k}. \end{cases} \tag{9.14}$$

By means of Theorem 1.1.2 we deduce the following auxiliary result.

Lemma 9.3.2 *For all $k \geq k_0$, problem (9.13) has at least one solution $y_k \in C^2(0,1) \cap C[0,1]$. Moreover, there exists $m > 0$, which is independent of k such that*

$$mt(1-t) \leq y_k(t) \leq M \quad \textit{for all } 0 \leq t \leq 1. \tag{9.15}$$

Proof For fixed $k \geq k_0$ and $0 < \lambda < 1$, let $y \in C^2(0,1) \cap C[0,1]$ be a solution of the auxiliary problem

$$\begin{cases} -y''(t) = \lambda q(t)G_k(y(t)) & 0 < t < 1, \\ y(t) > 0 & 0 < t < 1, \\ y(0) = y(1) = \dfrac{1}{k}. \end{cases} \tag{9.16}$$

Then y is concave and $y \geq 1/k$ in $[0,1]$. Denote by $t_{\lambda,k} \in (0,1)$ the maximum point of y. We also have $y' \geq 0$ in $(0, t_{\lambda,k})$ and $y' \leq 0$ in $(t_{\lambda,k}, 1)$. Because $G_k \leq g$ in $(0,\infty)$, from (9.16) we derive

$$-y''(t) \leq q(t)g(y(t)) \quad \text{for all } 0 < t < 1. \tag{9.17}$$

Integrating over $[t, t_{\lambda,k}]$ $(0 < t \leq t_{\lambda,k})$ in (9.17) we obtain

$$y'(t) \leq g(y(t)) \int_t^{t_{\lambda,k}} q(s)ds \quad \text{for all } 0 < t \leq t_{\lambda,k}.$$

An integration over $[0, t_{\lambda,k}]$ in the last estimate yields

$$\int_0^{t_{\lambda,k}} \frac{y'(s)}{g(y(s))} ds \leq \int_0^{t_{\lambda,k}} \int_t^{t_{\lambda,k}} q(s)ds.$$

That is,

$$\int_{1/k}^{y(t_{\lambda,k})} \frac{ds}{g(s)} \leq \int_0^{t_{\lambda,k}} sq(s)ds.$$

Because $k\varepsilon > 1$ for all $k \geq k_0$, the last inequality also implies that

$$\int_{\varepsilon}^{y(t_{\lambda,k})} \frac{ds}{g(s)} \leq \int_0^{t_{\lambda,k}} sq(s)ds.$$

Hence,

$$\int_{\varepsilon}^{y(t_{\lambda,k})} \frac{ds}{g(s)} \leq \frac{1}{1-t_{\lambda k}} \int_0^{t_{\lambda,k}} t(1-t)q(t)dt. \tag{9.18}$$

In the same manner, integrating in (9.17) over $[t_{\lambda k}, t]$ $(t_{\lambda,k} \leq t \leq 1)$, we have

$$\int_{\varepsilon}^{y(t_{\lambda,k})} \frac{ds}{g(s)} \leq \frac{1}{t_{\lambda,k}} \int_{t_{\lambda,k}}^{1} t(1-t)q(t)dt. \tag{9.19}$$

Combining (9.18) with (9.19) we obtain

$$\int_{\varepsilon}^{y(t_{\lambda,k})} \frac{ds}{g(s)} \leq 2\int_0^1 t(1-t)q(t)dt.$$

Using (9.12) we derive that $\|y\|_\infty = y(t_{\lambda,k}) < M$. Hence, by Theorem 1.1.2, problem (9.13) has a solution $y_k \in C^2(0,1) \cap C[0,1]$ such that

$$\frac{1}{k} \leq y_k(t) \leq M \quad \text{in } (0,1). \tag{9.20}$$

Furthermore, by (9.20) and the definition of G_k it follows that $G_k(y_k) = g(y_k)$ in $(0,1)$. Hence, for all $k \geq k_0$, y_k satisfies

$$\begin{cases} -y_k''(t) = q(t)g(y_k(t)) & 0 < t < 1\,, \\ y_k(t) > 0 & 0 < t < 1, \\ y_k(0) = y_k(1) = \dfrac{1}{k}. \end{cases} \tag{9.21}$$

It remains to prove the first part of the inequality in (9.15). By Green's representation we have

$$y_k(t) = \frac{1}{k} + (1-t)\int_0^t sq(s)g(y_k(s))ds + t\int_t^1 (1-s)q(s)g(y_k(s))ds,$$

for all $0 \leq t \leq 1$. Because g is decreasing, by (9.20) we obtain

$$y_k(t) \geq Q(t) \quad \text{for all } 0 \leq t \leq 1, \tag{9.22}$$

where

$$Q(t) = g(M)\bigg((1-t)\int_0^t sq(s)ds + t\int_t^1 (1-s)q(s)ds\bigg).$$

Remark that $Q(0) = Q(1) = 0$ and

$$Q'(t) = g(M)\bigg(-\int_0^t sq(s)ds + \int_t^1 (1-s)q(s)ds\bigg) \qquad \text{for all } 0 < t < 1.$$

In view of (9.10) it follows that $Q'(0)$ is finite. Also notice that $Q'(t)$ is positive in a neighborhood of zero. Hence, there exist $c_1 > 0$ and $\eta_1 > 0$ such that $Q(t) \geq c_1 t$ for all $0 \leq t \leq \eta_1$. Similarly, taking into account the fact that $Q'(t) < 0$ near $t = 1$, we can find $c_2 > 0$ and $\eta_2 > 0$ small enough such that $Q(t) \geq c_2(1-t)$ for all $1 - \eta_2 \leq t \leq 1$. Because $Q(t)/(t(1-t))$ is bounded away from zero on $[\eta_1, 1-\eta_2]$, we can find $m > 0$ such that

$$Q(t) \geq mt(1-t) \qquad \text{for all } 0 \leq t \leq 1. \tag{9.23}$$

Now the second inequality in (9.15) follows from (9.22) and (9.23). This finishes the proof. □

Lemma 9.3.3 *The sequence $(y_k)_{k\geq k_0}$ is bounded and equicontinuous.*

Proof Let $t_k \in (0,1)$ be the maximum point of y_k. Because y_k is concave, it follows that $y_k' \geq 0$ in $(0, t_k]$ and $y_k' \leq 0$ in $[t_k, 1)$. Integrating between t and t_k in (9.21) we derive

$$\frac{|y_k'(t)|}{g(y_k(t))} \leq \left|\int_t^{t_k} q(s)ds\right| \qquad \text{for all } 0 < t < 1. \tag{9.24}$$

We claim that

$$0 < \inf_{k\geq k_0} t_k \leq \sup_{k\geq k_0} t_k < 1. \tag{9.25}$$

Assume by contradiction that (9.25) does not hold. If $\inf_{k\geq k_0} t_k = 0$, then, up to a subsequence, we have $t_k \to 0$ as $k \to \infty$. Integrating over $[0, t_k]$ in (9.24) we have

$$\int_0^{t_k} \frac{|y_k'(t)|}{g(y_k(t))}dt \leq \int_0^{t_k}\int_t^{t_k} q(s)dsdt = \int_0^{t_k} sq(s)ds.$$

This yields

$$\int_0^{y_k(t_k)} \frac{ds}{g(s)} \leq \int_0^{t_k} sq(s)ds + \int_0^{1/k} \frac{ds}{g(s)}.$$

Because $t_k \to 0$ as $k \to \infty$, from the previous inequality we deduce $\|y_k\|_\infty = y_k(t_k) \to 0$ as $k \to \infty$. This means that $y_k \to 0$ in $C[0,1]$ as $k \to \infty$, which contradicts (9.15). Therefore $\inf_{k\geq k_0} t_k > 0$, and similarly $\sup_{k\geq k_0} t_k > 1$.

Let c_1, $c_2 > 0$ be such that

$$0 < c_1 < \inf_{k \geq k_0} t_k \leq \sup_{k \geq k_0} t_k < c_2 < 1.$$

By virtue of (9.24) and (9.25) we deduce that

$$\frac{|y_k'(t)|}{g(y_k(t))} \leq z(t) \qquad \text{for all } 0 < t < 1, \tag{9.26}$$

where

$$z(t) = \int_{\min\{c_1,t\}}^{\max\{c_2,t\}} q(s)ds \qquad \text{for all } 0 \leq t \leq 1.$$

Let Φ be the mapping defined in (9.11). According to (9.26) we obtain

$$|\Phi(y_k(t)) - \Phi(y_k(s))| = \left| \int_{y_k(s)}^{y_k(t)} \frac{d\tau}{g(\tau)} \right| = \left| \int_s^t \frac{y'(\tau)}{g(y_k(\tau))} d\tau \right| \leq \int_s^t z(\tau)d\tau.$$

Now, $z \in L^1(0,1)$ and the uniform continuity of Φ^{-1} on $[0, \Phi(M)]$ implies that $(y_k)_{k \geq k_0}$ is equicontinuous. This finishes the proof of our lemma. □

The result in Lemma 9.3.3 enables us to apply the Arzelà–Ascoli theorem. Therefore, there exists $y \in C[0,1]$ such that, up to a subsequence, we have $y_k \to y$ in $C[0,1]$ as $k \to \infty$. From (9.21) and Lemma 9.3.2 we deduce $y(0) = y(1) = 0$ and $mt(1-t) \leq y(t) \leq M$ for all $0 \leq t \leq 1$. In particular, we have $y > 0$ in $(0,1)$. Fix $t \in (0,1)$. Without losing the generality, we may assume that $t \neq 1/2$. Remark that y_k satisfies the integral equation

$$y_k(x) = y_k\left(\frac{1}{2}\right) + y_k'\left(\frac{1}{2}\right)\left(x - \frac{1}{2}\right) + \int_{1/2}^{x} (s-x)q(s)g(y_k(s))ds, \tag{9.27}$$

for all $0 < x < 1$. Taking, for instance, $x = 2/3$ in the previous relation and using (9.15) we deduce that the sequence $(y_k'(1/2))_{k \geq k_0}$ is bounded. Let $r \in \mathbb{R}$ be such that, up to a subsequence, $y_k'(1/2) \to r$ as $k \to \infty$. Now, letting $x = t$ in (9.27) and passing to the limit with $k \to \infty$ we have

$$y(t) = y\left(\frac{1}{2}\right) + r\left(t - \frac{1}{2}\right) + \int_{1/2}^{t} (s-t)q(s)g(y_k(s))ds.$$

Because $t \in (0,1)$ was arbitrary, the last equality implies $y''(t) + q(t)g(y(t)) = 0$ for all $0 < t < 1$—that is, $y \in C^2(0,1) \cap C[0,1]$ is a solution of (9.9). The proof is now complete. □

Condition (9.10) was introduced by Taliaferro [186]. In our setting, this condition reads

$$\int_0^1 tp(t)dt < \infty. \tag{9.28}$$

We have seen so far that condition (9.28) is necessary to have a classical solution for (9.1). In one dimension, Theorem 9.3.1 shows that the previous condition is also sufficient. In the next section we will argue that (9.28) suffices to have classical solutions in higher dimensions provided that the parameters λ and μ belong to a certain range.

9.4 Existence results in the sublinear case

Our aim in this section is to supply existence results for problem (9.1) in case f is sublinear. Recall that this means that f satisfies the hypotheses

$(f1)$ the mapping $(0,\infty) \ni t \longmapsto \dfrac{f(x,t)}{t}$ is nonincreasing for all $x \in \overline{\Omega}$;

$(f2)$ $\displaystyle\lim_{t\searrow 0} \frac{f(x,t)}{t} = \infty$ and $\displaystyle\lim_{t\to\infty} \frac{f(x,t)}{t} = 0$ uniformly for $x \in \overline{\Omega}$.

We also assume that $p : (0,\infty) \to (0,\infty)$ is a Hölder continuous function such that p is nonincreasing.

Nevertheless, we prove that condition (9.28) suffices to guarantee the existence of a classical solution for λ belonging to a certain range.

In this case the existence of a solution is strongly dependent on the exponent of the nonlinear gradient term $|\nabla u|^a$. To understand this dependence better, we assume $\mu = 1$, but the same results hold for any $\mu > 0$ (note only that the bifurcation point λ^* in the following results is dependent on μ).

Theorem 9.4.1 *Assume that* $0 < a < 1$, $\mu = 1$, *and conditions* $(f1)$, $(f2)$, $(g1)$, *and (9.28) are fulfilled. Then problem (9.1) has at least one classical solution for all* $\lambda \in \mathbb{R}$.

Proof CASE $\lambda > 0$. By Theorem 1.2.5 there exists a classical solution ζ of the problem

$$\begin{cases} -\Delta\zeta = f(x,\zeta) & \text{in } \Omega, \\ \zeta > 0 & \text{in } \Omega, \\ \zeta = 0 & \text{on } \partial\Omega. \end{cases} \tag{9.29}$$

Using the regularity of f we have $\zeta \in C^2(\overline{\Omega})$. Because $\lambda > 0$, it follows that ζ is a subsolution of (9.1). We now focus on finding a supersolution $\overline{u}_\lambda$ of (9.1) such that $\zeta \leq \overline{u}_\lambda$ in Ω.

According to Theorem 9.3.1, there exists $H \in C^2(0,1) \cap C[0,1]$ such that

$$\begin{cases} -H''(t) = p(t)g(H(t)) & 0 < t < 1, \\ H'(t) > 0 & 0 < t < 1, \\ H(0) = H(1) = 0. \end{cases} \tag{9.30}$$

Because H is concave, there exists $H'(0+) \in (0,\infty]$. We fix $0 < b < 1$ such that $H' > 0$ in $[0,b]$. Then H is increasing on $[0,b]$. Multiplying by H' in (9.30) and integrating on $[t,b]$, $(0 < t \leq b)$ we find

$$\begin{aligned}(H')^2(t) - (H')^2(b) &= 2\int_t^b p(s)g(H(s))H'(s)ds \\ &\leq 2p(t)\int_{H(t)}^{H(b)} g(\tau)d\tau,\end{aligned} \tag{9.31}$$

for all $0 < t \leq b$. Using the monotonicity of g it follows that

$$(H')^2(t) \leq 2H(b)p(t)g(H(t)) + (H')^2(b) \quad \text{for all } 0 < t \leq b. \tag{9.32}$$

Hence, there exist $c_1, c_2 > 0$ such that

$$H'(t) \leq c_1 p(t)g(H(t)) \quad \text{for all } 0 < t \leq b \tag{9.33}$$

and

$$(H')^2(t) \leq c_2 p(t)g(H(t)) \quad \text{for all } 0 < t \leq b. \tag{9.34}$$

The existence of a supersolution of (9.1) is obtained in the following result.

Lemma 9.4.2 *There exist two positive constants $M > 0$ (which depends on λ) and $c > 0$ such that $\overline{u}_\lambda := MH(c\varphi_1)$ is a supersolution of (9.1).*

Proof Let us first consider $c > 0$ such that

$$c\varphi_1 \leq \min\{b, d(x)\} \quad \text{in } \Omega. \tag{9.35}$$

By the strong maximum principle we have $\partial\varphi_1/\partial n < 0$ on $\partial\Omega$. Hence, there exist $\omega \subset\subset \Omega$ and $\delta > 0$ such that

$$|\nabla\varphi_1| > \delta \quad \text{in } \Omega \setminus \omega. \tag{9.36}$$

Moreover, because

$$\lim_{d(x)\searrow 0} \left\{c^2 p(c\varphi_1)g(H(c\varphi_1))|\nabla\varphi_1|^2 - 3f(x, H(c\varphi_1))\right\} = \infty,$$

we can assume that

$$c^2 p(c\varphi_1)g(H(c\varphi_1))|\nabla\varphi_1|^2 \geq 3f(x, H(c\varphi_1)) \quad \text{in } \Omega \setminus \omega. \tag{9.37}$$

Let $M > 1$ be such that

$$Mc^2\delta^2 > 3. \tag{9.38}$$

Using the fact that $H'(0+) > 0$ and $0 < a < 1$, we can choose $M > 1$ with

$$M\frac{(c\delta)^2}{c_1}H'(c\varphi_1) \geq 3\lambda(McH'(c\varphi_1)|\nabla\varphi_1|)^a \quad \text{in } \Omega \setminus \omega,$$

where c_1 is the constant from (9.33). By (9.33), (9.35), (9.36), and (9.38) we derive

$$Mc^2p(c\varphi_1)g(H(c\varphi_1))|\nabla\varphi_1|^2 \geq 3\lambda(McH'(c\varphi_1)|\nabla\varphi_1|)^a \quad \text{in } \Omega \setminus \omega. \tag{9.39}$$

Because g is decreasing and $H'(c\varphi_1) > 0$ in $\overline{\omega}$, there exists $M > 0$ such that

$$Mc\lambda_1\varphi_1 H'(c\varphi_1) \geq 3p(d(x))g(H(c\varphi_1)) \quad \text{in } \omega. \tag{9.40}$$

In the same manner, using $(f2)$ and the fact that $\varphi_1 > 0$ in $\overline{\omega}$, we can choose $M > 1$ large enough such that

$$Mc\lambda_1\varphi_1 H'(c\varphi_1) \geq 3\lambda(MH'(c\varphi_1)|\nabla\varphi_1|)^a \quad \text{in } \omega \tag{9.41}$$

and

$$Mc\lambda_1\varphi_1 H'(c\varphi_1) \geq 3f(x, MH(c\varphi_1)) \quad \text{in } \omega. \tag{9.42}$$

For M satisfying (9.38) through (9.42), we claim that

$$\overline{u}_\lambda(x) := MH(c\varphi_1(x)) \quad \text{for all } x \in \Omega \tag{9.43}$$

is a supersolution of (9.1). We have

$$-\Delta\overline{u}_\lambda = Mc^2p(c\varphi_1)g(H(c\varphi_1))|\nabla\varphi_1|^2 + Mc\lambda_1\varphi_1 H'(c\varphi_1) \quad \text{in } \Omega. \tag{9.44}$$

We first show that in $\Omega \setminus \omega$ there holds

$$Mc^2p(c\varphi_1)g(H(c\varphi_1))|\nabla\varphi_1|^2 \geq p(d(x))g(\overline{u}_\lambda) + \lambda|\nabla\overline{u}_\lambda|^a + f(x, \overline{u}_\lambda). \tag{9.45}$$

Indeed, by (9.35), (9.36), and (9.38) we have

$$\begin{aligned} \frac{M}{3}c^2p(c\varphi_1)g(H(c\varphi_1))|\nabla\varphi_1|^2 &\geq p(d(x))g(H(c\varphi_1)) \\ &\geq p(d(x))g(MH(c\varphi_1)) \\ &= p(d(x))g(\overline{u}_\lambda) \quad \text{in } \Omega \setminus \omega. \end{aligned} \tag{9.46}$$

The assumption $(f1)$ and (9.37) produce

$$\begin{aligned} \frac{M}{3}c^2p(c\varphi_1)g(H(c\varphi_1))|\nabla\varphi_1|^2 &\geq Mf(x, H(c\varphi_1)) \\ &\geq f(x, MH(c\varphi_1)) \\ &= f(x, \overline{u}_\lambda) \quad \text{in } \Omega \setminus \omega. \end{aligned} \tag{9.47}$$

From (9.36) and (9.39) we obtain

$$\begin{aligned} \frac{M}{3}c^2p(c\varphi_1)g(H(c\varphi_1))|\nabla\varphi_1|^2 &\geq \lambda(McH'(c\varphi_1)|\nabla\varphi_1|)^a \\ &= \lambda|\nabla\overline{u}_\lambda|^a \quad \text{in } \Omega \setminus \omega. \end{aligned} \tag{9.48}$$

Now, estimate (9.45) follows by (9.46), (9.47), and (9.48).

Next we prove that

$$Mc\lambda_1\varphi_1 H'(c\varphi_1) \geq p(d(x))g(\overline{u}_\lambda) + \lambda|\nabla\overline{u}_\lambda|^a + f(x,\overline{u}_\lambda) \quad \text{in } \omega. \tag{9.49}$$

From (9.40) and (9.41) we have

$$\begin{aligned} \frac{M}{3}c\lambda_1\varphi_1 H'(c\varphi_1) &\geq p(d(x))g(H(c\varphi_1)) \\ &\geq p(d(x))g(MH(c\varphi_1)) \\ &= p(d(x))g(\overline{u}_\lambda) \quad \text{in } \omega \end{aligned} \tag{9.50}$$

and

$$\begin{aligned} \frac{M}{3}c\lambda_1\varphi_1 H'(c\varphi_1) &\geq \lambda(McH'(c\varphi_1)|\nabla\varphi_1|)^a \\ &= \lambda|\nabla\overline{u}_\lambda|^a \quad \text{in } \omega. \end{aligned} \tag{9.51}$$

Finally, from (9.42) we derive

$$\frac{M}{3}c\lambda_1\varphi_1 H'(c\varphi_1) \geq f(x, MH(c\varphi_1)) = f(x,\overline{u}_\lambda) \quad \text{in } \omega. \tag{9.52}$$

Now, relation (9.49) follows from (9.50), (9.51), and (9.52). Combining (9.44) with (9.45) and (9.49) we conclude that $\overline{u}_\lambda$ is a supersolution of (9.1). This ends the proof. □

Let us come back to the proof of Theorem 9.4.1. So far we have constructed a subsolution ζ and a supersolution $\overline{u}_\lambda$ such that

$$\begin{gathered} \Delta\overline{u}_\lambda + f(x,\overline{u}_\lambda) \leq \Delta\zeta + f(x,\zeta) \quad \text{in } \Omega, \\ \overline{u}_\lambda, \zeta > 0 \quad \text{in } \Omega, \\ \overline{u}_\lambda = \zeta = 0 \quad \text{on } \partial\Omega. \end{gathered}$$

Because $\Delta\zeta \in L^1(\Omega)$ (note that $\zeta \in C^2(\overline{\Omega})$), by Theorem 1.3.17 we obtain $\zeta \leq \overline{u}_\lambda$ in Ω. The conclusion in this case follows now by the sub- and supersolution method for the ordered pair $(\zeta, \overline{u}_\lambda)$.

Case $\lambda \leq 0$. We fix $\nu > 0$ and let $u_\nu \in C^2(\Omega) \cap C(\overline{\Omega})$ be a solution of (9.1) for $\lambda = \nu$. Then u_ν is a supersolution of (9.1) for all $\lambda \leq 0$. Set

$$m := \inf_{(x,t)\in\Omega\times(0,\infty)} \Big(p(d(x))g(t) + f(x,t)\Big). \tag{9.53}$$

Because $g(0+) = \infty$ and the mapping $(0,\infty) \ni t \longmapsto \min_{x\in\overline{\Omega}} f(x,t)$ is positive and nondecreasing, we deduce that m is a positive. Consider the problem

$$\begin{cases} -\Delta v = m + \lambda|\nabla v|^a & \text{in } \Omega, \\ v = 0 & \text{on } \partial\Omega. \end{cases} \tag{9.54}$$

Clearly, zero is a subsolution of (9.54). Because $\lambda \le 0$, the solution w of the problem

$$\begin{cases} -\Delta w = m & \text{in } \Omega, \\ w = 0 & \text{on } \partial\Omega \end{cases}$$

is a supersolution of (9.54). Hence, (9.54) has at least one solution $v \in C^2(\Omega) \cap C(\overline{\Omega})$. We claim that $v > 0$ in Ω. Indeed, if not, we deduce that $\min_{x\in\overline{\Omega}} v$ is achieved at some point $x_0 \in \Omega$. Then $\nabla v(x_0) = 0$ and

$$-\Delta v(x_0) = m + \lambda |\nabla v(x_0)|^a = m > 0,$$

which is a contradiction. Therefore, $v > 0$ in Ω. It is easy to see that v is a subsolution of (9.1) and $-\Delta v \le m \le -\Delta u_\nu$ in Ω, which yields $v \le u_\nu$ in Ω. Again by the sub- and supersolution method we conclude that (9.1) has at least one classical solution $u_\lambda \in C^2(\Omega) \cap C(\overline{\Omega})$. This completes the proof. □

In the case $1 < a \le 2$, the complete description is given in the following result.

Theorem 9.4.3 *Assume that* $1 < a \le 2$, $\mu = 1$, *and conditions* $(f1)$, $(f2)$, $(g1)$, *and (9.28) are fulfilled. Then there exists* $\lambda^* > 0$ *such that (9.1) has at least one classical solution for all* $\lambda < \lambda^*$, *and no solutions exist if* $\lambda > \lambda^*$.

Proof We proceed in the same manner as in the proof of Theorem 9.4.1. The only difference is that (9.39) and (9.41) are no longer valid for any $\lambda > 0$. The main difficulty when dealing with estimates like (9.39) is that $H'(c\varphi_1)$ may blow up at the boundary. However, combining the assumption $1 < a \le 2$ with (9.34), we can choose $\lambda > 0$ small enough such that (9.39) and (9.41) hold. This implies that problem (9.1) has a classical solution provided $\lambda > 0$ is sufficiently small.

Set

$$A := \{\lambda > 0 : \text{ problem (9.1) has at least one classical solution}\}.$$

From the previous arguments, A is nonempty. Let $\lambda^* := \sup A > 0$. We first claim that if $\lambda \in A$, then $(0, \lambda) \subseteq A$. To this aim, let $\lambda_1 \in A$ and $0 < \lambda_2 < \lambda_1$. If u_{λ_1} is a solution of (9.1) with $\lambda = \lambda_1$, then u_{λ_1} is a supersolution of (9.1) with $\lambda = \lambda_2$, whereas ζ defined in (9.29) is a subsolution. Using Theorem 1.3.17 once more, we derive that $\zeta \le u_{\lambda_1}$ in Ω so that problem (9.1) has at least one classical solution for $\lambda = \lambda_2$. This proves the claim. Because $\lambda_1 \in A$ was arbitrary, we conclude that $(0, \lambda^*) \subseteq A$.

Next we prove that $\lambda^* < \infty$. This claim will be achieved in a more general framework. We have the following proposition.

Proposition 9.4.4 *Let* $F : [0, \infty) \to [0, \infty)$ *be a* C^1 *convex function such that*

$(F1)$ $F(0) = F'(0) = 0$;

$(F2)$ $\displaystyle\int_1^\infty \frac{ds}{F(s)} < \infty.$

Then there exists a positive number $\bar{\sigma}$ such that the problem

$$\begin{cases} -\Delta u = F(|\nabla u|) + \sigma & \text{in } \Omega, \\ u > 0 & \text{in } \Omega, \\ v = 0 & \text{on } \partial\Omega \end{cases} \tag{9.55}$$

has no classical solutions for $\sigma > \bar{\sigma}$.

Proof Let $v \in C^2(\Omega) \cap C(\overline{\Omega})$ be a solution of (9.55) and fix $\phi \in C_0^\infty(\Omega)$. Multiplying by ϕ in (9.55) and then integrating over Ω we obtain

$$\begin{aligned} \sigma \int_\Omega \phi dx &= \int_\Omega \Big(\nabla u \nabla \phi - \phi F(|\nabla u|)\Big) dx \\ &\leq \int_\Omega \phi \left(|\nabla u| \frac{|\nabla \phi|}{\phi} - F(|\nabla u|)\right) dx. \end{aligned} \tag{9.56}$$

Consider now the convex conjugate F^* of F defined as

$$F^*(t) = \sup_{\alpha \in \mathbb{R}} \{\alpha t - F(\alpha)\} \qquad \text{for all } t \in \mathbb{R}.$$

Now, from (9.56) we deduce

$$\sigma \int_\Omega \phi dx \leq \int_\Omega \phi F^* \left(\frac{|\nabla \phi|}{\phi}\right) dx \qquad \text{for all } \phi \in C_0^\infty(\Omega). \tag{9.57}$$

By density, the previous estimate holds for any $\phi \in W^{1,\infty}(\Omega)$. Hence, if we construct $\phi \in W^{1,\infty}(\Omega)$ such that $\phi > 0$ in Ω and $\int_\Omega \phi F^*(|\nabla \phi|/\phi) dx < \infty$, then, by (9.57), it follows that σ is finite. Without loss of generality, we may assume that Ω is the unit ball in $\mathbb{R}^N$. Let

$$\Phi(t) = \int_0^t \frac{ds}{F(s) + M} \qquad \text{for all } t \geq 0.$$

Using the assumption $(F2)$ we can choose $M > 0$ large enough such that $\ell := \lim_{t \to \infty} \Phi(t) \leq 1$. Then $\Phi : [0, \infty) \to [0, \ell)$ is bijective and let $\Gamma : [0, \ell) \to [0, \infty)$ be the inverse of Φ. Define now $\psi : [0, 1] \to [0, \infty)$ by

$$\psi(t) = \begin{cases} \dfrac{1}{F(\Gamma(t)) + M} & 0 \leq t \leq \ell, \\ 0 & \ell < t \leq 1. \end{cases}$$

We consider $\phi(x) = \psi(|x|)$, $x \in \overline{\Omega}$. Then

$$\psi'(t) = -\psi(t) F'(\Gamma(t)),$$

$$I := \int_\Omega \phi F^* \left(\frac{|\nabla \phi|}{\phi}\right) dx = \omega_N \int_0^\ell t^{N-1} \psi(t) F^*(F'(\Gamma(t))) dt.$$

On the other hand, a straightforward computation yields

$$F^*\big(F'(\Gamma(t))\big) = \Gamma(t)F'(\Gamma(t)) - F(\Gamma(t)) \quad \text{for all } t \geq 0.$$

Therefore, we have

$$\begin{aligned} I &\leq \omega_N \ell^{N-1} \int_0^\ell \Big(-\psi'(t)\Gamma(t) - \psi(t)F(\Gamma(t))\Big)dt \\ &= \omega_N \ell^{N-1} \left(-\psi(t)\Gamma(t)\Big|_0^\ell + \int_0^\ell \psi(t)\Big(\Gamma'(t) - F(\Gamma(t))\Big)dt \right) \\ &= \omega_N \ell^{N-1} M \int_0^\ell \psi(t)dt < \infty. \end{aligned}$$

This completes the proof. □

Consider $\lambda \in A$ and let u_λ be a classical solution of (9.1). Then

$$-\Delta u_\lambda \geq \lambda |\nabla u_\lambda|^a + m \quad \text{ in } \Omega,$$

where $m > 0$ is the infimum in (9.53). Because $1 < a \leq 2$, it follows that $v_\lambda := \lambda^{1/(a-1)} u_\lambda$ verifies

$$-\Delta v_\lambda \geq |\nabla v_\lambda|^a + m\lambda^{1/(a-1)} \quad \text{ in } \Omega.$$

Hence, v_λ is a supersolution of

$$\begin{cases} -\Delta v = |\nabla v|^a + m\lambda^{1/(a-1)} & \text{in } \Omega, \\ v = 0 & \text{on } \partial\Omega. \end{cases} \tag{9.58}$$

Because zero is a subsolution, it follows that problem (9.58) has a classical solution that, in view of the maximum principle, is positive. According to Proposition 9.4.4 with $F(t) = t^a$, we obtain $m\lambda^{1/(a-1)} \leq \bar{\sigma}$—that is, $\lambda \leq (\bar{\sigma}/m)^{a-1}$. This means that $\lambda^* = \inf A \leq (\bar{\sigma}/m)^{a-1} < \infty$. Hence, λ^* is finite. The existence of a solution in the case $\lambda \leq 0$ can be achieved exactly in the same way as in Theorem 9.4.1. This finishes the proof of Theorem 9.4.3. □

The case $a = 2$ is a special one, because by the change of variable (often called the *Gelfand transform*) $v = e^{\lambda u} - 1$, we obtain a new singular problem without a gradient term. If $f(x, u)$ depends on u, this change of variable does not preserve either the sublinearity conditions $(f1)$ and $(f2)$ on f, or the monotonicity of g. In turn, if $f(x, u)$ does not depend on u, this approach can be successfully used. This may help us to understand better the dependence between λ and μ in problem (9.1).

Let

$$m := \lim_{t \to \infty} g(t) \in [0, \infty).$$

Theorem 9.4.5 *Assume that $a = 2$, $\lambda \geq 0$, $\mu > 0$ and $p \equiv 1$, $f \equiv 1$.*

(i) *Problem (9.1) has a solution if and only if $\lambda(m + \mu) < \lambda_1$.*

(ii) *Assume $\mu > 0$ is fixed and let $\lambda^* = \lambda_1/(m+\mu)$. Then (9.1) has a unique solution u_λ for every $0 \leq \lambda < \lambda^*$, and the sequence $(u_\lambda)_{0 \leq \lambda < \lambda^*}$ is increasing with respect to λ. Moreover, if $\limsup_{s \searrow 0} s^\alpha g(s) < \infty$, for some $\alpha \in (0,1)$, then the sequence of solutions $(u_\lambda)_{0 \leq \lambda < \lambda^*}$ has the following properties:*

(ii1) *There exist two positive constants c_1, c_2 depending on λ such that $c_1 d(x) \leq u_\lambda \leq c_2 d(x)$ in Ω.*

(ii2) $u_\lambda \in C^{1,1-\alpha}(\overline{\Omega}) \cap C^2(\Omega)$.

(ii3) $\lim_{\lambda \nearrow \lambda^*} u_\lambda = \infty$ *uniformly on compact subsets of Ω.*

The situations described in Theorem 9.4.5 are depicted in the following bifurcation diagrams. Figure 9.1 corresponds to (i) and $a = 0$ (respectively $a > 0$), whereas Figure 9.2 is related to (ii), $\lambda > 0$, and $\mu =$ fixed.

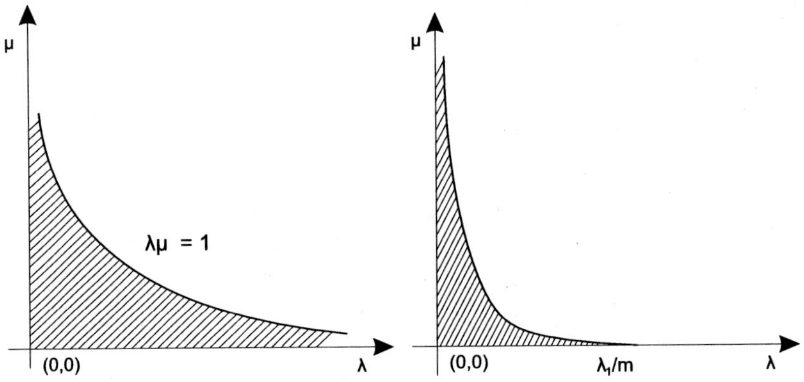

FIGURE 9.1. The bifurcation diagrams in Theorem 9.4.5 (i).

Proof If $\lambda = 0$, the existence follows by Theorem 1.2.5. Next we assume that $\lambda > 0$ and let us fix $\mu > 0$. With the change of variable $v = e^{\lambda u} - 1$, problem (9.1) becomes

$$\begin{cases} -\Delta v = \Phi_\lambda(v) & \text{in } \Omega, \\ v > 0 & \text{in } \Omega, \\ v = 0 & \text{on } \partial\Omega, \end{cases} \tag{9.59}$$

where

$$\Phi_\lambda(t) = \lambda(t+1)g\left(\frac{1}{\lambda}\ln(t+1)\right) + \lambda\mu(t+1) \quad \text{for all } t > 0.$$

Obviously, Φ_λ is not monotone, but we still have that the mapping $(0, \infty) \ni t \longmapsto \Phi_\lambda(t)/t$ is decreasing and

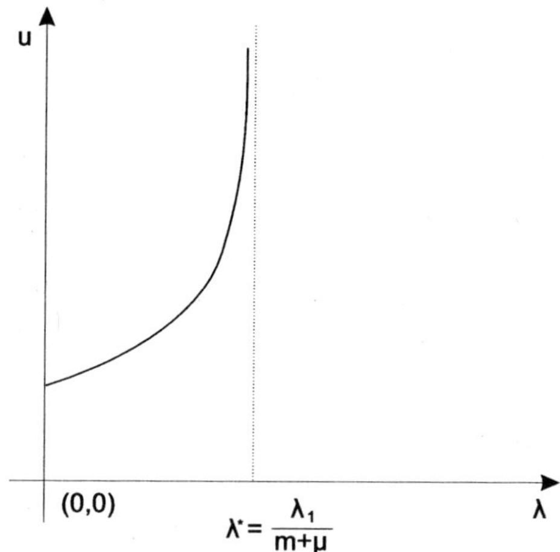

FIGURE 9.2. The bifurcation diagram in Theorem 9.4.5 (ii).

$$\lim_{t\to\infty} \frac{\Phi_\lambda(t)}{t} = \lambda(m+\mu) \quad \text{and} \quad \lim_{t\searrow 0} \frac{\Phi_\lambda(t)}{t} = \infty,$$

for all $\lambda > 0$. We first remark that Φ_λ satisfies the hypotheses of Theorem 1.2.5 provided that $\lambda(m+\mu) < \lambda_1$. Hence, problem (9.59) has at least one solution.

On the other hand, because $g \geq m$ on $(0, \infty)$, we obtain

$$\Phi_\lambda(t) \geq \lambda(m+\mu)(t+1) \quad \text{for all } \lambda, t \in (0,\infty). \tag{9.60}$$

This implies that (9.59) has no classical solutions if $\lambda(m+\mu) \geq \lambda_1$. Indeed, if $u_{\lambda\mu}$ would be a solution of (9.59) with $\lambda(m+\mu) \geq \lambda_1$, then $u_{\lambda\mu}$ is a supersolution of the problem

$$\begin{cases} -\Delta u = A(u+1) & \text{in } \Omega, \\ u = 0 & \text{on } \partial\Omega, \end{cases} \tag{9.61}$$

where $A = \lambda(m+\mu)$. Clearly zero is a subsolution of (9.61), so there exists a classical solution u of (9.61) such that $u \leq u_{\lambda\mu}$ in Ω. By the maximum principle and elliptic regularity, it follows that u is positive in Ω and $u \in C^2(\overline{\Omega})$. To raise a contradiction, we multiply by φ_1 in (9.61) and then integrate over Ω. We find

$$-\int_\Omega \varphi_1 \Delta u dx = A \int_\Omega u\varphi_1 dx + A \int_\Omega \varphi_1 dx.$$

This implies $\lambda_1 \int_\Omega u\varphi_1 dx = A \int_\Omega u\varphi_1 dx + A \int_\Omega \varphi_1 dx$, which is a contradiction, because $A \geq \lambda_1$. The proof of the first part in Theorem 9.4.5 is therefore complete. (ii) Follows exactly in the same manner as in Theorem 5.6.1. This concludes the proof. □

In what follows we discuss the case $a = 1$. Note that the method used in Theorem 9.4.1 does not apply here for large values of λ.

Assume that $\Omega = B_R$ for some $R > 0$, where B_R denotes the open ball centered at the origin, with a radius $R > 0$. In this case, and with $\mu = 1$, problem (9.1) reads

$$\begin{cases} -\Delta u = p(R - |x|)g(u) + \lambda|\nabla u| + f(x, u) & |x| < R, \\ u > 0 & |x| < R, \\ u = 0 & |x| = R. \end{cases} \tag{9.62}$$

Theorem 9.4.6 *Assume that $\Omega = B_R$ for some $R > 0$, $a = 1$, $\mu = 1$, and conditions $(f1)$, $(f2)$, and $(g1)$ hold. Then problem (9.62) has at least one solution for all $\lambda \in \mathbb{R}$.*

Proof The proof in the case $\lambda \leq 0$ is the same as in Theorem 9.4.1. In what follows we assume that $\lambda > 0$. By virtue of Theorem 9.4.1, there exists $\underline{u} \in C^2(\Omega) \cap C(\overline{\Omega})$ such that

$$\begin{cases} -\Delta \underline{u} = p(R - |x|)g(\underline{u}) & |x| < R, \\ \underline{u} > 0 & |x| < R, \\ \underline{u} = 0 & |x| = R. \end{cases}$$

It is obvious that $\underline{u}$ is a subsolution of (9.62) for all $\lambda > 0$. To provide a supersolution of (9.62) we consider the problem

$$\begin{cases} -\Delta u = p(R - |x|)g(u) + \lambda|\nabla u| + 1 & |x| < R, \\ u > 0 & |x| < R, \\ u = 0 & |x| = R. \end{cases} \tag{9.63}$$

We need the following auxiliary result.

Lemma 9.4.7 *Problem (9.63) has at least one classical solution.*

Proof We are looking for radially symmetric solutions u of (9.63)—that is,

$$u(x) = u(r) \quad \text{for all } 0 \leq r = |x| \leq R.$$

In this case, problem (9.63) reads

$$\begin{cases} -u'' - \dfrac{N-1}{r}u'(r) = p(R - r)g(u(r)) + \lambda|u'(r)| + 1 & 0 \leq r < R, \\ u > 0 & 0 \leq r < R, \\ u(R) = 0. \end{cases} \tag{9.64}$$

The first equality in (9.64) implies

$$-\left(r^{N-1}u'(r)\right)' \geq 0 \quad \text{for all } 0 \leq r < R.$$

This yields $u'(r) \le 0$ for all $0 \le r < R$. Again by (9.64) we find

$$-\left(u'' + \frac{N-1}{r}u'(r) + \lambda u'(r)\right) = p(R-r)g(u(r)) + 1 \quad 0 \le r < R.$$

The previous relation may be written as

$$-(e^{\lambda r} r^{N-1} u'(r))' = e^{\lambda r} r^{N-1} \psi(r, u(r)) \quad 0 \le r < R, \tag{9.65}$$

where

$$\psi(r,t) = p(R-r)g(t) + 1 \quad (r,t) \in [0,R) \times (0,\infty).$$

From (9.65) we have

$$u(r) = u(0) - \int_0^r e^{-\lambda t} t^{-N+1} \int_0^t e^{\lambda s} s^{N-1} \psi(s, u(s)) ds dt \quad 0 \le r < R. \tag{9.66}$$

On the other hand, in view of Theorem 9.4.1 and using the fact that g is decreasing, there exists a unique solution $w \in C^2(B_R) \cap C(\overline{B}_R)$ of the problem

$$\begin{cases} -\Delta w = p(R-|x|)g(w) + 1 & |x| < R, \\ w > 0 & |x| < R, \\ w = 0 & |x| = R. \end{cases} \tag{9.67}$$

Clearly, w is a subsolution of (9.63). As a result of the uniqueness and the symmetry of the domain, w is radially symmetric, so $w(x) = w(r)$ for all $0 \le r = |x| \le R$. As noted earlier, we obtain

$$w(r) = w(0) - \int_0^r t^{-N+1} \int_0^t s^{N-1} \psi(s, w(s)) ds dt \quad 0 \le r < R. \tag{9.68}$$

We claim that there exists a solution $v \in C^2[0,R) \cap C[0,R]$ of (9.66) such that $v > 0$ in $[0,R)$. Let $A = w(0)$ and define the sequence $(v_k)_{k \ge 0}$ by $v_0 = w$ and

$$v_k(r) = A - \int_0^r e^{-\lambda t} t^{-N+1} \int_0^t e^{\lambda s} s^{N-1} \psi(s, v_{k-1}(s)) ds dt, \tag{9.69}$$

for all $0 \le r < R$ and $k \ge 1$. Note that v_k is decreasing in $[0,R)$ for all $k \ge 0$. From (9.68) and (9.69) it is easy to see that $v_1 \ge v_0$ in $[0,R)$. A further induction argument yields $v_k \ge v_{k-1}$ in $[0,R)$ for all $k \ge 1$. Hence,

$$w = v_0 \le v_1 \le \cdots \le v_k \le \cdots \le A \quad \text{in } B_R.$$

Thus, for all $0 \le r < R$ there exists $v(r) := \lim_{k\to\infty} v_k(r)$ and, clearly, $v > 0$ in $[0,R)$. We now can pass to the limit in (9.69) to obtain

$$v(r) = A - \int_0^r e^{-\lambda t} t^{-N+1} \int_0^t e^{\lambda s} s^{N-1} \psi(s, v(s)) ds dt,$$

for all $0 \le r < R$. It follows that $v \in C^2[0,R) \cap C[0,R]$ is a solution of (9.66). This proves the claim.

We have obtained a supersolution v of (9.63) such that $v \geq w$ in B_R. So, problem (9.63) has at least one solution and the proof of lemma is now complete. □

Let u be a solution to problem (9.63). For $M > 1$ we have

$$\begin{aligned} -\Delta(Mu) &= Mp(R - |x|)g(u) + \lambda|\nabla(Mu)| + M \\ &\geq p(R - |x|)g(Mu) + \lambda|\nabla(Mu)| + M \quad \text{in } B_R. \end{aligned} \tag{9.70}$$

Because f is sublinear, we can choose $M > 1$ such that

$$M \geq f(x, M\|u\|_\infty) \quad \text{in } B_R.$$

Then, $\overline{u}_\lambda := Mu$ satisfies

$$-\Delta\overline{u}_\lambda \geq p(R - |x|)g(\overline{u}_\lambda) + \lambda|\nabla\overline{u}_\lambda| + f(x, \overline{u}_\lambda) \quad \text{in } B_R.$$

It follows that $\overline{u}_\lambda$ is a supersolution of (9.62). Because g is decreasing, we easily deduce $\underline{u} \leq \overline{u}_\lambda$ in B_R. Hence, problem (9.1) has at least one solution.

The proof of Theorem 9.4.6 is now complete. □

If p is bounded in the neighborhood of the origin, the previous procedure can be applied for general bounded domains. Let Ω be a smooth bounded domain and $R > 0$ be such that $\Omega \subseteq B_R$. According to Theorem 9.4.6, there exists $v \in C^2(B_R) \cap C(\overline{B}_R)$ such that

$$\begin{cases} -\Delta v = Mg(v) + \lambda|\nabla v| + f(x, v) & |x| < R, \\ v > 0 & |x| < R, \\ v = 0 & |x| = R, \end{cases}$$

where $M = \max_{x \in \overline{\Omega}} p(d(x))$. Obviously v is a supersolution of (9.1) with $a = 1$, whereas $w \in C^2(\Omega) \cap C(\overline{\Omega})$ defined by

$$\begin{cases} -\Delta w = p(d(x))g(w) & \text{in } \Omega, \\ w > 0 & \text{in } \Omega, \\ w = 0 & \text{on } \partial\Omega \end{cases}$$

is a subsolution of (9.1). It is easy to see that $w \leq v$ in Ω and, by Theorem 1.2.3, we deduce that problem (9.1) has at least one classical solution.

9.5 Existence results in the linear case

In this section we study problem (9.1) in which we drop the sublinearity assumptions $(f1)$ and $(f2)$ on f, but we require, in turn, that f is linear. More precisely, we assume that $f(x, t) = t$ for all $(x, t) \in \overline{\Omega} \times [0, \infty)$ and consider the problem

$$\begin{cases} -\Delta u = p(d(x))g(u) + \lambda|\nabla u|^a + \mu u & \text{in } \Omega, \\ u > 0 & \text{in } \Omega, \\ u = 0 & \text{on } \partial\Omega, \end{cases} \tag{9.71}$$

where $\lambda \geq 0$, $\mu > 0$ and p, g verify $(g1)$ and (9.28). We shall be concerned in this section with the case $0 < a < 1$.

Note that the existence result in Theorem 1.2.5 does not apply here because the mapping

$$\Psi(x,t) = p(d(x))g(t) + \mu t \quad (x,t) \in \Omega \times (0,\infty)$$

is not defined on $\partial\Omega \times (0,\infty)$.

Theorem 9.5.1 *Assume that* $0 < a < 1$ *and conditions* $(g1)$, *(9.28) are fulfilled. Then for all* $\lambda \geq 0$, *problem (9.71) has solutions if and only if* $\mu < \lambda_1$.

Proof Fix $\mu \in (0, \lambda_1)$ and $\lambda \geq 0$. By Theorem 9.4.1 there exists $u \in C^2(\Omega) \cap C(\overline{\Omega})$ such that

$$\begin{cases} -\Delta u = p(d(x))g(u) + \lambda|\nabla u|^a & \text{in } \Omega, \\ u > 0 & \text{in } \Omega, \\ u = 0 & \text{on } \partial\Omega. \end{cases}$$

Obviously, $\underline{u}_{\lambda\mu} := u$ is a subsolution of (9.71). Because $\mu < \lambda_1$, there exists $v \in C^2(\overline{\Omega})$ such that

$$\begin{cases} -\Delta v = \mu v + 2 & \text{in } \Omega, \\ v > 0 & \text{in } \Omega, \\ v = 0 & \text{on } \partial\Omega. \end{cases} \tag{9.72}$$

Using the fact that $0 < a < 1$, we can choose $M > 0$ large enough such that

$$M > \mu\|u\|_\infty \text{ and } M > \lambda(M|\nabla v|)^a \text{ in } \Omega. \tag{9.73}$$

From (9.72) and (9.73) we note that $w := Mv$ satisfies

$$-\Delta w \geq \lambda|\nabla w|^a + \mu(u+w) \quad \text{in } \Omega.$$

We claim that $\overline{u}_{\lambda\mu} := u + w$ is a supersolution of (9.71). Indeed, we have

$$-\Delta\overline{u}_{\lambda\mu} \geq p(d(x))g(u) + \lambda|\nabla u|^a + \lambda|\nabla w|^a + \mu\overline{u}_{\lambda\mu} \quad \text{in } \Omega. \tag{9.74}$$

Using the assumption $0 < a < 1$, we can easily deduce

$$t_1^a + t_2^a \geq (t_1 + t_2)^a \quad \text{for all } t_1, t_2 \geq 0.$$

Hence,

$$|\nabla u|^a + |\nabla w|^a \geq (|\nabla u| + |\nabla w|)^a \geq |\nabla(u+w)|^a \quad \text{in } \Omega. \tag{9.75}$$

Combining (9.74) with (9.75) we obtain

$$-\Delta\overline{u}_{\lambda\mu} \geq p(d(x))g(\overline{u}_{\lambda\mu}) + \lambda|\nabla\overline{u}_{\lambda\mu}|^a + \mu\overline{u}_{\lambda\mu} \quad \text{in } \Omega.$$

Hence, $(\underline{u}_{\lambda\mu}, \overline{u}_{\lambda\mu})$ is an ordered pair of sub- and supersolutions of (9.71), and thus problem (9.71) has a classical solution $u_{\lambda\mu}$ provided $\lambda \geq 0$ and $0 < \mu < \lambda_1$.

If $\mu \geq \lambda_1$, using the same method as in the proof of Theorem 9.4.5, we deduce that problem (9.71) has no classical solutions. This finishes the proof. □

9.6 Boundary estimates of the solution

In this section we are concerned with the asymptotic analysis of solutions of problem (9.1) in the special case $p(t) = t^{-\alpha}$, $g(t) = t^{-\beta}$, where α, $\beta > 0$. Hence, we study the boundary behavior of classical solutions to

$$\begin{cases} -\Delta u = d(x)^{-\alpha} u^{-\beta} + \lambda |\nabla u|^a + f(x,u) & \text{in } \Omega, \\ u > 0 & \text{in } \Omega, \\ u = 0 & \text{on } \partial\Omega, \end{cases} \tag{9.76}$$

where $0 < a \leq 2$, $\lambda > 0$, and f satisfies $(f1)$ and $(f2)$.

Recall that if $\int_0^1 tp(t)dt < \infty$ and λ belongs to a certain range, then Theorems 9.4.1 and 9.4.3 assert that (9.76) has at least one classical solution u_λ satisfying $u_\lambda \leq MH(c\varphi_1)$ in Ω, for some $M, c > 0$. Here, H is the solution of

$$\begin{cases} -H''(t) = t^{-\alpha} H^{-\beta}(t) & \text{for all } 0 < t \leq b < 1, \\ H, H' > 0 & \text{in } (0, b], \\ H(0) = 0. \end{cases} \tag{9.77}$$

With the same idea as in the proof of Theorem 9.4.1, we can show that there exists $m > 0$ small enough such that $v := mH(c\varphi_1)$ satisfies

$$-\Delta v \leq d(x)^{-\alpha} v^{-\beta} \quad \text{in } \Omega. \tag{9.78}$$

Indeed, we have

$$-\Delta v = m\Big(c^{2-\alpha} |\nabla \varphi_1|^2 \varphi_1^{-\alpha} H^{-\beta}(c\varphi_1) + \lambda_1 c\varphi_1 H'(c\varphi_1)\Big) \quad \text{in } \Omega.$$

Using the properties of φ_1 and (9.33), there exist two positive constants $c_1, c_2 > 0$ such that

$$-\Delta v \leq m\Big(c_1 |\nabla \varphi_1|^2 + c_2 \varphi_1\Big) d(x)^{-\alpha} H^{-\beta}(c\varphi_1) \quad \text{in } \Omega.$$

Clearly, (9.78) holds if we choose $0 < m < 1$ small enough such that

$$m\Big(c_1 |\nabla \varphi_1|^2 + c_2 \varphi_1\Big) < 1 \quad \text{in } \Omega.$$

Furthermore, from (9.78) we obtain that v is a subsolution of (9.76), for all $\lambda > 0$. It is easy to see that $v \leq u_\lambda$ in Ω. Therefore the solution u_λ satisfies

$$mH(c\varphi_1) \leq u_\lambda \leq MH(c\varphi_1) \quad \text{in } \Omega. \tag{9.79}$$

Now, a careful analysis of (9.77) together with (9.79) provides boundary estimates for the solutions of problem (9.76), as stated in the following result.

Theorem 9.6.1 *Assume that* $\lambda > 0$, $0 < a \leq 2$ *and conditions* $(f1)$ *and* $(f2)$ *hold.*

(i) *If* $\alpha \geq 2$, *then problem (9.76) has no classical solutions.*

(ii) *If* $\alpha < 2$, *then there exists* $0 < \lambda^* \leq \infty$ *(with* $\lambda^* = \infty$ *if* $0 < a < 1$*) such that for all* $0 < \lambda < \lambda^*$, *problem (9.76) has at least one classical solution. Moreover, there exist* $0 < \delta < 1$ *and* $C_1, C_2 > 0$ *such that* u_λ *satisfies the following:*

(ii1) *If* $\alpha + \beta > 1$, *then for all* $x \in \Omega$ *we have*

$$C_1 d(x)^{(2-\alpha)/(1+\beta)} \leq u_\lambda(x) \leq C_2 d(x)^{(2-\alpha)/(1+\beta)}. \tag{9.80}$$

(ii2) *If* $\alpha + \beta = 1$, *then for all* $x \in \Omega$ *with* $d(x) < \delta$ *we have*

$$C_1 d(x)(-\ln d(x))^{1/(2-\alpha)} \leq u_\lambda(x) \leq C_2 d(x)(-\ln d(x))^{1/(2-\alpha)}. \tag{9.81}$$

(ii3) *If* $\alpha + \beta < 1$, *then for all* $x \in \Omega$ *we have*

$$C_1 d(x) \leq u_\lambda(x) \leq C_2 d(x). \tag{9.82}$$

Proof The existence and nonexistence parts follow directly from Theorems 9.2.1, 9.4.1, and 9.4.3. We next establish the boundary estimates (9.80) through (9.82).

(ii1) Remark that

$$H(t) = \left(\frac{(1+\beta)^2}{(2-\alpha)(\alpha+\beta-1)}\right)^{1/(1+\beta)} t^{(2-\alpha)/(1+\beta)} \quad t > 0$$

is a solution to (9.77) provided that $\alpha+\beta > 1$. The conclusion in this case follows from (9.79).

(ii2) Note that in this case problem (9.77) becomes

$$\begin{cases} -H''(t) = t^{-\alpha} H^{\alpha-1}(t) & \text{for all } 0 < t \leq b < 1, \\ H > 0 & \text{in } (0, b], \\ H(0) = 0. \end{cases} \tag{9.83}$$

Because H is concave, it follows that

$$H(t) > tH'(t) \quad \text{for all } 0 < t \leq b. \tag{9.84}$$

Relations (9.83) and (9.84) yield

$$-H''(t) < \frac{(H'(t))^{\alpha-1}}{t} \quad \text{for all } 0 < t \leq b.$$

Hence,

$$-H''(t)(H'(t))^{1-\alpha} \le \frac{1}{t} \quad \text{for all } 0 < t \le b. \tag{9.85}$$

Integrating in (9.85) over $[t, b]$ we obtain

$$(H')^{2-\alpha}(t) - (H')^{2-\alpha}(b) \le (2-\alpha)(\ln b - \ln t) \quad \text{for all } 0 < t \le b.$$

Thus, there exist $c_1 > 0$ and $\delta_1 \in (0, b)$ such that

$$H'(t) \le c_1(-\ln t)^{1/(2-\alpha)} \quad \text{for all } 0 < t \le \delta_1. \tag{9.86}$$

Fix $t \in (0, \delta_1]$. Integrating in (9.86) over $[\varepsilon, t]$, $0 < \varepsilon < t$, we have

$$H(t) - H(\varepsilon) \le c_1 t(-\ln t)^{1/(2-\alpha)} + \frac{c_1}{2-\alpha}\int_\varepsilon^t (-\ln s)^{(\alpha-1)/(2-\alpha)} ds. \tag{9.87}$$

Note that

$$\int_0^t (-\ln s)^{(\alpha-1)/(2-\alpha)} ds < \infty \quad \text{and} \quad \lim_{t\searrow 0} \frac{\int_0^t (-\ln s)^{(\alpha-1)/(2-\alpha)} ds}{t(-\ln t)^{1/(2-\alpha)}} = 0. \tag{9.88}$$

Hence, taking $\varepsilon \to 0$ in (9.87), there exist $c_2 > 0$ and $\delta_2 \in (0, \delta_1)$ such that

$$H(t) \le c_2 t(-\ln t)^{1/(2-\alpha)} \quad \text{for all } 0 < t \le \delta_2. \tag{9.89}$$

From (9.83) and (9.89) we then obtain

$$-H''(t) \ge c_2^{\alpha-1} \frac{(-\ln t)^{(\alpha-1)/(2-\alpha)}}{t} \quad \text{for all } 0 < t \le \delta_2.$$

Integrating over $[t, \delta_2]$ in the previous inequality we find

$$H'(t) \ge (2-\alpha)c_2^{\alpha-1}\left[(-\ln t)^{1/(2-\alpha)} - (-\ln \delta_2)^{1/(2-\alpha)}\right] \quad \text{for all } 0 < t \le \delta_2.$$

Therefore, there exist $c_3 > 0$ and $\delta_3 \in (0, \delta_2)$ such that

$$H'(t) \ge c_3(-\ln t)^{1/(2-\alpha)} \quad \text{for all } 0 < t \le \delta_3.$$

Using the same arguments as in (9.86) through (9.89) we obtain $c_4 > 0$ and $\delta_4 \in (0, \delta_3)$ such that

$$H(t) \ge c_4 t(-\ln t)^{1/(2-\alpha)} \quad \text{for all } 0 < t \le \delta_4. \tag{9.90}$$

The conclusion of (ii) in Theorem 9.6.1 follows now from (9.89), (9.90), and (9.79).

(ii3) Using the inequality (9.84) and the fact that $H'(0+) > 0$, there exists $c > 0$ such that

$$H(t) > ct \qquad \text{for all } 0 < t \le b.$$

This yields

$$-H''(t) \le c^{-\beta} t^{-(\alpha+\beta)} \qquad \text{for all } 0 < t \le b.$$

Because $\alpha + \beta < 1$, it follows that $H'(0+) < \infty$—that is, $H \in C^1[0, b]$. Thus, there exist $c_1, c_2 > 0$ such that

$$c_1 t \le H(t) \le c_2 t \qquad \text{for all } 0 < t \le b. \tag{9.91}$$

The conclusion in Theorem 9.6.1 (iii) follows directly from (9.91) and (9.79).
This completes the proof of Theorem 9.6.1. □

9.7 The case of a negative singular potential

This section is devoted to the study of (9.1) in case p is negative. More exactly, we shall be concerned with an equivalent form of (9.1)—namely,

$$\begin{cases} -\Delta u + p(d(x))g(u) = \lambda f(x, u) + \mu|\nabla u|^a & \text{in } \Omega, \\ u > 0 & \text{in } \Omega, \\ u = 0 & \text{on } \partial\Omega, \end{cases} \tag{9.92}$$

where $0 < a \le 2$, $\lambda > 0$, $\mu \in \mathbb{R}$, g satisfies $(g1)$, and $p : (0, \infty) \to (0, \infty)$ is a nonincreasing Hölder continuous function possibly singular at the origin. In this framework, the existence of a classical solution for problem (9.92) depends on both singular terms $p(d(x))$ and unbounded nonlinearity g. More precisely, we show that a necessary condition to have a classical solution is

$$\int_0^1 p(t)g(t)dt < \infty. \tag{9.93}$$

In case f is sublinear—that is, f fulfills the hypotheses $(f1)$ and $(f2)$—condition (9.93) is also sufficient for the existence of classical solutions provided that λ and μ belong to a certain range. Obviously, relation (9.93) implies the following Keller–Osserman–type condition around the origin that have been already encountered in Chapter 4:

$$\int_0^1 \left(\int_0^s \Phi(t)dt \right)^{-1/2} ds < \infty, \tag{9.94}$$

where $\Phi(t) = p(t)g(t)$, $t > 0$.

9.7.1 *A nonexistence result*

We establish here the following nonexistence result related to problem (9.92).

Theorem 9.7.1 *Assume that $\int_0^1 p(t)g(t)dt = \infty$ and let $\Phi : \overline{\Omega} \times [0,\infty) \to [0,\infty)$ be a continuous function such that $\Phi \not\equiv 0$. Then the problem*

$$\begin{cases} -\Delta u + p(d(x))g(u) \le \Phi(x,u) & \text{in } \Omega, \\ u > 0 & \text{in } \Omega, \\ u = 0 & \text{on } \partial\Omega \end{cases} \tag{9.95}$$

has no classical solutions.

Proof Assume that problem (9.95) has a classical solution u and fix $C > \max_{\overline{\Omega}} \Phi(x,u)$. Let $v \in C^2(\overline{\Omega})$ be the unique solution of

$$\begin{cases} -\Delta v = C & \text{in } \Omega, \\ v > 0 & \text{in } \Omega, \\ v = 0 & \text{on } \partial\Omega. \end{cases} \tag{9.96}$$

Moreover, there exist $c_1, c_2 > 0$ such that

$$c_1 d(x) \le v \le c_2 d(x) \quad \text{for all } x \in \Omega. \tag{9.97}$$

By the weak maximum principle, it follows that $u \le v$ in Ω. Next we consider the perturbed problem

$$\begin{cases} -\Delta u + p(d(x)+\varepsilon)g(u+\varepsilon) = C & \text{in } \Omega, \\ u > 0 & \text{in } \Omega, \\ u = 0 & \text{on } \partial\Omega. \end{cases} \tag{9.98}$$

Obviously, u and v are, respectively, sub- and supersolution of (9.98). By standard arguments, there exists $u_\varepsilon \in C^2(\overline{\Omega})$ a solution of (9.98) such that $u \le u_\varepsilon \le v$ in Ω. Integrating in (9.98) we obtain

$$-\int_\Omega \Delta u_\varepsilon dx + \int_\Omega p(d(x)+\varepsilon)g(u_\varepsilon+\varepsilon)dx = C|\Omega|,$$

which yields

$$-\int_{\partial\Omega} \frac{\partial u_\varepsilon}{\partial n} d\sigma(x) + \int_\Omega p(d(x)+\varepsilon)g(u_\varepsilon+\varepsilon)dx \le M, \tag{9.99}$$

where M is a positive constant. Taking into account the fact that $\partial u_\varepsilon/\partial n \le 0$ on $\partial\Omega$, from (9.99) we find

$$\int_\Omega p(d(x)+\varepsilon)g(u_\varepsilon+\varepsilon)dx \le M.$$

Because g is decreasing and $u_\varepsilon \le v$ in $\overline{\Omega}$, it follows that

$$\int_\omega p(d(x)+\varepsilon)g(v+\varepsilon)dx \le M,$$

for any compact subset $\omega \subset\subset \Omega$. Passing to the limit with $\varepsilon \to 0$ in the previous estimate we obtain $\int_\omega p(d(x))g(v)dx \le M$, for all $\omega \subset\subset \Omega$. Therefore

$$\int_\Omega p(d(x))g(v)dx \le M. \tag{9.100}$$

On the other hand, by (9.97) and $\int_0^1 p(t)g(t)ds = \infty$, it follows that

$$M \ge \int_\Omega p(d(x))g(v)dx \ge \int_\Omega p(d(x))g(c_2 d(x))dx = \infty,$$

which is a contradiction. Hence, problem (9.95) has no classical solutions. This completes the proof. □

Corollary 9.7.2 *Assume that $\int_0^1 p(t)g(t)dt = \infty$. Then, for all $\mu \le 0$, problem (9.92) has no classical solutions.*

9.7.2 *Existence result*

Theorem 9.7.3 *Assume that $\int_0^1 p(t)g(t)dt < \infty$ and conditions $(f1)$, $(f2)$, and $(g1)$ are satisfied.*

(i) *If $\mu = -1$ and $0 < a \le 2$, then there exists $\lambda^* > 0$ such that problem (9.92) has at least one classical solution if $\lambda > \lambda^*$, and no solutions exist if $0 < \lambda < \lambda^*$.*

(ii) *If $\mu = 1$ and $0 < a < 1$, then there exists $\lambda^* > 0$ such that problem (9.92) has at least one classical solution for all $\lambda > \lambda^*$, and no solutions exist if $0 < \lambda < \lambda^*$.*

Proof (i) We split the proof into three steps.
Step 1: Existence of a solution for large λ. By virtue of Theorem 1.3.17, for all $\lambda > 0$ the problem

$$\begin{cases} -\Delta U = \lambda f(x,U) & \text{in } \Omega, \\ U > 0 & \text{in } \Omega, \\ U = 0 & \text{on } \partial\Omega \end{cases} \tag{9.101}$$

has at least one classical solution U_λ. Using the regularity of f it follows that $U_\lambda \in C^2(\overline{\Omega})$ and there exist $c_1, c_2 > 0$ depending on λ such that

$$c_1 d(x) \le U_\lambda(x) \le c_2 d(x) \quad \text{in } \Omega. \tag{9.102}$$

Fix $\lambda > 0$ and observe that U_λ is a supersolution of (9.92). The main point is to find a subsolution $\underline{u}_\lambda$ of (9.92) such that $\underline{u}_\lambda \le U_\lambda$ in Ω. For this purpose, let $\Phi(t) = p(t)g(t)$, $t > 0$, and define

$$\Psi : [0, \infty) \to [0, \infty), \quad \Psi(t) = \int_0^t \left(\int_0^s \Phi(\tau) d\tau \right)^{-1/2} ds.$$

As in the proof of Theorem 4.5.1, we obtain that Ψ is a bijective map. Let $h : [0, \infty) \to [0, \infty)$ be the inverse of Ψ. Then h satisfies

$$\begin{cases} h > 0 & \text{in } (0, \infty), \\ h'(t) = \sqrt{2 \int_0^{h(t)} \Phi(s) ds} & \text{in } (0, \infty), \\ h''(t) = \Phi(h(t)) & \text{in } (0, \infty), \\ h(0) = h'(0) = 0. \end{cases} \tag{9.103}$$

Thus, $h \in C^2(0, \infty) \cap C^1[0, \infty)$.

The key result for this part of the proof is the following lemma.

Lemma 9.7.4 *There exist two positive constants $c > 0$ and $M > 0$ such that $\underline{u}_\lambda := Mh(c\varphi_1)$ is a subsolution of (9.92) provided $\lambda > 0$ is large enough.*

Proof Because $h \in C^1[0, \infty)$ and $h(0) = 0$, we can take $c > 0$ small enough such that

$$h(c\varphi_1) \le d(x) \quad \text{in } \Omega. \tag{9.104}$$

By the strong maximum principle, there exist $\delta > 0$ and $\omega \subset\subset \Omega$ such that $|\nabla \varphi_1| \ge \delta$ in $\Omega \setminus \omega$. Let

$$M = \max\{1, 2(c\delta)^{-2}\}. \tag{9.105}$$

Because

$$\lim_{d(x) \searrow 0} \Big\{ -p(d(x))g(h(c\varphi_1)) + Mc\lambda_1\varphi_1 h'(c\varphi_1) + (Mch'(c\varphi_1)|\nabla\varphi_1|)^a \Big\} = -\infty,$$

we can assume that in $\Omega \setminus \omega$ there holds

$$-p(d(x))g(h(c\varphi_1)) + Mc\lambda_1\varphi_1 h'(c\varphi_1) + (Mch'(c\varphi_1)|\nabla\varphi_1|)^a < 0. \tag{9.106}$$

We are now able to show that $\underline{u}_\lambda := Mh(c\varphi_1)$ is a subsolution of (9.92), provided $\lambda > 0$ is sufficiently large. Indeed, we have

$$-\Delta \underline{u}_\lambda = -Mc^2 p(h(c\varphi_1))g(h(c\varphi_1))|\nabla\varphi_1|^2 + Mc\lambda_1\varphi_1 h'(c\varphi_1).$$

By the monotonicity of g and (9.104) we obtain

$$\begin{aligned} -\Delta \underline{u}_\lambda + p(d(x))g(\underline{u}_\lambda) + |\nabla \underline{u}_\lambda|^a \le & \, p(d(x))g(h(c\varphi_1))(1 - Mc^2|\nabla\varphi_1|^2) \\ & + Mc\lambda_1\varphi_1 h'(c\varphi_1) \\ & + (Mch'(c\varphi_1)|\nabla\varphi_1|)^a. \end{aligned} \tag{9.107}$$

Next, by the definition of M and (9.106), for all $x \in \Omega \setminus \omega$ we find

$$\begin{aligned} -\Delta \underline{u}_\lambda + p(d(x))g(\underline{u}_\lambda) + |\nabla \underline{u}_\lambda|^a \leq & - p(d(x))g(h(c\varphi_1)) \\ & + Mc\lambda_1\varphi_1 h'(c\varphi_1) \\ & + (Mch'(c\varphi_1)|\nabla\varphi_1|)^a \\ < & \, 0. \end{aligned} \tag{9.108}$$

On the other hand, from (9.107) and for all $x \in \omega$ we have

$$\begin{aligned} -\Delta \underline{u}_\lambda + p(d(x))g(\underline{u}_\lambda) + |\nabla \underline{u}_\lambda|^a \leq & \, p(d(x))g(h(c\varphi_1)) + Mc\lambda_1\varphi_1 h'(c\varphi_1) \\ & + (Mch'(c\varphi_1)|\nabla\varphi_1|)^a. \end{aligned} \tag{9.109}$$

Because $\varphi_1 > 0$ in $\overline{\omega}$ and f is positive on $\overline{\omega} \times (0, \infty)$, we may choose $\lambda > 0$ large enough such that

$$\begin{aligned} \lambda f(x, Mh(c\varphi_1)) \geq & \, p(d(x))g(h(c\varphi_1)) + Mc\lambda_1\varphi_1 h'(c\varphi_1) \\ & + (Mch'(c\varphi_1)|\nabla\varphi_1|)^a \quad \text{for all } x \in \overline{\omega}. \end{aligned} \tag{9.110}$$

From (9.109) and (9.110) we deduce

$$-\Delta \underline{u}_\lambda + p(d(x))g(\underline{u}_\lambda) + |\nabla \underline{u}_\lambda|^a \leq \lambda f(x, \underline{u}_\lambda) \quad \text{in } \omega. \tag{9.111}$$

Now, relations (9.108) and (9.111) show that $\underline{u}_\lambda = Mh(c\varphi_1)$ is a subsolution of problem (9.92) provided that $\lambda > 0$ satisfies (9.110). This finishes the proof of the lemma. □

By virtue of Theorem 1.3.17, it follows that $\underline{u}_\lambda \leq U_\lambda$ in Ω and, thus, from Theorem 1.2.3 we obtain a classical solution u_λ of (9.92) such that $\underline{u}_\lambda \leq u_\lambda \leq U_\lambda$ in Ω.

Step 2: Nonexistence for small $\lambda > 0$. Let us remark that

$$\lim_{t \searrow 0}(f(x,t) - p(d(x))g(t)) = -\infty \quad \text{uniformly for } x \in \Omega.$$

Hence, there exists $t_0 > 0$ such that

$$f(x,t) - p(d(x))g(t) < 0 \quad \text{for all } (x,t) \in \Omega \times (0, t_0). \tag{9.112}$$

On the other hand, assumption $(f1)$ yields

$$\frac{f(x,t) - p(d(x))g(t)}{t} \leq \frac{f(x,t)}{t} \leq \frac{f(x,t_0)}{t_0}, \tag{9.113}$$

for all $(x,t) \in \Omega \times [t_0, \infty)$. Let $m = \max_{x \in \overline{\Omega}} f(x, t_0)/t_0$. Combining (9.112) with (9.113) we find

$$f(x,t) - p(d(x))g(t) < mt \quad \text{for all } (x,t) \in \Omega \times (0, \infty). \tag{9.114}$$

Set $\lambda_0 = \min\{1, \lambda_1/2m\}$. We claim that problem (9.92) has no classical solution for $0 < \lambda \leq \lambda_0$. Indeed, assume by contradiction that u_0 is a classical solution of (9.92) with $0 < \lambda \leq \lambda_0$. Then, according to (9.114), u_0 is a subsolution of

$$\begin{cases} -\Delta u = \dfrac{\lambda_1}{2} u & \text{in } \Omega, \\ u > 0 & \text{in } \Omega, \\ u = 0 & \text{on } \partial\Omega. \end{cases} \tag{9.115}$$

By Theorem 1.3.17 we have $u_0 \le U_\lambda$ in Ω. Furthermore, from (9.102) we have $cu_0 \le \varphi_1$ in Ω for some positive constant $0 < c < 1$. Note that cu_0 is still a subsolution of (9.115) whereas φ_1 is a supersolution of (9.115). Hence, problem (9.115) has a solution $u \in C^2(\overline{\Omega})$. Multiplying by φ_1 in (9.115) and then integrating over Ω we have

$$-\int_\Omega \varphi_1 \Delta u dx = \frac{\lambda_1}{2} \int_\Omega u\varphi_1 dx.$$

That is,

$$\lambda_1 \int_\Omega u\varphi_1 dx = -\int_\Omega u\Delta\varphi_1 dx = \frac{\lambda_1}{2} \int_\Omega u\varphi_1 dx.$$

The previous equality yields $\int_\Omega u\varphi_1 dx = 0$, but this is clearly a contradiction, because both u and φ_1 are positive in Ω. It follows that (9.92) has no classical solutions for $0 < \lambda \le \lambda_0$.

Step 3: Dependence on $\lambda > 0$. Set

$$A := \{\lambda > 0 : \text{ problem (9.92) has at least one classical solution}\}.$$

From the previous arguments we deduce that A is nonempty and $\lambda^* := \inf A$ is positive. We show that if $\lambda \in A$, then $(\lambda, \infty) \subseteq A$. To this aim, let $\lambda_1 \in A$ and $\lambda_2 > \lambda_1$. If u_{λ_1} is a solution of (9.92) with $\lambda = \lambda_1$, then u_{λ_1} is a subsolution of (9.92) with $\lambda = \lambda_2$ whereas U_{λ_2} defined by (9.101) for $\lambda = \lambda_2$ is a supersolution. Furthermore, we have

$$\Delta U_{\lambda_2} + \lambda_2 f(x, U_{\lambda_2}) \le 0 \le \Delta u_{\lambda_1} + \lambda_2 f(x, u_{\lambda_1}) \quad \text{in } \Omega,$$

$$U_{\lambda_2}, u_{\lambda_1} > 0 \quad \text{in } \Omega,$$

$$U_{\lambda_2} = u_{\lambda_1} = 0 \quad \text{on } \partial\Omega,$$

$$\Delta U_{\lambda_2} \in L^1(\Omega).$$

Again by Theorem 1.3.17 we find $u_{\lambda_1} \le U_{\lambda_2}$ in Ω. Therefore, problem (9.92) with $\lambda = \lambda_2$ has at least one classical solution. Because $\lambda \in A$ was arbitrary, we conclude that $(\lambda^*, \infty) \subseteq A$. This completes the proof of (i).

(ii) *Step 1: Existence of a solution for large λ.*

According to Lemma 9.7.4, there exists $\lambda^* > 0$ such that (9.92) has a subsolution $\underline{u}_\lambda$ for $\lambda > \lambda^*$ and $\mu = -1$. Then $\underline{u}_\lambda$ is also a subsolution in case $\mu = 1$, provided that $\lambda > \lambda^*$. Let us now construct a supersolution. By Theorem 1.2.5, for all $\lambda > \lambda^*$ there exists $v_\lambda \in C^2(\overline{\Omega})$ a solution of

$$\begin{cases} -\Delta v = \lambda f(x,v) + 1 & \text{in } \Omega, \\ v > 0 & \text{in } \Omega, \\ v = 0 & \text{on } \partial\Omega. \end{cases}$$

Because $0 < a < 1$, we can choose $M = M(\lambda) > 1$ large enough such that

$$M > M^a |\nabla v_\lambda|^a \quad \text{in } \Omega.$$

Then, using $(f1)$ we obtain

$$-\Delta(Mv_\lambda) = \lambda M f(x, v_\lambda) + M \geq \lambda f(x, Mv_\lambda) + |\nabla(Mv_\lambda)|^a \quad \text{in } \Omega.$$

Hence, $\overline{u}_\lambda := Mv_\lambda \in C^2(\overline{\Omega})$ is a supersolution of (9.92) for all $\lambda > \lambda^*$. On the other hand, because $\Delta\overline{u}_\lambda + \lambda f(x, \overline{u}_\lambda) \leq 0 \leq \Delta\underline{u}_\lambda + \lambda f(x, \underline{u}_\lambda)$ in Ω, by Theorem 1.3.17 we obtain $\underline{u}_\lambda \leq \overline{u}_\lambda$, and finally problem (9.92) has at least one solution for all $\lambda > \lambda^*$.

Step 2: Nonexistence for small $\lambda > 0$. We first extend Theorem 1.3.17 in the following way.

Lemma 9.7.5 *Let* $0 < a < 1$ *and* $\Psi : \overline{\Omega} \times (0, \infty) \to \mathbb{R}$ *be a Hölder continuous function such that the mapping* $(0, \infty) \ni t \longmapsto \Psi(x,t)/t$ *is decreasing for all* $x \in \Omega$. *Assume that there exist* $v, w \in C^2(\Omega) \cap C(\overline{\Omega})$ *such that*

(a) $\Delta w + \Psi(x, w) + |\nabla w|^a \leq 0 \leq \Delta v + \Psi(x, v) + |\nabla v|^a$ *in* Ω;

(b) $v, w > 0$ *in* Ω *and* $v < w$ *on* $\partial\Omega$.

Then $v \leq w$ *in* Ω.

Proof Assume by contradiction that the inequality $v \leq w$ does not hold in Ω and let $\zeta = v/w$. Clearly, $\zeta < 1$ on $\partial\Omega$ and

$$-\nabla \cdot \left(w^2 \nabla \zeta\right) = -w\Delta v + v\Delta w \quad \text{in } \Omega.$$

Let $x_0 \in \Omega$ denote a maximum point of ζ. In particular,

$$\nabla\zeta(x_0) = 0, \quad -\Delta\zeta(x_0) \geq 0,$$

which yields

$$0 \leq (-w\Delta v + v\Delta w)(x_0). \tag{9.116}$$

Because $w(x_0) < v(x_0)$, from assumption (a), the property of Ψ, and (9.116), it follows that

$$0 < (w|\nabla v|^a - v|\nabla w|^a)(x_0).$$

From $\nabla\zeta(x_0) = 0$ we also have

$$w(x_0)|\nabla v|(x_0) = v(x_0)|\nabla w|(x_0).$$

Using the previous two relations we find

$$0 < \left[\left(\frac{v}{w}\right)^a w - v\right] |\nabla w|^a(x_0) = v^a\left(w^{1-a} - v^{1-a}\right)|\nabla w|^a(x_0),$$

which contradicts $w(x_0) < v(x_0)$. This concludes the proof of our lemma. □

Assume by contradiction that there exists a sequence of solutions $(u_n)_{n\geq 1}$ of problem (9.92) associated with a sequence of parameters $(\lambda_n)_{n\geq 1} \subset (0,\infty)$ such that $\lambda_n \to 0$ as $n \to \infty$. A simple computation shows that $w(x) := A(R^2 - |x|^2)$ is positive and satisfies the inequality

$$\Delta w + f(x,w) + |\nabla w|^a \leq 0 \quad \text{in } \Omega,$$

where $A, R > 0$ are large constants. In particular, it follows from Lemma 9.7.5 that $0 < u_n \leq w$ whenever $\lambda_n \leq 1$. Let $x_n \in \Omega$ be a maximum point of u_n. Then $\nabla u_n(x_n) = 0$ and $-\Delta u_n(x_n) \geq 0$. Letting $d_n = d(x_n)$, $M_n = u_n(x_n)$, it follows from (9.92) that

$$p(d_n)g(M_n) \leq \lambda_n f(x_n, M_n) \leq \lambda_n \max_{x\in\overline{\Omega}} f(x, w(x)) \leq C\lambda_n,$$

which yields a contradiction as $n \to \infty$. Hence, problem (9.92) has no classical solutions when $\lambda > 0$ is small.

The third step concerning the dependence on λ follows in exactly the same way as in the case $\mu = -1$. This finishes the proof of Theorem 9.7.3. □

9.8 Ground-state solutions of singular elliptic problems with gradient term

We continue here the study of singular elliptic problems in unbounded domains started in Section 4.7. We shall be concerned in the sequel with the corresponding problem to (9.1) in the case $\Omega = \mathbb{R}^N$, $N \geq 3$. More precisely, we consider

$$\begin{cases} -\Delta u = p(x)(g(u) + f(u) + |\nabla u|^a) & \text{in } \mathbb{R}^N, \\ u > 0 & \text{in } \mathbb{R}^N, \\ u(x) \to 0 & \text{as } |x| \to \infty, \end{cases} \tag{9.117}$$

where $p : \mathbb{R}^N \to (0,\infty)$ is a Hölder continuous function and $0 < a < 1$. We also assume that f fulfills $(f1)$ and $(f2)$, and $g \in C^1(0,\infty)$ satisfies $g > 0$, $g' < 0$ in $(0,\infty)$ and $(g1)$.

As we have argued in Section 4.7, solutions of such problems are called *ground-state solutions*, especially for the prescribed condition at infinity.

For $f \equiv 0$ and $a = 0$, we saw in Section 4.7 that a necessary condition have a solution is

$$\int_1^\infty t\psi(t)dt < \infty, \tag{9.118}$$

where $\psi(r) = \min_{|x|=r} p(x)$, $r \geq 0$. As in Section 4.7 we require that p satisfies

$$\int_1^\infty t\phi(t)dt < \infty, \tag{9.119}$$

where $\phi(r) = \max_{|x|=r} p(x)$, $r \geq 0$. In this framework we have the following theorem.

Theorem 9.8.1 *Assume that (f1), (f2), (g1), and (9.119) are fulfilled. Then, problem (9.117) has at least one solution.*

Proof The approach is similar to that in Theorem 4.7.1. The construction of the monotone sequence with a pointwise limit that is the solution of (9.117) relies on Theorem 9.4.1. More precisely, let $B_n := \{x \in \mathbb{R}^N : |x| < n\}$. According to Theorem 9.4.1, for all $n \geq 1$ there exists $u_n \in C^2(B_n) \cap C(\overline{B_n})$ such that

$$\begin{cases} -\Delta u_n = p(x)(g(u_n) + f(u_n) + |\nabla u_n|^a) & \text{in } B_n, \\ u_n > 0 & \text{in } B_n, \\ u_n = 0 & \text{on } \partial B_n. \end{cases}$$

Set $\Psi(t) := g(t) + f(t)$, $t > 0$. Then we have

$$\Delta u_{n+1} + p(x)(\Psi(u_{n+1}) + |\nabla u_{n+1}|^a) \leq 0 \leq \Delta u_n + p(x)(\Psi(u_n) + |\nabla u_n|^a) \quad \text{in } B_n,$$

$$u_n = 0 < u_{n+1} \quad \text{on } \partial B_n.$$

With the same proof as in Lemma 9.7.5 (note that the only difference here is the presence of the positive potential $p(x)$), we find $u_n \leq u_{n+1}$ in B_n, for all $n \geq 1$.

Extending u_n by zero outside of B_n, this means that

$$0 \leq u_1 \leq \cdots \leq u_n \leq u_{n+1} \leq \cdots \quad \text{in } \mathbb{R}^N.$$

The main point is to find an upper bound for the sequence $(u_n)_{n\geq 1}$.

Lemma 9.8.2 *There exists $v \in C^2(\mathbb{R}^N)$ such that*

$$\begin{cases} -\Delta v \geq p(x)(g(v) + f(v) + |\nabla v|^a) & \text{in } \mathbb{R}^N, \\ v > 0 & \text{in } \mathbb{R}^N, \\ v(x) \to 0 & \text{as } |x| \to \infty. \end{cases} \tag{9.120}$$

Proof Let ζ be defined by (4.56) and fix $k > 2$ such that

$$k^{1-a} \geq 2 \max_{r \geq 0} \zeta^a(r). \tag{9.121}$$

In view of Lemma 2.2.4 we can define

$$\xi(x) := k \int_{|x|}^\infty \zeta(t)dt \quad \text{for all } x \in \mathbb{R}^N.$$

Then ξ satisfies

$$\begin{cases} -\Delta\xi = k\phi(|x|) & \text{in } \mathbb{R}^N, \\ \xi > 0 & \text{in } \mathbb{R}^N, \\ \xi(x) \to 0 & \text{as } |x| \to \infty. \end{cases}$$

To proceed further, we implicitly define the mapping $w : \mathbb{R}^N \to (0, \infty)$ by

$$\int_0^{w(x)} \frac{dt}{g(t)+1} = \xi(x) \quad \text{for all } x \in \mathbb{R}^N.$$

It is easy to see that $w \in C^2(\mathbb{R}^N)$ and $w(x) \to 0$ as $|x| \to \infty$. Moreover, we have

$$|\nabla w| = |\nabla\xi|(g(w)+1) = k\zeta(|x|)(g(w)+1) \quad \text{in } \mathbb{R}^N \tag{9.122}$$

and

$$\begin{aligned} -\Delta w &= -(g(w)+1)\Delta\xi - g'(w)(g(w)+1)|\nabla\xi|^2 \\ &\geq k\phi(|x|)(g(w)+1) \\ &\geq \phi(|x|)(g(w)+1) + \frac{k}{2}\phi(|x|)(g(w)+1). \end{aligned}$$

By (9.121) and (9.122) we deduce

$$\frac{k}{2}\phi(|x|)(g(w)+1) \geq \phi(|x|)k^a(g(w)+1)^a\zeta^a(|x|) \geq p(x)|\nabla w|^a \quad \text{in } \mathbb{R}^N.$$

Hence,

$$\begin{cases} -\Delta w \geq p(x)(g(w)+1+|\nabla w|^a) & \text{in } \mathbb{R}^N, \\ w > 0 & \text{in } \mathbb{R}^N, \\ w(x) \to 0 & \text{as } |x| \to \infty. \end{cases} \tag{9.123}$$

Using the assumption $(f1)$ and the fact that w is bounded in $\mathbb{R}^N$, we can find $M > 1$ large enough such that $M > f(Mw)$ in $\mathbb{R}^N$. Multiplying by M in (9.123) we deduce that $v := Mw$ satisfies (9.120), and the proof of Lemma 9.8.2 is now complete. □

Let us come back to the proof of Theorem 9.8.1. With the same arguments used earlier, we obtain $u_n \leq v$ in B_n, for all $n \geq 1$. This implies

$$0 \leq u_1 \leq \cdots \leq u_n \leq \cdots \leq v \quad \text{in } \mathbb{R}^N.$$

Thus, there exists $u(x) := \lim_{n\to\infty} u_n(x)$, for all $x \in \mathbb{R}^N$ and $u_n \leq u \leq v$ in $\mathbb{R}^N$. Because $v(x) \to 0$ as $|x| \to \infty$, we deduce that $u(x) \to 0$ as $|x| \to \infty$. A standard bootstrap argument implies that $u_n \to u$ in $C^{2,\gamma}_{\text{loc}}(\mathbb{R}^N)$ for some $0 < \gamma < 1$ and that u is a solution of problem (9.117).

This ends the proof of Theorem 9.8.1. □

9.9 Comments and historical notes

We have discussed in this chapter the influence of a gradient term in singular elliptic problems. As remarked by many authors (see, for instance, Serrin [181], Choquet–Bruhat and Leray [45], Kazdan and Warner [114]), the requirement that the nonlinearity $|\nabla u|^a$ grows at most quadratically is natural to apply the maximum principle. Elliptic equations involving a gradient term appear in many fields. For instance, Bellman's dynamic programming principle arising in optimal stochastic control problems indicates that the *value function* (also called *Bellman function*) u, which minimizes the *cost functional*, is also a solution of the nonlinear elliptic equation

$$-\frac{1}{2}\,\Delta u + \frac{1}{a}\,|\nabla u|^a + \lambda u = f(x,u) \qquad \text{in } \Omega,$$

where $\Omega \subset \mathbb{R}^N$ is a bounded domain, $0 < a \leq 2$, $\lambda > 0$ denotes the *discount factor*, and f is a smooth or singular nonlinearity.

Special attention has been paid to the case when $p > 0$ in problem (9.1). In this sense, we saw that if the exponent a is inferior to 1, then the presence of the gradient term does not change in a significant way the results presented in previous chapters. More precisely, if the nonlinear gradient term has a sublinear growth, then it does not perturb the structure of the problems encountered so far in this book.

A different situation occurs when $1 < a \leq 2$—that is, the gradient term has a superlinear behavior. A key result in the study of problem (9.1) is played by Proposition 9.4.4, which is a work by Alaa and Pierre [3].

Another major role in our analysis was played by the potential $p(d(x))$. We emphasize the different levels at which the presence of the (possibly) singular potential p affects the study of existence in this chapter. First, we saw that the existence of a classical solution to problem (9.1) is related to the growth of $tp(t)$ near zero. In turn, the existence of a classical solution to problem (9.92), is closely related to the behavior near the origin of both singular terms p and g. The *competition* between all these regular or singular nonlinearities or potential terms, combined with the presence of the bifurcation parameters λ and μ, decides whether a solution exists.

The study of problems (9.1) and (9.92) in smooth bounded domains $\Omega \subset \mathbb{R}^N$, $N \geq 2$, is a result of the work by Dupaigne, Ghergu, and Rădulescu [72] (see also [88], [89]). Problem (9.117) was considered in [90]. We emphasized the importance of the growth condition (9.28) supplied first by Taliaferro [186] (see also [2]) for similar one-dimensional problems. The nonexistence result in Theorem 9.2.1 generalizes a result by Zhang and Cheng [202].

10

SINGULAR GIERER–MEINHARDT SYSTEMS

> To myself I am only a child playing on the beach, while vast oceans of truth lie undiscovered before me.
>
> Sir Isaac Newton (1642–1727)

10.1 Introduction

In this chapter we extend the study of single singular elliptic equations to the following singular elliptic system:

$$\begin{cases} -\Delta u + \alpha u = \dfrac{f(u)}{g(v)} + \rho(x),\ u > 0 & \text{in } \Omega, \\ -\Delta v + \beta v = \dfrac{h(u)}{k(v)},\ v > 0 & \text{in } \Omega, \\ u = 0,\ v = 0 & \text{on } \partial\Omega. \end{cases} \tag{$\mathcal{S}$}$$

Here, $\Omega \subset \mathbb{R}^N$, $(N \geq 1)$ is a smooth bounded domain, $\alpha, \beta > 0$, $\rho \in C^{0,\gamma}(\Omega)$, $(0 < \gamma < 1)$, $\rho \geq 0$, $\rho \not\equiv 0$, and $f, g, h, k \in C^{0,\gamma}[0, \infty)$ are nonnegative and nondecreasing functions such that $g(0) = k(0) = 0$. This last assumption on g and k, together with the Dirichlet boundary conditions on $\partial\Omega$, makes the system singular at the boundary.

In the particular case of pure powers in nonlinearities, system $(\mathcal{S})$ reads

$$\begin{cases} -\Delta u + \alpha u = \dfrac{u^p}{v^q} + \rho(x),\ u > 0 & \text{in } \Omega, \\ -\Delta v + \beta v = \dfrac{u^r}{v^s},\ v > 0 & \text{in } \Omega, \\ u = 0,\ v = 0 & \text{on } \partial\Omega. \end{cases} \tag{10.1}$$

The associated parabolic system to (10.1) subject to Neumann boundary conditions is known as the Gierer–Meinhardt system and it occurs in morphogenesis and cellular differentiation. These problems arise in the study of biological pattern formation by auto- and cross-catalysis, and are related to known biochemical processes and cellular properties. The unknowns u and v represent the concentrations of two morphogens called *activator* and *inhibitor*, respectively.

The main difficulties in solving system $(\mathcal{S})$ are the result of the presence of singular nonlinearities on the one hand and the noncooperative (that is, non-quasimonotone) character of our system on the other. We are mainly interested in the case when the activator and inhibitor have different source terms—that is, the mappings $t \longmapsto f(t)/h(t)$ and $t \longmapsto g(t)/k(t)$ are not constant on $(0, \infty)$.

10.2 A nonexistence result

In this section we present a nonexistence result for classical solutions to $(\mathcal{S})$. The main idea is to speculate the asymptotic behavior of v from the second equation of $(\mathcal{S})$. This will be then used in the first equation of the system and, by a similar argument to that in Theorem 9.2.1, we obtain the desired nonexistence result. Special attention will be paid to the case of pure powers in nonlinearities. In this sense we obtain some relations between the exponents p, q, r, and s for which the system (10.1) has no classical solutions.

Several times in this chapter we shall apply the following comparison result, which is a direct consequence of the weak maximum principle stated in Theorem 1.3.2.

Lemma 10.2.1 *Let $k \in C(0, \infty)$ be a positive nondecreasing function and let a_1, $a_2 \in C(\Omega)$ be such that $0 < a_2 \le a_1$ in Ω.*

Assume there exist $\beta > 0$ and v_1, $v_2 \in C^2(\Omega) \cap C(\overline{\Omega})$ such that

(i) $\Delta v_1 - \beta v_1 + \dfrac{a_1(x)}{k(v_1)} \le 0 \le \Delta v_2 - \beta v_2 + \dfrac{a_2(x)}{k(v_2)}$ *in Ω;*

(ii) v_1, $v_2 > 0$ *in Ω, $v_1 \ge v_2$ on $\partial\Omega$.*

Then $v_1 \ge v_2$ in Ω.

Let $\Phi : [0, 1) \to [0, \infty)$ defined by

$$\Phi(t) = \int_0^t \left(2 \int_\tau^1 1/k(\theta)d\theta \right)^{-1/2} d\tau, \quad 0 \le t < 1. \tag{10.2}$$

Set $a := \lim_{t \nearrow 1} \Phi(t)$ and let $\Psi : [0, a) \to [0, 1)$ be the inverse of Φ.

The main result of this section is the following nonexistence property.

Theorem 10.2.2 *Assume that*

$$\int_0^a \frac{tf(mt)}{g(M\Psi(t))} dt = \infty, \tag{10.3}$$

for all $0 < m < 1 < M$. Then system $(\mathcal{S})$ has no classical solutions.

Proof Assume by contradiction that there exists a classical solution (u, v) of system $(\mathcal{S})$ and let φ_1 be the normalized first eigenfunction of $(-\Delta)$ in $H_0^1(\Omega)$. Also let ζ be the unique solution of the problem

$$\begin{cases} -\Delta\zeta + \alpha\zeta = \rho(x) & \text{in } \Omega, \\ \zeta = 0 & \text{on } \partial\Omega. \end{cases} \tag{10.4}$$

By standard elliptic arguments and the strong maximum principle, we deduce that $\zeta \in C^2(\overline{\Omega})$ and $\zeta > 0$ in Ω. Furthermore, taking into account the regularity of the domain, there exist c_1, $c_2 > 0$ such that

$$c_1 d(x) \leq \varphi_1, \zeta \leq c_2 d(x) \quad \text{in } \Omega, \tag{10.5}$$

where $d(x) = \operatorname{dist}(x, \partial\Omega)$.

Because

$$\begin{cases} -\Delta(u-\zeta) + \alpha(u-\zeta) \geq 0 & \text{in } \Omega, \\ u - \zeta = 0 & \text{on } \partial\Omega, \end{cases}$$

by the weak maximum principle (see Theorem 1.3.2) we have $u \geq \zeta$ in Ω. Hence, by (10.5), it follows that

$$u(x) \geq m d(x) \text{ in } \Omega, \tag{10.6}$$

for some $m > 0$ small enough. Set $C := \max_{x\in\overline{\Omega}} h(u(x)) > 0$. Then v satisfies

$$\begin{cases} -\Delta v + \beta v \leq \dfrac{C}{k(v)} & \text{in } \Omega, \\ v > 0 & \text{in } \Omega, \\ v = 0 & \text{on } \partial\Omega. \end{cases} \tag{10.7}$$

Let $c > 0$ be such that

$$c\varphi_1 \leq \min\{a, d(x)\} \quad \text{in } \Omega. \tag{10.8}$$

We need the following auxiliary result.

Lemma 10.2.3 *There exists $M > 1$ large enough such that $\overline{v} := M\Psi(c\varphi_1)$ satisfies*

$$-\Delta\overline{v} + \beta\overline{v} \geq \frac{C}{k(\overline{v})} \quad \text{in } \Omega. \tag{10.9}$$

Proof Because Ψ is the inverse of Φ defined in (10.2), we have $\Psi(0) = 0$ and $\Psi \in C^1(0, a)$ with

$$\Psi'(t) = \sqrt{2\int_{\Psi(t)}^{1} \frac{1}{k(\tau)} d\tau} \quad \text{for all } 0 < t < a. \tag{10.10}$$

This yields

$$\begin{cases} -\Psi''(t) = \dfrac{1}{k(\Psi(t))} & \text{for all } 0 < t < a, \\ \Psi'(t), \Psi(t) > 0 & \text{for all } 0 < t < a, \\ \Psi(0) = 0. \end{cases} \tag{10.11}$$

By the strong maximum principle, there exist $\omega \subset\subset \Omega$ and $\delta > 0$ such that

$$|\nabla\varphi_1| > \delta \quad \text{in } \Omega \setminus \omega \quad \text{and} \quad \varphi_1 > \delta \quad \text{in } \omega. \tag{10.12}$$

Fix $M > 1$ large enough such that

$$M(c\delta)^2 > C \quad \text{and} \quad Mc\lambda_1\delta\Psi'(c\|\varphi_1\|_\infty) > \frac{C}{\min_{x\in\overline{\omega}} k(\Psi(c\varphi_1))}. \tag{10.13}$$

We have

$$-\Delta\overline{v} = \frac{Mc^2}{k(\Psi(c\varphi_1))}|\nabla\varphi_1|^2 + Mc\lambda_1\varphi_1\Psi'(c\varphi_1) \quad \text{in } \Omega.$$

By (10.12) and (10.13) we obtain

$$\begin{aligned} -\Delta\overline{v} &\geq Mc\lambda_1\varphi_1\Psi'(c\varphi_1) \geq Mc\lambda_1\delta\Psi'(c\|\varphi_1\|_\infty) \geq \frac{C}{k(\overline{v})} && \text{in } \omega, \\ -\Delta\overline{v} &\geq \frac{Mc^2}{k(\Psi(c\varphi_1))}|\nabla\varphi_1|^2 \geq \frac{C}{k(\Psi(c\varphi_1))} \geq \frac{C}{k(\overline{v})} && \text{in } \Omega\setminus\omega. \end{aligned}$$

The last two inequalities imply that $\overline{v}$ satisfies

$$-\Delta\overline{v} \geq \frac{C}{k(\overline{v})} \quad \text{in } \Omega.$$

Hence, $\overline{v}$ fulfills (10.9). This ends the proof of the lemma. □

By virtue of Lemma 10.2.1, from (10.7) and (10.9) we find $v \leq \overline{v}$ in Ω. Using (10.6) we then obtain

$$\frac{f(u)}{g(v)} \geq \frac{f(md(x))}{g(M\Psi(c\varphi_1))} \quad \text{in } \Omega.$$

Furthermore, u satisfies

$$\begin{cases} -\Delta u + \alpha u \geq \dfrac{f(md(x))}{g(M\Psi(c\varphi_1))} & \text{in } \Omega, \\ u > 0 & \text{in } \Omega, \\ u = 0 & \text{on } \partial\Omega. \end{cases} \tag{10.14}$$

To avoid the singularities in (10.14) near the boundary, we consider the approximated problem

$$\begin{cases} -\Delta w + \alpha w = \dfrac{f(md(x))}{g(M\Psi(c\varphi_1)) + \varepsilon} & \text{in } \Omega, \\ w = 0 & \text{on } \partial\Omega. \end{cases} \tag{10.15}$$

Clearly, $\overline{w} := u$ is a supersolution of (10.15) whereas $\underline{w} = 0$ is a subsolution. By the sub- and supersolution method and the strong maximum principle, problem (10.15) has a unique solution $w_\varepsilon \in C^2(\overline{\Omega})$ such that $0 < w_\varepsilon \leq u$ in Ω. To raise a contradiction, we multiply by φ_1 in (10.15) and then we integrate over Ω. We obtain

$$(\alpha+\lambda_1)\int_\Omega w_\varepsilon \varphi_1 dx = \int_\Omega \varphi_1 \frac{f(md(x))}{g(M\Psi(c\varphi_1))+\varepsilon}dx.$$

Because $w_\varepsilon \le u$ in Ω, we find

$$(\alpha+\lambda_1)\int_\Omega u\varphi_1 dx \ge \int_\omega \varphi_1 \frac{f(md(x))}{g(M\Psi(c\varphi_1))+\varepsilon}dx \quad \text{for all } \omega \subset\subset \Omega.$$

Let $\widetilde{C} := (\alpha+\lambda_1)\int_\Omega u\varphi_1 dx$. Passing to the limit with $\varepsilon \to 0$ in the previous inequality we obtain

$$\int_\omega \varphi_1 \frac{f(md(x))}{g(M\Psi(c\varphi_1))}dx \le \widetilde{C} < \infty \quad \text{for all } \omega \subset\subset \Omega.$$

Hence,

$$\int_\Omega \varphi_1 \frac{f(md(x))}{g(M\Psi(c\varphi_1))}dx \le \widetilde{C} < \infty.$$

Now let $\Omega_0 := \{x \in \Omega : d(x) < a\}$. The previous estimate combined with (10.8) produces

$$\int_{\Omega_0} d(x)\frac{f(md(x))}{g(M\Psi(d(x)))}dx < \infty,$$

but this clearly contradicts (10.3). Hence, system $(\mathcal{S})$ has no positive classical solutions. This finishes the proof. □

If $k(t) = t^s$, $s > 0$, condition (10.3) can be written more explicitly by describing the asymptotic behavior of Ψ. We have the following corollary.

Corollary 10.2.4 *Assume that $k(t) = t^s$, $s > 0$, and for all $0 < m < 1 < M$ one of the following conditions hold:*

(i) $s > 1$ *and* $\int_0^a tf(mt)/g(Mt^{2/(1+s)})dt = \infty$.

(ii) $s = 1$ *and* $\int_0^{\min\{a,1/2\}} tf(mt)/g(Mt\sqrt{-\ln t})dt = \infty$.

(iii) $0 < s < 1$ *and* $\int_0^a tf(mt)/g(Mt)dt = \infty$.

Then, system $(\mathcal{S})$ has no positive classical solutions.

Proof The main idea is to describe the asymptotic behavior of Ψ near the origin. Notice that in our setting the mapping $\Psi : [0,a) \to [0,1)$ satisfies

$$\begin{cases} -\Psi''(t) = \Psi^{-s}(t) & \text{for all } 0 < t < a, \\ \Psi'(t), \Psi(t) > 0 & \text{for all } 0 < t < a, \\ \Psi(0) = 0. \end{cases} \tag{10.16}$$

The asymptotic behavior of Ψ was studied in a more general framework concerning problem (9.77) in Chapter 9. Using the arguments in the proof of Theorem 9.6.1, there exist c_1, $c_2 > 0$ such that

$$
\begin{aligned}
c_1 t^{2/(1+s)} \le \Psi(t) \le c_2 t^{2/(1+s)} \qquad & \text{if } s>1,\ 0<t<a,\\
c_1 t\sqrt{-\ln t} \le \Psi(t) \le c_2 t\sqrt{-\ln t} \qquad & \text{if } s=1,\ 0<t<\min\{1/2,a\},\\
c_1 t \le \Psi(t) \le c_2 t \qquad & \text{if } 0<s<1,\ 0<t<a.
\end{aligned}
$$

This completes the proof. □

In case of pure powers in the nonlinearities, we have the following nonexistence result for system (10.1).

Corollary 10.2.5 *Let* $p,\ q,\ r,\ s>0$ *be such that one of the following conditions hold:*

(i) $s>1$ *and* $2q \ge (s+1)(p+2)$.
(ii) $s=1$ *and* $q>p+2$.
(iii) $0<s<1$ *and* $q \ge p+2$.

Then, system (10.1) has no positive classical solutions.

Proof The proofs of (i) and (iii) are simple exercices of calculus. For (ii), by Corollary 10.2.4 we have that (10.1) has no classical solutions provided $s=1$ and

$$\int_0^{1/2} t^{1+p-q}(-\ln t)^{-q/2}dt = \infty. \tag{10.17}$$

On the other hand, for $a, b \in \mathbb{R}$ we have $\int_0^{1/2} t^a(-\ln t)^b dt < \infty$ if and only if $a>-1$ or $a=-1$ and $b<-1$. Now condition (10.17) reads $q>p+2$. This concludes the proof. □

10.3 Existence results

In this section we provide existence results for classical solutions to $(\mathcal{S})$ under the additional hypothesis $\beta \le \alpha$. The existence is obtained without assuming any growth condition on ρ near the boundary, because we are able to establish general bounds for the regularized system associated with $(\mathcal{S})$. In particular, we obtain that problem (10.1) has solutions provided that $r-p=s-q\ge 0$ and $q>p-1$.

For all $t_1,\ t_2>0$ define

$$A(t_1,t_2) := \frac{f(t_1)}{h(t_1)} - \frac{g(t_2)}{k(t_2)}.$$

In this section we suppose that A fulfills

(A_1) $\quad A(t_1,t_2)\le 0$ for all $t_1 \ge t_2 > 0$.

We also assume that $k \in C^1(0,\infty)$ is a nonnegative and nondecreasing function such that

(A_2) $\quad \lim_{t\to\infty} \frac{K(t)}{h(t+c)} = \infty$ for all $c>0$, where $K(t)=\int_0^t k(\tau)d\tau$.

Here are some examples of nonlinearities that fulfill $(A1)$ and $(A2)$:

(i) $f(t) = t^p$; $g(t) = t^q$; $h(t) = t^r$; $k(t) = t^s$; $t \geq 0$; $p, q, r, s > 0$; $r - p = s - q \geq 0$; and $p - q < 1$.
(ii) $f(t) = \ln(1 + t^p)$; $g(t) = e^{t^q} - 1$; $h(t) = t^p$; and $k(t) = t^q$, $t \geq 0$, $p, q > 0$, $p - q < 1$.
(iii) $f(t) = \log(1 + at)$; $g(t) = \log(1 + t)$; $h(t) = at$; and $k(t) = t$; $t \geq 0$, $a \geq 1$.

In the following we provide a general method to construct nonlinearities f, g, h, k that verify hypotheses $(A1)$ and $(A2)$. Let $f, g, h, k : [0, \infty) \to [0, \infty)$ be nondecreasing functions such that k and h verify $(A2)$ and one of the following assumptions hold:

(a) $fk = gh$ and the mapping $(0, \infty) \ni t \longmapsto f(t)/h(t)$ is nonincreasing.
(b) There exists $m > 0$ such that $f(t)/h(t) \leq m \leq g(t)/k(t)$, for all $t > 0$.

Then mapping A verifies $(A1)$.

For instance, the mappings in example (i) satisfy condition (a) whereas the mappings in (ii) verify condition (b).

The first result of this section concerns the existence of classical solutions for the general system $(\mathcal{S})$.

Theorem 10.3.1 *Assume that the hypotheses* $(A1)$ *and* $(A2)$ *are fulfilled. Then system* $(\mathcal{S})$ *has at least one classical solution.*

The existence of a solution to $(\mathcal{S})$ is obtained by considering the regularized system

$$\begin{cases} -\Delta u + \alpha u = \dfrac{f(u+\varepsilon)}{g(v+\varepsilon)} + \rho(x) & \text{in } \Omega, \\ -\Delta v + \beta v = \dfrac{h(u+\varepsilon)}{k(v+\varepsilon)} & \text{in } \Omega, \\ u = 0, \; v = 0 & \text{on } \partial\Omega. \end{cases} \qquad (\mathcal{S})_\varepsilon$$

First, we need to provide some a priori estimates for a classical solution of $(\mathcal{S})_\varepsilon$. We start with the following auxiliary result.

Lemma 10.3.2 *There exists* $M > 0$ *that is independent of* ε *such that any solution* $(u_\varepsilon, v_\varepsilon)$ *of system* $(\mathcal{S})_\varepsilon$ *with* $u_\varepsilon, v_\varepsilon \in C^2(\Omega) \cap C(\overline{\Omega})$ *satisfies*

$$\max\{\|u_\varepsilon\|_\infty, \|v_\varepsilon\|_\infty\} \leq M. \qquad (10.18)$$

Proof Let $w_\varepsilon := u_\varepsilon - v_\varepsilon$ and $\omega := \{x \in \Omega : w_\varepsilon > 0\}$. To establish (10.18), it suffices to provide a uniform upper bound for v_ε and w_ε. Subtracting the two equations in $(S)_\varepsilon$ we have

$$\begin{aligned} -\Delta w_\varepsilon + \alpha w_\varepsilon &= (\beta - \alpha) w_\varepsilon + \frac{f(u_\varepsilon + \varepsilon)}{g(v_\varepsilon + \varepsilon)} - \frac{h(u_\varepsilon + \varepsilon)}{gk(v_\varepsilon + \varepsilon)} + \rho(x) \\ &= (\beta - \alpha) v_\varepsilon + \frac{h(u_\varepsilon + \varepsilon)}{g(v_\varepsilon + \varepsilon)} A(u_\varepsilon + \varepsilon, v_\varepsilon + \varepsilon) + \rho(x) \quad \text{in } \Omega. \end{aligned}$$

Notice that $A(u_\varepsilon + \varepsilon, v_\varepsilon + \varepsilon) \geq 0$ in ω and $w_\varepsilon = 0$ on $\partial\omega$. This yields

$$-\Delta w_\varepsilon + \alpha w_\varepsilon \leq \rho(x) \quad \text{in } \omega.$$

Let $\zeta \in C^2(\overline{\Omega})$ be the unique solution of (10.4). Then,

$$\begin{cases} -\Delta(w_\varepsilon - \zeta) + \alpha(w_\varepsilon - \zeta) \geq 0 & \text{in } \omega, \\ w_\varepsilon - \zeta \leq 0 & \text{on } \partial\omega. \end{cases}$$

Furthermore, by the weak maximum principle, $w_\varepsilon \leq \zeta$ in ω. Because $w_\varepsilon \leq 0$ in $\Omega \setminus \omega$, it follows that

$$w_\varepsilon \leq \zeta \quad \text{in } \Omega. \tag{10.19}$$

We next multiply by $k(v_\varepsilon)$ in the second equation of $(S)_\varepsilon$ and obtain

$$-k(v_\varepsilon)\Delta v_\varepsilon + \beta v_\varepsilon k(v_\varepsilon) = \frac{k(v_\varepsilon)}{k(v_\varepsilon + \varepsilon)} h(u_\varepsilon + \varepsilon) \quad \text{in } \Omega. \tag{10.20}$$

On the other hand,

$$-k(v_\varepsilon)\Delta v_\varepsilon = -\Delta K(v_\varepsilon) + k'(v_\varepsilon)|\nabla v_\varepsilon|^2 \quad \text{in } \Omega. \tag{10.21}$$

Because k is nondecreasing, we also have

$$K(v_\varepsilon) = \int_0^{v_\varepsilon} k(t)dt \leq v_\varepsilon k(v_\varepsilon) \quad \text{in } \Omega. \tag{10.22}$$

Using (10.21) and (10.22) in (10.20), we deduce

$$-\Delta K(v_\varepsilon) + k'(v_\varepsilon)|\nabla v_\varepsilon|^2 + \beta K(v_\varepsilon) \leq \frac{k(v_\varepsilon)}{k(v_\varepsilon + \varepsilon)} h(u_\varepsilon + \varepsilon) \quad \text{in } \Omega.$$

Hence,

$$-\Delta K(v_\varepsilon) + \beta K(v_\varepsilon) \leq h(u_\varepsilon + \varepsilon) \quad \text{in } \Omega. \tag{10.23}$$

By Theorem 1.3.3, there exists a positive constant $C > 1$ that depends only on Ω such that

$$\sup_{\overline{\Omega}} K(v_\varepsilon) \leq C \sup_{\overline{\Omega}} h(u_\varepsilon + \varepsilon) \leq C \sup_{\overline{\Omega}} h(v_\varepsilon + \|\zeta\|_\infty + 1).$$

Using assumption $(A2)$, we deduce that $(v_\varepsilon)_{\varepsilon>0}$ is uniformly bounded—that is, $\|v_\varepsilon\|_\infty \leq m$ for some $m > 0$ independent of ε. In view of (10.19), this yields

$$u_\varepsilon = v_\varepsilon + w_\varepsilon \leq m + \|\zeta\|_\infty \quad \text{in } \Omega,$$

and the proof of Lemma 10.3.2 is now complete. □

Lemma 10.3.3 *For all $0 < \varepsilon < 1$ there exists a solution $(u_\varepsilon, v_\varepsilon) \in C^2(\overline{\Omega}) \times C^2(\overline{\Omega})$ of system $(S)_\varepsilon$.*

Proof We use topological degree arguments. Consider the set

$$\mathcal{U} := \left\{ (u,v) \in C^2(\overline{\Omega}) \times C^2(\overline{\Omega}) : \begin{array}{r} \|u\|_\infty, \|v\|_\infty \le M+1 \\ u, v \ge 0 \text{ in } \Omega, u|_{\partial\Omega} = v|_{\partial\Omega} = 0 \end{array} \right\},$$

where $M > 0$ is the constant in (10.18). Define

$$\Phi_t : \mathcal{U} \to \mathcal{U}, \quad \Phi_t(u,v) = (\Phi_t^1(u,v), \Phi_t^2(u,v)),$$

by

$$\Phi_t^1(u,v) = u - t(-\Delta + \alpha)^{-1} \left(\frac{f(u+\varepsilon)}{g(v+\varepsilon)} + \rho \right),$$
$$\Phi_t^2(u,v) = v - t(-\Delta + \beta)^{-1} \left(\frac{h(u+\varepsilon)}{k(v+\varepsilon)} \right).$$

Using Lemma 10.3.2, we have $\Phi_t(u,v) \neq (0,0)$ on $\partial\mathcal{U}$, for all $0 \le t \le 1$. Therefore, by the invariance at homotopy of the topological degree we have

$$\deg(\Phi_1, \mathcal{U}, (0,0)) = \deg(\Phi_0, \mathcal{U}, (0,0)) = 1.$$

Hence, there exists $(u,v) \in \mathcal{U}$ such that $\Phi_1(u,v) = (0,0)$. This means that system $(\mathcal{S})_\varepsilon$ has at least one classical solution. The proof is now complete. □

Let us come back to the proof of Theorem 10.3.1. Let $(u_\varepsilon, v_\varepsilon) \in C^2(\overline{\Omega}) \times C^2(\overline{\Omega})$ be a solution of $(\mathcal{S})_\varepsilon$. Then,

$$\begin{cases} \Delta(u_\varepsilon - \zeta) - \alpha(u_\varepsilon - \zeta) \le 0 & \text{in } \Omega, \\ u_\varepsilon - \zeta = 0 & \text{on } \partial\Omega, \end{cases}$$

where ζ is the unique solution of (10.4). Hence $\zeta \le u_\varepsilon$ in Ω. By (10.19) it follows that

$$w_\varepsilon \le \zeta \le u_\varepsilon \quad \text{in } \Omega. \tag{10.24}$$

Let $\xi \in C^2(\overline{\Omega})$ be the unique positive solution of the boundary value problem

$$\begin{cases} -\Delta\xi + \beta\xi = \dfrac{h(\zeta)}{k(\xi+1)} & \text{in } \Omega, \\ \xi = 0 & \text{on } \partial\Omega. \end{cases} \tag{10.25}$$

In view of Lemma 10.2.1 we have $\xi \le v_\varepsilon$ in Ω, so that, by Lemma 10.3.2, the following estimates hold:

$$\begin{cases} \zeta(x) \le u_\varepsilon(x) \le M & \text{in } \Omega, \\ \xi(x) \le v_\varepsilon(x) \le M & \text{in } \Omega. \end{cases} \tag{10.26}$$

Now, standard Hölder and Schauder estimates imply that $\{(u_\varepsilon, v_\varepsilon)\}_{0<\varepsilon<1}$ converges (up to a subsequence) in $C^2_{\text{loc}}(\Omega) \times C^2_{\text{loc}}(\Omega)$ to (u,v). Passing to the limit in (10.26) we find

$$\begin{cases} \zeta(x) \le u(x) \le M & \text{in } \Omega, \\ \xi(x) \le v(x) \le M & \text{in } \Omega. \end{cases} \tag{10.27}$$

It remains only to obtain an upper bound near $\partial\Omega$ for $(u_\varepsilon, v_\varepsilon)$, which will lead us to the continuity up to the boundary of the solution (u, v). This will be done by combining standard arguments with the estimate (10.24). First, by (10.23) we have

$$\Delta K(v_\varepsilon) + h(M+1) \ge 0 \quad \text{in } \Omega. \tag{10.28}$$

Fix $x_0 \in \partial\Omega$. Because $\partial\Omega$ is smooth (more precisely, Ω satisfies the exterior sphere condition), there exist $y \in \mathbb{R}^N \setminus \Omega$ and a ball B centered at y, having radius $R > 0$ such that $\overline{\Omega} \cap \overline{B} = \partial\Omega \cap \overline{B} = \{x_0\}$.

Let $\delta(x) := |x-y| - R$ and $\Omega_0 := \{x \in \Omega : 4(N-1)\delta(x) < R\}$.

Consider $\psi \in C^2(0,\infty)$ such that $\psi' > 0$ and $\psi'' < 0$ on $(0,\infty)$, and set $\phi(x) := \psi(\delta(x))$, $x \in \Omega_0$. Then

$$\begin{aligned} \Delta\phi(x) &= \psi'(\delta(x))\Delta\delta(x) + \psi''(\delta(x))|\nabla\delta(x)|^2 \\ &= \frac{N-1}{|x-y|}\psi'(\delta(x)) + \psi''(\delta(x)) \\ &\le \frac{N-1}{R}\psi'(\delta(x)) + \psi''(\delta(x)) \quad \text{in } \Omega_0. \end{aligned}$$

Let us now choose $\psi(t) = C\sqrt{t}$, $t > 0$, where $C > 0$. Therefore,

$$\Delta\phi(x) \le \frac{C}{4}\delta^{-3/2}(x)\left[\frac{2(N-1)\delta(x)}{R} - 1\right] \le -\frac{C}{8}\delta^{-3/2}(x) < 0 \quad \text{in } \Omega_0.$$

We choose $C > 0$ large enough such that

$$\Delta\phi(x) \le -h(M+1) \quad \text{in } \Omega_0 \tag{10.29}$$

and

$$\phi(x) > K(M) \ge \sup_{\overline{\Omega}_0} K(v_\varepsilon) \quad \text{on } \partial\Omega_0 \setminus \partial\Omega. \tag{10.30}$$

Furthermore, by relations (10.28), (10.29), and (10.30), we obtain

$$\begin{cases} -\Delta(\phi(x) - K(v_\varepsilon)) \ge 0 & \text{in } \Omega_0, \\ \phi(x) - K(v_\varepsilon) \ge 0 & \text{on } \partial\Omega_0. \end{cases}$$

This implies $\phi(x) \ge K(v_\varepsilon)$ in Ω_0—that is,

$$0 \le v_\varepsilon(x) \le K^{-1}(\phi(x)) \quad \text{in } \Omega_0.$$

Passing to the limit with $\varepsilon \to 0$ in the last inequality, we have $0 \le v \le K^{-1}(\phi(x))$ in Ω_0. Hence,

$$0 \le \lim_{x\to x_0} v(x) \le \lim_{x \to x_0} K(\phi(x)) = 0.$$

Because $x_0 \in \partial\Omega$ was arbitrarily chosen, it follows that $v \in C(\overline{\Omega})$. Using the fact that $u_\varepsilon = w_\varepsilon + v_\varepsilon \le \zeta + v_\varepsilon$ in Ω, in the same manner we conclude $u \in C(\overline{\Omega})$. This finishes the proof of Theorem 10.3.1. □

The next result concerns the following singular system:

$$\begin{cases} -\Delta u + \alpha u = \dfrac{u^p}{v^q} + \rho(x) & \text{in } \Omega, \\ -\Delta v + \beta v = \dfrac{u^{p+\sigma}}{v^{q+\sigma}} & \text{in } \Omega, \\ u = v = 0 & \text{on } \partial\Omega, \end{cases} \tag{10.31}$$

where $\sigma \geq 0$ is a nonnegative real number.

Theorem 10.3.4 *Assume that p, $q \geq 0$ satisfy $p - q < 1$. Then the following properties are valid:*

(i) *System (10.31) has solutions for all $\sigma \geq 0$.*

(ii) *For any solution (u, v) of (10.31), there exist c_1, $c_2 > 0$ such that*

$$c_1 d(x) \leq u, v \leq c_2 d(x) \quad \text{in } \Omega. \tag{10.32}$$

Moreover, the following properties hold:

(ii1) *If $-1 < p - q < 0$, then $u, v \in C^2(\Omega) \cap C^{1,1+p-q}(\overline{\Omega})$.*

(ii2) *If $0 \leq p - q < 1$, then $u, v \in C^2(\overline{\Omega})$.*

Proof (i) Follows directly from Theorem 10.3.1, because both conditions $(A1)$ and $(A2)$ are fulfilled.
(ii) Recall that from (10.5) and (10.27) we have $u \geq \zeta \geq \overline{c}\varphi_1$ in Ω. From the second equation in (10.31) we deduce

$$-\Delta v + \beta v \geq \overline{c}^{p+\sigma} \frac{\varphi_1^{p+\sigma}}{v^{q+\sigma}} \quad \text{in } \Omega.$$

Because $p - q < 1$, we deduce that $\underline{v} := \underline{c}\varphi_1$ satisfies

$$-\Delta \underline{v} + \beta \underline{v} \geq \overline{c}^{p+\sigma} \frac{\varphi_1^{p+\sigma}}{\underline{v}^{q+\sigma}} \quad \text{in } \Omega,$$

for some $\underline{c} > 0$ small enough. Therefore, by virtue of Lemma 10.2.1, we obtain $v \geq \underline{c}\varphi_1$ in Ω.

Let us now prove the second inequality in (10.32). To this aim, set $w := u - v$. With the same idea as in the proof of Lemma 10.3.2, we find

$$-\Delta w + \alpha w \leq \rho(x) \quad \text{in the set } \{x \in \Omega : w(x) > 0\},$$

which yields

$$w \leq \zeta \leq c\varphi_1 \quad \text{in } \Omega. \tag{10.33}$$

Let $w^+ := \max\{w, 0\}$. Then v satisfies

$$\begin{cases} -\Delta v + \beta v \leq \dfrac{(w^+ + v)^{p+\sigma}}{v^{q+\sigma}} & \text{in } \Omega, \\ v = 0 & \text{on } \partial\Omega. \end{cases}$$

Consider now the problem

$$\begin{cases} -\Delta z + \beta z = 2^{p+\sigma} z^{p-q} & \text{in } \Omega, \\ z > 0 & \text{in } \Omega, \\ z = 0 & \text{on } \partial\Omega. \end{cases} \tag{10.34}$$

The existence of a classical solution to (10.34) follows from Theorem 1.2.5. Moreover, if $0 \le p - q < 1$, then $z \in C^2(\overline{\Omega})$, and if $-1 < p - q < 0$ by Theorem 4.3.2, we have $z \in C^2(\Omega) \cap C^{1,1+p-q}(\overline{\Omega})$ and $\Delta z \in L^1(\Omega)$. Furthermore, $z \le m\varphi_1$ in Ω for some $m > 0$. On the other hand, $\tilde{c}\varphi_1$ is a subsolution of (10.34) provided that $\tilde{c} > 0$ is small enough. Therefore, by Lemma 1.3.17 we derive $z \ge \tilde{c}\varphi_1$ in Ω. This last inequality, together with (10.33), allows us to choose $M > 1$ large enough such that $Mz \ge w^+$ in Ω. Hence,

$$\begin{aligned} -\Delta(Mz) + \beta(Mz) &= M(-\Delta z + \beta z) = 2^{p+\sigma} M z^{p-q} \\ &\ge 2^{p+\sigma}(Mz)^{p-q} \\ &\ge \frac{(w^+ + Mz)^{p+\sigma}}{(Mz)^{q+\sigma}} \quad \text{in } \Omega. \end{aligned}$$

This means that $\overline{v} := Mz$ verifies

$$-\Delta\overline{v} + \beta\overline{v} \ge \frac{(w^+ + \overline{v})^{p+\sigma}}{\overline{v}^{q+\sigma}} \quad \text{in } \Omega \quad \text{and} \quad \overline{v} = 0 \quad \text{on } \partial\Omega.$$

Remark now that

$$\Psi(x,t) := -\beta t + \frac{(w^+(x) + t)^{p+\sigma}}{t^{q+\sigma}} \qquad (x,t) \in \overline{\Omega} \times (0,\infty)$$

satisfies the hypotheses in Lemma 1.3.17, because $p - q < 1$. Furthermore, we have

$$\Delta\overline{v} + \Psi(x,\overline{v}) \le 0 \le \Delta v + \Psi(x,v) \quad \text{in } \Omega,$$

$$\overline{v}, v > 0 \quad \text{in } \Omega, \quad \overline{v} = v = 0 \quad \text{on } \partial\Omega, \quad \text{and } \Delta\overline{v} \in L^1(\Omega).$$

Hence, by Lemma 1.3.17, we obtain

$$v \le \overline{v} \le \tilde{c}\varphi_1 \quad \text{in } \Omega. \tag{10.35}$$

Combining (10.33) and (10.35) we derive $u = w + v \le C\varphi_1$ in Ω, for some $C > 0$. This completes the proof of (ii). As a consequence, there exists $M > 1$ such that

$$0 \le \frac{u^p}{v^q}, \ \frac{u^{p+\sigma}}{v^{q+\sigma}} \le M\varphi_1^{p-q} \quad \text{in } \Omega.$$

If $0 \le p - q < 1$, then by classical regularity arguments we have $u, v \in C^2(\overline{\Omega})$. If $-1 < p - q < 0$, then the same method as in the proof of Theorem 4.3.2 can be used to obtain $u, v \in C^2(\Omega) \cap C^{1,1+p-q}(\overline{\Omega})$.

This finishes the proof of Theorem 10.3.4. □

10.4 Uniqueness of the solution in one dimension

The uniqueness of the solution is a delicate matter. Actually, this issue seems to be a particular feature of the Dirichlet boundary conditions because in the case of Neumann boundary conditions, the Gierer–Meinhardt system does not have a unique solution.

In this section we are concerned with the uniqueness of the solution associated with the one-dimensional system:

$$\begin{cases} u'' - \alpha u + \dfrac{u^p}{v^q} + \rho(x) = 0 & \text{in } (0,1), \\ v'' - \beta v + \dfrac{u^{p+\sigma}}{v^{q+\sigma}} = 0 & \text{in } (0,1), \\ u(0) = u(1) = 0, \\ v(0) = v(1) = 0. \end{cases} \tag{10.36}$$

Our approach originates in the methods developed by Choi and McKenna [43], where a C^2 regularity of the solution up to the boundary is needed. So, we restrict our attention to the case $0 < q \le p \le 1$. Thus, by virtue of Theorem 10.3.4, any solution of (10.36) belongs to $C^2[0,1] \times C^2[0,1]$. By the strong maximum principle we also have $u'(0) > 0$, $v'(0) > 0$, $u'(1) < 0$, and $v'(1) < 0$ for any solution (u, v) of system (10.36).

The main result of this section is the following.

Theorem 10.4.1 *Assume that* $0 < q \le p \le 1$, $\sigma \ge 0$. *Then system (10.36) has a unique solution* $(u,v) \in C^2(\overline{\Omega}) \times C^2(\overline{\Omega})$.

Proof The existence part follows from Theorem 10.3.4. We prove here only the uniqueness. Suppose that there exist two distinct solutions $(u_1, v_1), (u_2, v_2) \in C^2[0,1] \times C^2[0,1]$ of system (10.36).

First we claim that we cannot have $u_2 \ge u_1$ or $v_2 \ge v_1$ in [0,1]. Indeed, let us assume that $u_2 \ge u_1$ in [0,1]. Then,

$$v_2'' - \beta v_2 + \frac{u_2^{p+\sigma}}{v_2^{q+\sigma}} = 0 = v_1'' - \beta v_1 + \frac{u_1^{p+\sigma}}{v_1^{q+\sigma}} \quad \text{in } (0,1)$$

and, by Lemma 10.2.1, we further obtain $v_2 \ge v_1$ in $[0,1]$. On the other hand,

$$u_1'' - \alpha u_1 + \frac{u_1^p}{v_2^q} + \rho(x) \le 0 = u_2'' - \alpha u_2 + \frac{u_2^p}{v_2^q} + \rho(x) \quad \text{in } (0,1). \tag{10.37}$$

Because $p < 1$, the mapping

$$\Psi(x,t) = -\alpha t + \frac{t^p}{v_2(x)^q} + \rho(x) \quad (x,t) \in (0,1) \times (0,\infty)$$

satisfies the hypotheses in Lemma 1.3.17. Hence, $u_2 \le u_1$ in $[0,1]$—that is $u_1 \equiv u_2$. This also implies $v_1 \equiv v_2$, which is a contradiction. Replacing u_1 with u_2 and

v_1 with v_2, we also deduce that the situation $u_1 \geq u_2$ or $v_1 \geq v_2$ in [0,1] is not possible.

Set $U := u_2 - u_1$ and $V := v_2 - v_1$. From the previous arguments, both U and V change sign in $(0,1)$. Moreover, we have the following proposition.

Proposition 10.4.2 *The functions U and V vanish only finitely many times in the interval $[0,1]$.*

Proof We write system (10.36) as

$$\begin{cases} \mathbf{W}''(x) + A(x)\mathbf{W}(x) = \mathbf{0} & \text{in } (0,1), \\ \mathbf{W}(0) = \mathbf{W}(1) = \mathbf{0}, \end{cases} \tag{10.38}$$

where $\mathbf{W} = (U,V)$ and $A(x) = (A_{ij}(x))_{1\leq i,j\leq 2}$ is a 2×2 matrix defined as

$$A_{11}(x) = -\alpha + \begin{cases} \dfrac{1}{v_2^q(x)}\cdot\dfrac{u_2^p(x)-u_1^p(x)}{u_2(x)-u_1(x)}, & u_1(x)\neq u_2(x), \\ p\dfrac{u_1^{p-1}(x)}{v_1^q(x)}, & u_1(x) = u_2(x), \end{cases}$$

$$A_{12}(x) = \begin{cases} -\dfrac{u_1^p(x)}{v_1^q(x)v_2^q(x)}\cdot\dfrac{v_2^q(x)-v_1^q(x)}{v_2(x)-v_1(x)}, & v_1(x)\neq v_2(x), \\ -q\dfrac{u_1^p(x)}{v_1^{q+1}(x)}, & v_1(x) = v_2(x), \end{cases}$$

$$A_{21}(x) = \begin{cases} \dfrac{1}{v_2^{q+\sigma}(x)}\cdot\dfrac{u_2^{p+\sigma}(x)-u_1^{p+\sigma}(x)}{u_2(x)-u_1(x)}, & u_1(x)\neq u_2(x), \\ (p+\sigma)\dfrac{u_1^{p+\sigma-1}(x)}{v_1^{q+\sigma}(x)}, & u_1(x) = u_2(x), \end{cases}$$

$$A_{22}(x) = -\beta - \begin{cases} \dfrac{u_1^{p+\sigma}(x)}{v_1^{q+\sigma}(x)v_2^{q+\sigma}(x)}\cdot\dfrac{v_2^{q+\sigma}(x)-v_1^{q+\sigma}(x)}{v_2(x)-v_1(x)}, & v_1(x)\neq v_2(x), \\ (q+\sigma)\dfrac{u_1^{p+\sigma}(x)}{v_1^{q+\sigma+1}(x)}, & v_1(x) = v_2(x). \end{cases}$$

Therefore, $A \in C(0,1)$ and $A_{12}(x) \neq 0$, $A_{21}(x) \neq 0$ for all $x \in (0,1)$. Moreover, $xA(x)$, $(1-x)A(x)$ are bounded in $L^\infty(0,1)$. Indeed, let us first notice that by (10.32) in Theorem 10.3.4, there exist c_1, $c_2 > 0$ such that

$$c_1 \leq \frac{u_i}{\min\{x,1-x\}}, \frac{v_i}{\min\{x,1-x\}} \leq c_2, \; (i=1,2) \quad \text{in } (0,1).$$

Then, by the mean value theorem we have

$$\begin{aligned} x|A_{12}(x)| &\leq qx\frac{u_1^p(x)}{v_1^q(x)v_2^q(x)}\max\{v_1^{q-1}(x), v_2^{q-1}(x)\} \\ &\leq qx^{p-q}\left(\frac{u_1(x)}{x}\right)^p \max\left\{\left(\frac{x}{v_1(x)}\right)^{q+1}, \left(\frac{x}{v_2(x)}\right)^{q+1}\right\} \\ &\leq cx^{p-q} \quad \text{for all } 0 < x \leq 1/2. \end{aligned}$$

We obtain similar estimates for xA_{11}, xA_{21}, and xA_{22}.

Basic to our approach are the following two general results.

Lemma 10.4.3 *Let $0 < a < b$ and $\mathbf{W} = (U, V) \in C^2[a, b] \times C^2[a, b]$ be such that $\mathbf{W} \not\equiv 0$ and*

$$\mathbf{W}''(x) + A(x)\mathbf{W}(x) = \mathbf{0} \quad \text{in} \quad [a, b], \tag{10.39}$$

where $A = (A_{ij})_{1 \le i,j \le 2}$ satisfies

(i) *$A_{ij} \in C[a, b]$, for all $1 \le i, j \le 2$;*
(ii) *$A_{12}(x) \neq 0$ and $A_{21}(x) \neq 0$, for all $x \in [a, b]$.*

Then neither U nor V can have infinitely many zeros in $[a, b]$.

Proof Supposing the contrary, let us assume that U has an infinite number of zeros in $[a, b]$. Hence, we can find a monotone sequence $(x_n)_{n \ge 1} \subset [a, b]$ such that $U(x_n) = 0$, for all $n \ge 1$. Without losing the generality, we may assume that $(x_n)_{n \ge 1}$ is decreasing and let $x_0 := \lim_{n \to \infty} x_n$. Then $U(x_0) = 0$ and, by Rolle's theorem, it follows that U' and U'' have infinitely many zeros in $(x_0, x_0 + \delta)$ for some $\delta > 0$ small enough. By continuity, we find $U'(x_0) = U''(x_0) = 0$. From (10.39) it follows that $A_{12}(x_0)V(x_0) = 0$. Because $A_{12}(x_0) \neq 0$, we have $V(x_0) = 0$.

If $V'(x_0) = 0$, then the uniqueness of the Cauchy problem

$$\begin{cases} \mathbf{W}''(x) + A(x)\mathbf{W}(x) = \mathbf{0} & \text{in} \quad (x_0, b], \\ \mathbf{W}(x_0) = \mathbf{W}'(x_0) = \mathbf{0} \end{cases}$$

implies $\mathbf{W} \equiv 0$ on $[x_0, b]$ and, similarly, $\mathbf{W} \equiv 0$ on $[a, x_0]$, which is a contradiction. Therefore, $V'(x_0) \neq 0$. Because A_{12} is bounded away from zero in $[a, b]$, we can find $\eta > 0$ small enough such that

$$|A_{12}(x)V(x)| > |A_{11}(x)U(x)| \qquad \text{for all } x \in (x_0, x_0 + \eta). \tag{10.40}$$

Combining now (10.39) with (10.40), we deduce that U'' has a constant sign on the interval $(x_0, x_0 + \eta)$. This clearly contradicts the fact that U vanishes infinitely many times in $(x_0, x_0 + \eta)$.

Similarly, if we assume that V has infinitely many zeros in $[a, b]$, we arrive at a contradiction using the fact that $A_{21}(x) \neq 0$, for all $x \in [a, b]$. This finishes the proof of our lemma. □

Lemma 10.4.4 *Let $\mathbf{W} \in C^2[0, a]$, $a > 0$, be such that*

$$\begin{cases} \mathbf{W}''(x) + A(x)\mathbf{W}(x) = \mathbf{0} & \text{in} \quad [0, a], \\ \mathbf{W}(0) = \mathbf{W}'(0) = \mathbf{0}. \end{cases}$$

Assume that $A = (A_{ij})_{1 \le i,j \le 2}$ satisfies

(i) *$A_{ij} \in C(0, a]$, for all $1 \le i, j \le 2$;*
(ii) *$xA_{ij}(x) \in L^\infty(0, a)$, for all $1 \le i, j \le 2$.*

Then $\mathbf{W} \equiv 0$ *in* $[0, a]$.

Proof Using the regularity of $\mathbf{W}$, a straightforward argument based on integration by parts leads us to

$$\mathbf{W}(x) = \int_0^x (t - x)A(t)\mathbf{W}(t)dt \quad \text{for all } 0 \le x \le a. \tag{10.41}$$

Define $B_{ij}(x) := xA_{ij}(x)$, $1 \le i, j \le 2$. Then $B_{ij} \in L^\infty(0, a)$. Set

$$M := \max_{1 \le i,j \le 2} \|B_{ij}\|_\infty, \quad k := \max_{1 \le i \le 2} \left\{ \frac{W_i(x)}{x} : 0 \le x \le a \right\}.$$

Notice that k is finite, because $\mathbf{W} \in C^1[0, a]$. From (10.41) we derive

$$\mathbf{W}(x) = \int_0^x (t - x)B(t)\frac{\mathbf{W}(t)}{t}dt \quad \text{for all } 0 \le t \le a. \tag{10.42}$$

It follows that

$$|\mathbf{W}(x)| \le Mk \int_0^x (x - t)dt = \frac{Mk}{2}x^2 \quad \text{for all } 0 \le x \le a. \tag{10.43}$$

Using (10.43) in (10.42) we obtain

$$|\mathbf{W}(x)| \le \frac{M^2k}{2} \int_0^x (x - t)tdt \le \frac{M^2k}{3!}x^3 \quad \text{for all } 0 \le x \le a.$$

By induction we deduce that for any $n \ge 1$ we have

$$|\mathbf{W}(x)| \le \frac{M^n k}{(n + 1)!}x^{n+1}, \quad \text{for all } 0 \le x \le a.$$

Now, we can pass to the limit with $n \to \infty$ in the last inequality to obtain the conclusion. □

Let us come back to the proof of Proposition 10.4.2. By Lemma 10.4.3, it follows that $\mathbf{W}$ vanishes for finitely many times in any interval $[a, b] \subset (0, 1)$. We next assume that U or V has infinitely many zeros near $x = 0$, the situation where U or V vanishes for infinitely many times near $x = 1$ being similar.

Without losing the generality, we assume that U has infinitely many zeros in a neighborhood of $x = 0$. Because $U \in C^2[0, 1]$, by Rolle's theorem it follows that both U' and U'' have infinitely many zeros near $x = 0$. As a consequence, we obtain $U'(0) = 0$—that is, $u_1'(0) = u_2'(0)$.

If $V'(0) = 0$, then $\mathbf{W}(0) = \mathbf{W}'(0) = \mathbf{0}$, and by Lemma 10.4.4 we deduce $\mathbf{W} \equiv \mathbf{0}$ in $[0, a]$ for any $0 < a < 1$. By continuity, we derive $\mathbf{W} \equiv \mathbf{0}$ in $[0, 1]$, which is a contradiction. Hence, $V'(0) \neq 0$. From (10.36) we have

$$u_1'' - \alpha u_1 + \frac{u_1^p}{v_1^q} + \rho(x) = 0 \quad \text{in } (0,1),$$

$$u_2'' - \alpha u_2 + \frac{u_2^p}{v_2^q} + \rho(x) = 0 \quad \text{in } (0,1).$$

Subtracting the previous relations we obtain

$$\begin{aligned} U''(x) &= \alpha U(x) + \frac{u_1^p(x)}{v_1^q(x)} - \frac{u_2^p(x)}{v_2^q(x)} \\ &= x^{p-q} \left\{ \alpha \frac{U(x)}{x^{p-q}} + \left(\frac{u_1(x)}{x} \right)^p \left(\frac{x}{v_1(x)} \right)^q - \left(\frac{u_2(x)}{x} \right)^p \left(\frac{x}{v_2(x)} \right)^q \right\}. \end{aligned}$$

On the other hand, because $0 \le p - q < 1$, $u_1'(0) = u_2'(0)$, and $v_1'(0) \neq v_2'(0)$, we have

$$\begin{aligned} &\lim_{x \searrow 0} \left\{ \alpha \frac{U(x)}{x^{p-q}} + \left(\frac{u_1(x)}{x} \right)^p \left(\frac{x}{v_1(x)} \right)^q - \left(\frac{u_2(x)}{x} \right)^p \left(\frac{x}{v_2(x)} \right)^q \right\} \\ &\quad = u_1'^p(0) \left(\frac{1}{v_1'^q(0)} - \frac{1}{v_1'^q(0)} \right) \neq 0. \end{aligned}$$

Therefore, U'' has a constant sign in a small neighborhood of $x = 0$, which contradicts the previous arguments.

This finishes the proof of Proposition 10.4.2. □

Let us complete the proof of Theorem 10.4.1. Set

$$\mathcal{I}^+ := \{x \in [0,1] : U(x) \ge 0\}, \quad \mathcal{I}^- := \{x \in [0,1] : U(x) \le 0\},$$

$$\mathcal{J}^+ := \{x \in [0,1] : V(x) \ge 0\}, \quad \mathcal{J}^- := \{x \in [0,1] : V(x) \le 0\}.$$

According to Proposition 10.4.2, the previous sets consist of finitely many disjoint closed intervals. Therefore, $\mathcal{I}^+ = \cup_{i=1}^m I_i^+$. For simplicity, let I^+ denote any interval I_i^+, and we will use similar notations for I^-, J^+, and J^- as well. We then have the following lemma.

Lemma 10.4.5 *For any intervals I^+, I^-, J^+, and J^- defined earlier, the following situations cannot occur:*

(i) $I^+ \subset J^+$.
(ii) $I^- \subset J^-$.
(iii) $J^+ \subset I^-$.
(iv) $J^- \subset I^+$.

Proof (i) Assume that $I^+ \subset J^+$. Because $v_2 \geq v_1$ in I^+, we deduce that inequality (10.37) holds in I^+. Using the fact that $u_2 = u_1$ on ∂I^+, by virtue of Lemma 1.3.17 we obtain $u_2 \leq u_1$ in I^+. Hence, $u_2 \equiv u_1$ in I^+, which contradicts Proposition 10.4.2. Similarly we prove statement (ii).

(iii) Assume that $J^+ \subset I^-$. Then $u_1^{p+\sigma}/v_1^{q+\sigma} \geq u_2^{p+\sigma}/v_2^{q+\sigma}$ in J^+. Notice that $V = v_2 - v_1$ satisfies

$$\begin{cases} -V'' + \beta V = \dfrac{u_2^{p+\sigma}}{v_2^{q+\sigma}} - \dfrac{u_1^{p+\sigma}}{v_1^{q+\sigma}} & \text{in } J^+, \\ V = 0 & \text{on } \partial J^+. \end{cases}$$

By the weak maximum principle, it follows that $V \leq 0$ in J^+—that is, $v_2 \leq v_1$ in J^+. This yields $v_2 \equiv v_1$ in J^+, which again contradicts Proposition 10.4.2. The proof of (iv) follows in the same manner. □

With no loss of generality, we may assume that $U > 0$ in a neighborhood of $x = 0$. We may also assume that $U \geq 0$ on $[0, a_1]$, $[a_2, a_3]$, ..., $[a_{2n}, a_{2n+1}]$, and $U \leq 0$ on $[a_1, a_2]$, $[a_3, a_4]$, ..., where $a_{2n+1} = 1$ or $a_{2n} = 1$ depending on the sign of U on the last interval on which U does not vanish.

Let us analyze the following distinct situations.

CASE 1: *$V > 0$ in a neighborhood of $x = 0$.* If V does not change the sign in $(0, a_1)$, then we arrive at a contradiction by Lemma 10.4.5 (i). Furthermore, if V changes sign more than once in $(0, a_1)$, then we again find a contradiction, by Lemma 10.4.5 (iv). Therefore V changes sign exactly once in $(0, a_1)$—that is, $V(a_1) < 0$.

CASE 2: *$V < 0$ in a neighborhood of $x = 0$.* If V changes sign in $(0, a_1]$, then we contradict Lemma 10.4.5 (iv). Hence, V has a constant sign on $(0, a_1]$—that is, $V(a_1) < 0$.

In both cases we have obtained $V(a_1) < 0$. A similar argument yields $V(a_2) > 0$ and, more generally, $V(a_k) < 0$ if k is odd and $V(a_k) > 0$ for all even values of k.

If the last interval is $[a_{2n}, 1]$, then $V(a_{2n}) > 0$. Using Lemma 10.4.5 (i), it follows that V vanishes on $(a_{2n}, 1)$, but this will contradict case Lemma 10.4.5 (iv).

The case in which the last interval where $U \leq 0$ is $[a_{2n+1}, 1]$, follows in the same manner.

This finishes the proof of Theorem 10.4.1. □

As a consequence of Theorem 10.3.1, solution (u, v) of system (10.31) can be approximated by the solutions of $(\mathcal{S})_\varepsilon$. Furthermore, the shooting method combined with the Broyden method (to avoid the derivatives) are suitable to approximate the solution of (10.31) numerically. We have considered $\alpha = 1$, $\beta = 0.5$, $p = q = 1$, $\varepsilon = 10^{-2}$, and $\rho(x) = \varphi_1(x) = \sin(\pi x)$. In the following figures, we have plotted solution (u, v) of (S_ε) for $\sigma = 0$ (Figure 10.1) and $\sigma = 2$ (Figure 10.2), respectively.

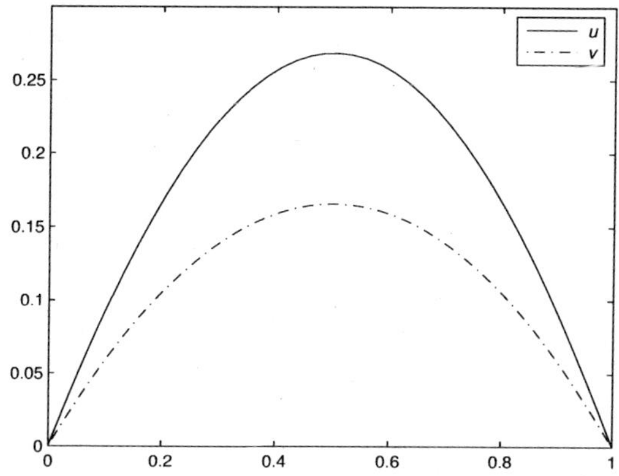

FIGURE 10.1. Solution (u, v) of system (S_ε) for $\sigma = 0$.

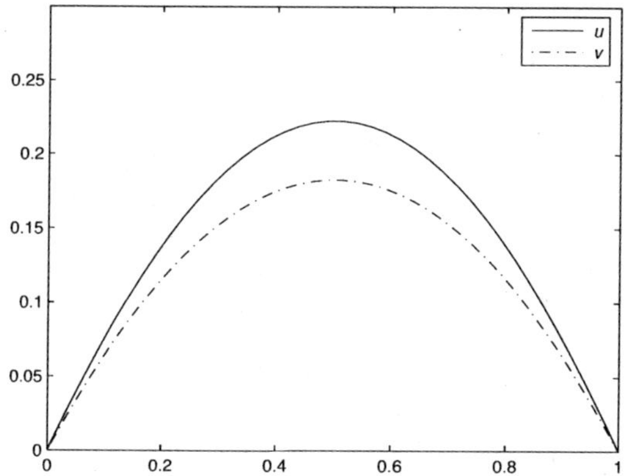

FIGURE 10.2. Solution (u, v) of system (S_ε) for $\sigma = 2$.

10.5 Comments and historical notes

Activator–inhibitor systems account for many important types of pattern formation and morphogenesis that rely on cell differentiation. A central question is how the cells, which carry identical genetic code, become different from each other. Spontaneous pattern formation is also common in inorganic systems. For instance, large sand dunes are formed despite the fact that the wind permanently redistributes the sand; sharply contoured and branching river systems (which are in fact quite similar to the branching patterns of a nerve) are formed as a result to erosion, despite the fact that the rain falls more or less homogeneously over the ground.

The main feature of these pattern-forming systems is that a deviation from homogeneity has a strong positive feedback on its further increase. Pattern formation requires, in addition, a longer ranging confinement of the locally self-enhancing process.

Turing [189] suggested in 1952 that, under certain conditions, chemicals can react and diffuse in such a way to produce steady-state heterogeneous spatial patterns of chemical or morphogen concentration.

In 1972 Gierer and Meinhardt [94] proposed a mathematical model for pattern formation of spatial tissue structures in morphogenesis, a biological phenomenon discovered by Trembley [188] in 1744. The influential activator–inhibitor mechanism suggested by Gierer and Meinhardt [94], [140] may be written as

$$\begin{cases} u_t = d_1 \Delta u - \alpha u + c\rho \dfrac{u^p}{v^q} + \rho_0 \rho & \text{in } \Omega \times (0,T), \\ v_t = d_2 \Delta v - \beta v + c'\rho' \dfrac{u^r}{v^s} & \text{in } \Omega \times (0,T), \\ \dfrac{\partial u}{\partial \nu} = 0, \ \dfrac{\partial u}{\partial \nu} = 0 & \text{on } \partial\Omega \end{cases} \tag{10.44}$$

in a smooth bounded region $\Omega \subset \mathbb{R}^N$ ($N \geq 1$). In system (10.44), the unknown u represents the concentration of a short-range autocatalytic substance (that is, activator), and v is the concentration of its long-range antagonist (that is, inhibitor). Also ρ and ρ' stand for the source distributions of the activator and inhibitor, respectively; d_1, d_2 are diffusion coefficients with $d_1 << d_2$; and α, β, c, c', ρ_0 are positive constants. The exponents p, q, r, $s > 0$ satisfy the relation $qr > (p-1)(s+1) > 0$.

System (10.44) is of a reaction–diffusion type and involves the determination of an activator and inhibitor concentration field. The basic idea behind Gierer–Meinhardt's system is the so-called *diffusion–driven instability*, originally the result of the work by Turing [189], which asserts that different diffusion rates could lead to nonhomogeneous distributions of the reactants. Also, the model introduced in [94] takes into account the classification between the concentration of activators and inhibitors on the one hand, and the densities of their sources on the other.

The model introduced by Gierer and Meinhardt has been used with satisfactory quantitative results for modeling the head regeneration process of hydra, an animal a few millimeters in length, consisting of 100,000 cells of about 15 different types and having a polar structure.

A similar model was proposed by Meinhardt [139] in 1976 that is applied both in geomorphology and in hydrology. For instance, the formation of leaf veins, blood vessels, and the fibers of the nervous system, but also the channels in which streams flow, are described by Meinhardt's model (see also Willgoose, Bras, and Rodriguez–Iturbe [196]). We refer to the recent monograph by Murray [145] for further details and references.

It has been shown that dynamics of system (10.44) exhibit various interesting behaviors such as periodic solutions, unbounded oscillating global solutions, and finite time blow-up solutions. We refer the interested reader to the paper by Ni, Suzuki, and Takagi [150] for the entire description of dynamics concerning system (10.44). Also a global existence results for a more general system than (10.44) is given in the recent work by Jiang [109].

Many recent works have been devoted to the study of the steady-states solutions of (10.44)—that is, solutions of the stationary system

$$\begin{cases} d_1\Delta u - \alpha u + c\rho\dfrac{u^p}{v^q} + \rho_0\rho = 0 & \text{in } \Omega, \\ d_2\Delta v - \beta v + c'\rho'\dfrac{u^r}{v^s} = 0 & \text{in } \Omega, \\ \dfrac{\partial u}{\partial \nu} = 0,\ \dfrac{\partial u}{\partial \nu} = 0 & \text{on } \partial\Omega. \end{cases} \tag{10.45}$$

System (10.45) is quite difficult to solve, because it does not have a variational structure. An attempt in solving (10.45) is to consider the *shadow system*, an idea due to Keener [116]. More exactly, dividing the second equation of (10.45) by d_2 and then letting $d_2 \to \infty$, we reduce system (10.45) to a single equation. The nonconstant solutions of such equation present interior or boundary peaks or spikes—that is, they exhibit a *point concentration* phenomenon. Among the great number of works in this direction, we refer the reader to [151], [192], [193], [194] and the references therein, as well as to the survey papers of Ni [148], [149]. For the study of instability of solutions to (10.45), we also mention here the works of Miyamoto [144] and Yanagida [198].

In the case $\Omega = \mathbb{R}^N$, $N = 1, 2$, it has been shown in del Pino et al. [63], [64] that there exist ground-state solutions of (10.1) with single or multiple bumps in the activator that, after a rescaling of u, approach a universal profile.

The study of system $(\mathcal{S})$ presented in this chapter is a result of the work by Ghergu and Rădulescu [91], [92] and is motivated by some questions addressed in recent research papers by Choi and McKenna [43], [44] and Kim [119], [120] concerning existence and nonexistence or even uniqueness of the classical solutions for model system (10.1). The existence of classical solutions for the general system $(\mathcal{S})$ was obtained in this chapter under the additional hypothesis $\beta \le \alpha$. In fact, this assumption is quite natural if we look at the steady-state system (10.45). We have only to divide the first equation by d_1, the second one by d_2, and then take into account the fact that $d_1 << d_2$.

In [43], [119] it is assumed that the activator and inhibitor have common sources and the approach relies on Schauder's fixed point theorem through a *decouplization* of the system. More precisely, if $p = r$ and $q = s$, then subtracting the two equations of (10.1) we obtain the following equivalent form:

$$\begin{cases} \Delta w - \alpha w + (\beta - \alpha) w v + \rho(x) = 0, & \text{in } \Omega, \\ \Delta v - \beta v + \dfrac{(v+w)^p}{v^s} = 0 & \text{in } \Omega, \\ v = w = 0 & \text{on } \partial\Omega. \end{cases} \tag{10.46}$$

Because the first equation in (10.46) is linear, we can easily obtain a priori estimates in order to control the map whose fixed points are solutions of (10.1).

In Choi and McKenna [44] it is obtained the existence of radially symmetric solutions of $(\mathcal{S})$ in the case $\Omega = (0,1)$ or $\Omega = B_1 \subset \mathbb{R}^2$ and $p = r > 1$, $q = 1$, $s = 0$. In [44] a priori bounds are obtained via sharp estimates of the associated Green's function. The method we used in proving the uniqueness of the solution to system (10.36) is a result of the work by Choi and McKenna [43], where it was initially obtained for $p = q = r = s = 1$ and $\rho \equiv 0$.

APPENDIX A

SPECTRAL THEORY FOR DIFFERENTIAL OPERATORS

> All intelligent thoughts have already been thought; what is necessary is only to try to think them again.
>
> Johann Wolfgang von Goethe (1749–1832)

A.1 Eigenvalues and eigenfunctions for the Laplace operator

Recall first the following theorem.

Theorem A.1.1 (Poincaré) *Let $\Omega \subset \mathbb{R}^N$ be a bounded domain (or, more generally, such that one of its projections to the coordinate axes is bounded). Then there exists a constant $c > 0$ such that for all $u \in H_0^1(\Omega)$,*

$$\|u\|_{H_0^1(\Omega)} \leq C\|\nabla u\|_{L^2(\Omega)}. \tag{A.1}$$

Consider the operator $(-\Delta)^{-1} : L^2(\Omega) \to H_0^1(\Omega)$ defined as follows: For all $f \in L^2(\Omega)$, $(-\Delta)^{-1}f$ is the unique $u \in H_0^1(\Omega)$ such that

$$\begin{cases} -\Delta u = f & \text{in } \Omega, \\ u = 0 & \text{on } \partial\Omega. \end{cases} \tag{A.2}$$

Remark that the operator $(-\Delta)^{-1}$ is continuous. Indeed, by (A.2) we have

$$\|u\|_{H_0^1(\Omega)}^2 \leq \int_\Omega fu dx \leq \|f\|_{L^2(\Omega)} \|u\|_{L^2(\Omega)}.$$

By Poincaré's inequality, we have $\|u\|_{H_0^1(\Omega)}^2 \leq C\|f\|_{L^2(\Omega)}\|\nabla u\|_{L^2(\Omega)}$, where C is the constant from (A.1). Then $\|u\|_{H_0^1(\Omega)} \leq c\|f\|_{L^2(\Omega)}$ for some $c > 0$ not dependent on f and u.

Theorem A.1.2 *The following properties hold true:*

(i) *The eigenvalue set of the operator $-\Delta$ subject to the Dirichlet boundary condition consists of a sequence*

$$0 < \lambda_1 < \lambda_2 < \cdots < \lambda_n < \cdots$$

and $\lambda_n \to \infty$ as $n \to \infty$.

(ii) *The associated eigenfunctions belong to* $C^2(\overline{\Omega})$.

Proof Let $i : H_0^1(\Omega) \to L^2(\Omega)$ be the canonical injection. By the Rellich–Kondrachov embedding theorem, i is a compact operator. Set $T = -\Delta \circ i : H_0^1(\Omega) \to H_0^1(\Omega)$. Then T is a compact, self-adjoint, and positive operator. The first part follows from the Riesz theory for compact operators in Hilbert spaces. The last part is a direct consequence of the Schauder and Hölder regularity theories for elliptic equations. □

Taking into account that the first eigenvalue φ_1 belongs to $C^2(\overline{\Omega})$, we deduce the following asymptotic behavior of the first eigenfunction near the boundary.

Corollary A.1.3 *The following properties hold true:*

(i) *There exist positive constants* c_1 *and* c_2 *such that* $c_1\, d(x) \leq \varphi_1(x) \leq c_2\, d(x)$ *in* Ω.

(ii) $\varphi_1^k \in L^1(\Omega)$ *if and only if* $k > -1$.

We point out that the existence of a countable family of eigenvalues for the Laplace operator under the Dirichlet boundary condition was established in 1894 by Poincaré [165]. This pioneering result is the beginning of the spectral theory, which has played a crucial role in the development of theoretical physics and functional analysis. Linear eigenvalue problems with nonconstant potential of the type

$$\begin{cases} -\Delta u = \lambda V(x) u & \text{in } \Omega, \\ u = 0 & \text{on } \partial\Omega \end{cases} \tag{A.3}$$

were studied by Bocher [28], Minakshisundaran and Pleijel [141], Pleijel [162], and Hess and Kato [107]. For instance, Minakshisundaran and Pleijel [141], [162] were concerned with the case when V is a variable potential such that $V \in L^\infty(\Omega)$, $V \geq 0$ in Ω, and $V > 0$ in $\Omega_0 \subset \Omega$ with $|\Omega_0| > 0$.

A.2 Krein–Rutman theorem

Definition A.2.1 *Let* X *be a real Banach space. A nonempty set* $\mathcal{C}$ *is said to be a cone if* $\mathcal{C}$ *is a closed convex set and* $\mathcal{C} \cap (-\mathcal{C}) = \{0\}$.

Definition A.2.2 *A real Banach space* X *is called ordered if there exists a cone* $\mathcal{C}$ *and a partial order relation "* $\leq$ *" such that* $x \leq y$ *if and only if* $x - y \in \mathcal{C}$.

A very useful tool in the spectral theory of differential operators is the following result.

Theorem A.2.3 (Krein–Rutman) *Let* X *be a real Banach space with an order cone* $\mathcal{C}$ *having a nonempty interior. Assume that* $T : X \to X$ *is a linear compact operator such that* $T(\mathcal{C}) \subset \mathcal{C}$. *Then the following properties hold true:*

(i) T *has exactly one eigenfunction* $x \in \mathcal{C}$. *The corresponding eigenvalue equals the spectral radius* $r(T)$ *of* T.

(ii) *For all complex numbers λ in the spectrum of T that are different from $r(T)$ we have $|\lambda| < r(T)$.*

Instead of using the Krein–Rutman theorem, we can establish by means of elementary arguments that the least eigenvalue of the Laplace operator in $H_0^1(\Omega)$ is simple. For this purpose, because eigenfunctions corresponding to higher eigenvalues λ_k $(k \geq 2)$ do necessarily change sign in Ω, it is enough to show that the solutions of the problem

$$\begin{cases} -\Delta u = \lambda u & \text{in } \Omega, \\ u > 0 & \text{in } \Omega, \\ u = 0 & \text{on } \partial\Omega \end{cases} \tag{A.4}$$

are unique, up to a multiplicative constant. For this purpose, we apply an idea introduced by Benguria, Brezis, and Lieb [21], which relies on the following facts:

(i) Any solution of problem (A.4) is a minimizer of the Dirichlet energy functional

$$E(v) := \int_\Omega |\nabla v|^2 dx$$

on the manifold

$$M := \left\{ v \in H_0^1(\Omega) : v > 0 \text{ in } \Omega \text{ and } \int_\Omega v^2 dx = 1 \right\}.$$

(ii) The energy functional $E(v)$ is convex in v^2 on the cone

$$\mathcal{C} := \left\{ v \in H_0^1(\Omega) : v > 0 \text{ in } \Omega \right\}.$$

Assuming that u_1 and u_2 are arbitrary solutions of (A.4), then both u_1 and u_2 minimize E on M. Set $v := (u_1^2+u_2^2)/2$ and $w := \sqrt{v}$. Thus, by (ii), we deduce that $w \in M$. A straightforward computation yields

$$\nabla w = \frac{u_1 \nabla u_1 + u_2 \nabla u_2}{2\sqrt{v}} \qquad \text{in } \Omega.$$

Hence,

$$\begin{aligned} E(w) &= \int_\Omega |\nabla w|^2 dx = \int_\Omega \frac{1}{2v} \left| u_1 \nabla u_1 + u_2 \nabla u_2 \right|^2 dx \\ &= \int_\Omega \frac{u_1^2+u_2^2}{2} \left| \frac{u_1^2}{u_1^2+u_2^2} \frac{\nabla u_1}{u_1} + \frac{u_2^2}{u_1^2+u_2^2} \frac{\nabla u_2}{u_2} \right|^2 dx \\ &\leq \int_\Omega \frac{u_1^2+u_2^2}{2} \left(\frac{u_1^2}{u_1^2+u_2^2} \frac{|\nabla u_1|^2}{u_1^2} + \frac{u_2^2}{u_1^2+u_2^2} \frac{|\nabla u_2|^2}{u_2^2} \right) dx \\ &= \frac{1}{2} \int_\Omega \left(|\nabla u_1|^2 + |\nabla u_2|^2 \right) dx \\ &= \frac{E(u_1) + E(u_2)}{2}. \end{aligned}$$

Because u_1 and u_2 are minimizers of E, the previous relation shows that w is a minimizer of E, too. Thus, $E(w) = E(u_1) = E(u_2)$. Hence, $\nabla u_1/u_1 = \nabla u_2/u_2$ in Ω. Therefore, $\nabla(u_1/u_2) = 0$ in Ω, which implies that u_1/u_2 is constant.

APPENDIX B

IMPLICIT FUNCTION THEOREM

> Mediocrity knows nothing higher than itself, but talent instantly recognizes genius.
>
> Arthur Conan Doyle (1859–1930)

The germs of the implicit function theorem appeared in the works of Newton, Leibniz, and Lagrange. In its simplest form, this theorem involves an equation of the form $F(u,\lambda)=0$ (we can assume without loss of generality that $F(0,0)=0$), where λ is a parameter and u is an unknown function. The basic assumption is that $F_u(u,\lambda)$ is invertible for u and λ "close" to their respective origins, but with some loss of derivatives when computing the inverse.

In his celebrated proof of the result that any compact Riemannian manifold may be isometrically embedded in some Euclidean space, Nash [147] developed a deep extension of the implicit function theorem. His ideas have been extended by various people to a technique that is now called the Nash–Moser theory. We recall that John F. Nash Jr. is an outstanding scientist who received the 1994 Nobel Prize in economics for his pioneering analysis of equilibria in the theory of noncooperative games.

Throughout these lecture notes we have used the following "standard" version of the implicit function theorem. This property is an important tool in nonlinear analysis and, roughly speaking, it asserts that the behavior of a nonlinear system is qualitatively determined by its linearized system around the zeros of the nonlinear system.

Theorem B.1 *Let X and Y be real Banach spaces and let $(u_0,\lambda_0)\in X\times\mathbb{R}$. Consider a mapping $F=F(u,\lambda):X\times\mathbb{R}\to Y$ of class C^1 such that*

(i) *$F(u_0,\lambda_0)=0$;*

(ii) *the linear mapping $F_u(u_0,\lambda_0):X\to Y$ is bijective.*

Then there exists a neighborhood U_0 of u_0 and a neighborhood V_0 of λ_0 such that for every $\lambda\in V_0$, there is a unique element $u(\lambda)\in U_0$ so that $F(u(\lambda),\lambda)=0$. Moreover, the mapping $V_0\ni\lambda\longmapsto u(\lambda)$ is of class C^1.

Proof Consider the mapping $\Phi(u,\lambda):X\times\mathbb{R}\to Y\times\mathbb{R}$ defined by $\Phi(u,\lambda)=(F(u,\lambda),\lambda)$. By our hypotheses, Φ is a mapping of class C^1. We apply to Φ the inverse function theorem. To conclude the proof, it remains to verify that $\Phi'(u_0,\lambda_0):X\times\mathbb{R}\to Y\times\mathbb{R}$ is bijective. Indeed, we have

$$\Phi(u_0 + tu, \lambda_0 + t\lambda) = (F(u_0 + tu, \lambda_0 + t\lambda), \lambda_0 + t\lambda) =$$
$$(F(u_0, \lambda_0) + F_u(u_0, \lambda_0) \cdot (tu) + F_\lambda(u_0, \lambda_0) \cdot (t\lambda) + o(1), \lambda_0 + t\lambda).$$

It follows that

$$F'(u_0, \lambda_0) = \begin{pmatrix} F_u(u_0, \lambda_0) & F_\lambda(u_0, \lambda_0) \\ 0 & \mathrm{I} \end{pmatrix},$$

which is a bijective operator, by our hypotheses. Thus, by the inverse function theorem, there exist a neighborhood $\mathcal{U}$ of the point (u_0, λ_0) and a neighborhood $\mathcal{V}$ of $(0, \lambda)$ such that the equation

$$\Phi(u, \lambda) = (f, \lambda_0)$$

has a unique solution, for every $(f, \lambda) \in \mathcal{V}$. Now, it is sufficient to take here $f = 0$ and our conclusion follows. □

With a similar proof we can establish the following global version of the implicit function theorem.

Theorem B.2 *Assume $F : X \times \mathbb{R} \to Y$ is a function of class C^1 such that*

(i) $F(0, 0) = 0$;
(ii) *the linear mapping $F_u(0, 0) : X \to Y$ is bijective.*

Then there exist an open neighborhood I of 0 and a mapping $I \ni \lambda \longmapsto u(\lambda)$ of class C^1 such that $u(0) = 0$ and $F(u(\lambda), \lambda) = 0$.

The following result has been of particular importance in our arguments in the study of bifurcation problems.

Theorem B.3 *Assume the same hypotheses on F as in Theorem B.2. Then there exists an open maximal interval I containing the origin and there exists a unique mapping $I \ni \lambda \longmapsto u(\lambda)$ of class C^1 such that the following hold:*

(i) $F(u(\lambda), \lambda) = 0$ *for every* $\lambda \in I$.
(ii) *The linear mapping $F_u(u(\lambda), \lambda)$ is bijective, for any $\lambda \in I$.*
(iii) $u(0) = 0$.

Proof Let u_1, u_2 be solutions and consider the corresponding open intervals I_1 and I_2 on which these solutions exist, respectively. It follows that $u_1(0) = u_2(0) = 0$ and

$$F(u_1(\lambda), \lambda) = 0 \quad \text{for every } \lambda \in I_1,$$
$$F(u_2(\lambda), \lambda) = 0 \quad \text{for every } \lambda \in I_2.$$

Moreover, the mappings $F_u(u_1(\lambda), \lambda)$ and $F_u(u_2(\lambda), \lambda)$ are bijective on I_1, respectively I_2. But, for λ sufficiently close to 0, we have $u_1(\lambda) = u_2(\lambda)$. We wish to show that we have global uniqueness. For this purpose, let

$$I := \{\lambda \in I_1 \cap I_2 : u_1(\lambda) = u_2(\lambda)\}.$$

Our aim is to show that $I = I_1 \cap I_2$. We first observe that $0 \in I$, so $I \neq \emptyset$. A standard argument implies that I is closed in $I_1 \cap I_2$. To prove that $I = I_1 \cap I_2$, it is sufficient to argue that I is an open set in $I_1 \cap I_2$. The proof of this statement follows by applying Theorem B.1 for λ instead of 0. Thus, $I = I_1 \cap I_2$.

Next, to justify the existence of a maximal interval I, we consider the C^1 curves $u_n(\lambda)$ defined on the corresponding open intervals I_n, such that $0 \in I_n$, $u_n(0) = u_0$, $F(u_n(\lambda), \lambda) = 0$ and $F_u(u_n(\lambda), \lambda)$ is an isomorphism, for any $\lambda \in I_n$. A standard argument enables us to construct a maximal solution on the set $\bigcup_n I_n$. This concludes the proof. □

We refer to a recent book [121] for various applications of the implicit function theorem.

APPENDIX C

EKELAND'S VARIATIONAL PRINCIPLE

> It is not enough that we do our best; sometimes we have to do what is required.
>
> Sir Winston Churchill (1874–1965)

C.1 Minimization of weak lower semicontinuous coercive functionals

An important topology in which many arguments can be carried out in reflexive Banach spaces (such as $L^p(\Omega)$ or $W^{k,p}(\Omega)$ for $1 < p < \infty$) is the weak topology, in which the unit ball is weakly compact. The weak lower semicontinuity is a major tool in reflexive Banach spaces. We recall that if E is a Banach space, then a functional $\Phi : E \to \mathbb{R}$ is said to be weak lower semicontinuous if $\Phi(u) \leq \liminf_{n\to\infty} \Phi(u_n)$ for any sequence (u_n) in E converging weakly to u. A basic sufficient condition for a continuous functional $\Phi : E \to \mathbb{R}$ to be weakly lower semicontinuous is that Φ is convex.

One of the deepest consequences is that weak lower semicontinuous coercive functionals attain their infimum on some suitable sets, as stated in the following basic result.

Theorem C.1.1 *Let E be a reflexive Banach space, $M \subset E$ be a weakly closed subset of E, and $\Phi : M \to \mathbb{R}$ such that*

(i) *Φ is coercive on M with respect to E—that is*

$$\Phi(u) \to \infty \qquad \text{as } \|u\| \to \infty.$$

(ii) *Φ is weakly lower semicontinuous on M.*

Then Φ is bounded from below and attains its infimum on M.

The proof of Theorem C.1.1 uses elementary arguments based on the definition of the weak lower semicontinuity. We refer to [27] for complete details and further comments.

C.2 Ekeland's variational principle

Ekeland's variational principle [75] was established in 1974 and is the nonlinear version of the Bishop–Phelps theorem [161], with its main feature of how to use the norm completeness and a partial ordering to obtain a point where a linear functional achieves its supremum on a closed bounded convex set. A major consequence of Ekeland's variational principle is that even if it is not always possible to minimize a nonnegative C^1 functional Φ on a Banach space; however, there is always a minimizing sequence $(u_n)_{n\geq 1}$ such that $\Phi'(u_n) \to 0$ as $n \to \infty$. We first state the original version of Ekeland's variational principle, which is valid in the general framework of complete metric spaces.

Theorem C.2.1 (Ekeland's variational principle) *Let (M, d) be a complete metric space and assume that $\Phi : M \to (-\infty, \infty]$, $\Phi \not\equiv \infty$, is a lower semicontinuous functional that is bounded from below.*

Then, for every $\varepsilon > 0$ and for any $z_0 \in M$, there exists $z \in M$ such that

(i) $\Phi(z) \leq \Phi(z_0) - \varepsilon\, d(z, z_0)$;
(ii) $\Phi(x) \geq \Phi(z) - \varepsilon\, d(x, z)$, *for any $x \in M$.*

Proof We may assume without loss of generality that $\varepsilon = 1$. Define the following binary relation on M:

$$y \leq x \qquad \text{if and only if} \qquad \Phi(y) - \Phi(x) + d(x, y) \leq 0\,.$$

Then "$\leq$" is a partial order relation—that is,
(a) $x \leq x$, for any $x \in M$;
(b) if $x \leq y$ and $y \leq x$ then $x = y$;
(c) if $x \leq y$ and $y \leq z$ then $x \leq z$.

For arbitrary $x \in M$, set

$$S(x) := \{y \in M : y \leq x\}.$$

Let $(\varepsilon_n)_{n\geq 1}$ be a sequence of positive numbers such that $\varepsilon_n \to 0$ and fix $z_0 \in M$. For any $n \geq 0$, let $z_{n+1} \in S(z_n)$ be such that

$$\Phi(z_{n+1}) \leq \inf_{S(z_n)} \Phi + \varepsilon_{n+1}.$$

The existence of z_{n+1} follows from the definition of $S(x)$. We prove that the sequence $(z_n)_{n\geq 1}$ converges to some element z, which satisfies (i) and (ii).

Let us first remark that $S(y) \subset S(x)$, provided that $y \leq x$. Hence, $S(z_{n+1}) \subset S(z_n)$. It follows that for any $n \geq 0$,

$$\Phi(z_{n+1}) - \Phi(z_n) + d(z_n, z_{n+1}) \leq 0,$$

which implies $\Phi(z_{n+1}) \leq \Phi(z_n)$. Because Φ is bounded from below, we deduce that the sequence $(\Phi(z_n))_{n\geq 1}$ converges.

We prove in what follows that $(z_n)_{n\geq 1}$ is a Cauchy sequence. Indeed, for any n and p we have

$$\Phi(z_{n+p}) - \Phi(z_n) + d(z_{n+p}, z_n) \leq 0. \tag{C.1}$$

Therefore,

$$d(z_{n+p}, z_n) \leq \Phi(z_n) - \Phi(z_{n+p}) \to 0 \quad \text{as } n \to \infty,$$

which shows that $(z_n)_{n\geq 1}$ is a Cauchy sequence, so it converges to some $z \in M$. Now, taking $n = 0$ in (C.1), we find

$$\Phi(z_p) - \Phi(z_0) + d(z_p, z_0) \leq 0.$$

So, as $p \to \infty$, we find (i).

To prove (ii), let us choose arbitrarily $x \in M$. We distinguish the following situations.

CASE 1: $x \in S(z_n)$, for any $n \geq 0$. It follows that $\Phi(z_{n+1}) \leq \Phi(x) + \varepsilon_{n+1}$, which implies that $\Phi(z) \leq \Phi(x)$.

CASE 2: There exists an integer $N \geq 1$ such that $x \notin S(z_n)$, for any $n \geq N$ or, equivalently,

$$\Phi(x) - \Phi(z_n) + d(x, z_n) > 0 \quad \text{for every } n \geq N.$$

Passing at the limit in this inequality as $n \to \infty$ we find (ii). □

The following consequence of Ekeland's variational principle has been of particular interest in our arguments. Roughly speaking, this property establishes the existence of *almost critical points* for bounded from below C^1 functionals defined on Banach spaces. In order words, Ekeland's variational principle can be viewed as a generalization of the Fermat theorem, which establishes that interior extrema points of a smooth functional are, necessarily, critical points of this functional.

Corollary C.2.2 *Let E be a Banach space and let $\Phi : E \to \mathbb{R}$ be a C^1 functional that is bounded from below. Then, for any $\varepsilon > 0$, there exists $z \in E$ such that*

$$\Phi(z) \leq \inf_E \Phi + \varepsilon \qquad \textit{and} \qquad \|\Phi'(z)\|_{E^\star} \leq \varepsilon.$$

Proof The first part of the conclusion follows directly from Theorem C.2.1. For the second part we have

$$\|\Phi'(z)\|_{E^\star} = \sup_{\|u\|=1} \langle \Phi'(z), u \rangle.$$

But,

$$\langle \Phi'(z), u \rangle = \lim_{\delta \to 0} \frac{\Phi(z + \delta u) - \Phi(z)}{\delta \|u\|}.$$

So, by Ekeland's variational principle,

$$\langle \Phi'(z), u \rangle \geq -\varepsilon.$$

Replacing now u with $-u$ we find

$$\langle \Phi'(z), u \rangle \leq \varepsilon,$$

which concludes our proof. □

Sullivan [185] observed that Ekeland's variational principle characterizes complete metric spaces in the following sense.

Theorem C.2.3 *Let* (M, d) *be a metric space. Then* M *is complete if and only if the following holds: For every application* $\Phi : M \to (-\infty, \infty]$, $\Phi \not\equiv \infty$, *which is bounded from below, and for every* $\varepsilon > 0$, *there exists* $z_\varepsilon \in M$ *such that*

(i) $\Phi(z_\varepsilon) \leq \inf_M \Phi + \varepsilon$;
(ii) $\Phi(x) > \Phi(z_\varepsilon) - \varepsilon\, d(x, z_\varepsilon)$, *for any* $x \in M \setminus \{z_\varepsilon\}$.

APPENDIX D

MOUNTAIN PASS THEOREM

> There are more things in heaven
> and earth, Horatio, than have
> been dreamt of in our philosophy.
>
> ---
>
> William Shakespeare (1564–1616)
> *Hamlet, Prince of Denmark.* Act I

D.1 Ambrosetti–Rabinowitz theorem

The mountain pass theorem was established by Ambrosetti and Rabinowitz in [8]. Their original proof relies on some deep deformation techniques developed by Palais and Smale [157], [158], who put the main ideas of the Morse theory into the framework of differential topology on infinite dimensional manifolds. In this way, Palais and Smale replaced the finite dimensionality assumption with an appropriate compactness hypothesis, which is stated in the following.

Definition D.1.1 (Palais–Smale condition) *Let E be a real Banach space. A functional $J : E \to \mathbb{R}$ of class C^1 satisfies the Palais–Smale condition if any sequence $(u_n)_{n\geq 1}$ in E is relatively compact, provided*

$$\sup_{n\geq 1} |J(u_n)| < \infty \quad \textit{and} \quad \lim_{n\to\infty} \|J'(u_n)\|_{E^\star} = 0\,. \tag{D.1}$$

A simple consequence of Ekeland's variational principle, in connection with the Palais–Smale condition is stated next.

Proposition D.1.2 *Let E be a real Banach space and assume that $\Phi : E \to \mathbb{R}$ is a functional of class C^1 that is bounded from below, and satisfies the Palais–Smale condition. Then the following properties hold true:*

(i) *Φ is coercive.*

(ii) *Any minimizing sequence of Φ has a convergent subsequence.*

The mountain pass theorem is a powerful tool for proving the existence of critical points of energy functionals, hence of weak solutions to wide classes of nonlinear problems. The precise statement of this result is the following.

Theorem D.1.3 (Mountain pass theorem) *Let E be a real Banach space and assume that $J : E \to \mathbb{R}$ is a C^1 functional that satisfies the following conditions: There exist positive constants α and R such that*

(i) *$J(0) = 0$ and $J(v) \geq \alpha$ for all $v \in E$ with $\|v\| = R$;*

(ii) $J(v_0) \leq 0$, *for some* $v_0 \in E$ *with* $\|v_0\| > R$.

Set

$$\Gamma := \{p \in C([0,1]; E) : p(0) = 0 \text{ and } p(1) = v_0\}$$

and

$$c := \inf_{p \in \Gamma} \max_{t \in [0,1]} J(p(t)).$$

Then there exists a sequence $(u_n)_{n \geq 1}$ *in* E *such that* $J(u_n) \to c$ *and* $J'(u_n) \to 0$ *as* $n \to \infty$. *Moreover, if* J *satisfies the Palais–Smale condition, then* c *is a nontrivial critical value of* J*—that is, there exists* $u \in E$ *such that* $J(u) = c$ *and* $J'(u) = 0$.

The elementary meaning of the mountain pass theorem is the following. The set Γ denotes the family of all continuous paths that join the "villages" 0 and v_0. Hypotheses (i) and (ii) imply $c \geq \alpha > \max\{J(0), J(v_0)\}$. In fact, in the proof of the mountain pass theorem, it is used only the basic assumption $c > \max\{J(0), J(v_0)\}$. We also observe that the geometric conditions of the mountain pass theorem imply the existence of two valleys around these villages, but also of a mountain range represented by the sphere centered at the origin and of radius R. Under these hypotheses, Theorem D.1.3 states that there exists a lowest mountain pass at the level c.

Proof Our arguments are the same as those developed by Brezis and Nirenberg [35], who provided a more elementary proof with respect to the original one of Ambrosetti and Rabinowitz. The arguments we give in what follows combine two basic tools: Ekeland's variational principle and the following pseudogradient lemma.

Lemma D.1.4 *Let* $F : [0,1] \to E^\star$ *be a continuous function. Then, for any* $\varepsilon > 0$, *there exists a continuous map* $v : [0,1] \to E$ *such that, for all* $t \in [0,1]$,

$$\|v(t)\| \leq 1 \qquad \text{and} \qquad \langle F(t), v(t)\rangle_{E^\star,E} \geq \|F(t)\|_{E^\star} - \varepsilon.$$

Proof Fix $t_0 \in [0,1]$. Then there exists $w(t_0) \in E$ such that

$$\|w(t_0)\| \leq 1 \qquad \text{and} \qquad \langle F(t_0), w(t_0)\rangle_{E^\star,E} \geq \|F(t_0)\|_{E^\star} - \varepsilon.$$

By continuity, there exists an open neighborhood $\mathcal{N}(t_0)$ of t_0 such that for any $t \in \mathcal{N}(t_0)$,

$$\langle F(t), w(t_0)\rangle_{E^\star,E} \geq \|F(t)\|_{E^\star} - \varepsilon.$$

The set of all these neighborhoods $\mathcal{N}(t)$ (for $t \in [0,1]$) covers $[0,1]$. So, there exist $t_1, \ldots, t_k \in [0,1]$ such that $[0,1] \subset \bigcup_{j=1}^k \mathcal{N}(t_j)$. Next, we consider the partition of unity associated with this finite covering. So, for any $1 \leq j \leq k$, we define the continuous map

$$\phi_j(t) := \frac{\text{dist}\,(t, [0,1] \setminus \mathcal{N}(t_j))}{\sum_{i=1}^k \text{dist}\,(t, [0,1] \setminus \mathcal{N}(t_i))} \qquad \text{for any } t \in [0,1].$$

Set

$$v(t) := \sum_{j=1}^k \phi_j(t) w(t_j) \qquad \text{for any } t \in [0,1].$$

We first observe that

$$\|v(t)\| \leq \sum_{j=1}^k \phi_j(t) \, \|w(t_j)\| \leq \sum_{j=1}^k \phi_j(t) = 1\,.$$

To conclude the proof, it remains to argue that for any $t \in [0,1]$,

$$\langle F(t), v(t) \rangle_{E^\star, E} = \sum_{j=1}^k \phi_j(t) \langle F(t), w(t_j) \rangle_{E^\star, E} \geq \|F(t)\|_{E^\star} - \varepsilon.$$

This follows after observing that for any $1 \leq j \leq k$ (t being assumed to be fixed), only one of the following situations may occur:

(i) If $t \in \mathcal{N}(t_j)$, then $\langle F(t), w(t_j) \rangle_{E^\star, E} \geq \|F(t)\|_{E^\star} - \varepsilon$.

(ii) If $t \notin \mathcal{N}(t_j)$, then $\phi(t_j) = 0$.

This concludes the proof of the lemma. □

We are now in the position to give the proof of the mountain pass theorem. Consider the complete metric space Γ endowed with the metric

$$d(p_1, p_2) := \max_{t \in [0,1]} \|p_1(t) - p_2(t)\| \qquad \text{for any } p_1,\, p_2 \in \Gamma.$$

Define the continuous mapping

$$\Phi(p) := \max_{t \in [0,1]} J(p(t)) \qquad \text{for any } p \in \Gamma.$$

Then Φ is bounded from below, because $\Phi(p) \geq \max\{J(0),\, J(v_0)\}$. Moreover, $\inf_{p \in \Gamma} \Phi(p) = c$. Fix a positive integer n. Thus, by Ekeland's variational principle, there exists $p_n \in \Gamma$ such that

$$c \leq \Phi(p_n) \leq c + \frac{1}{n} \tag{D.2}$$

and

$$\Phi(p) - \Phi(p_n) + \frac{1}{n}\, d(p, p_n) \geq 0 \qquad \text{for any } p \in \Gamma. \tag{D.3}$$

Because Φ is continuous, it follows that

$$M_n := \{t \in [0,1] : J(p_n(t)) = \Phi(p_n)\} \neq \emptyset\,.$$

Thus, to conclude the proof, it is enough to show that there exists $t_n \in M_n$ such that $\|J'(p_n(t_n))\|_{E^\star} \leq 2/n$. To this end, we apply Lemma D.1.4 to $F(t) =$

$J'(p_n(t))$. Hence, there exists a continuous function $v : [0,1] \to E$ such that for any $t \in [0,1]$,

$$\|v(t)\| \leq 1 \qquad \text{and} \qquad \langle J'(p_n(t)), v(t)\rangle \geq \|J'(p_n(t))\|_{E^\star} - \frac{1}{n}. \tag{D.4}$$

On the other hand, because $c > \max\{J(0), J(v_0)\}$, there exists a continuous function $\alpha : [0,1] \to [0,1]$ such that $\alpha(0) = \alpha(1) = 0$ and $\alpha \equiv 1$ on an open set $U \supset M_n$. As admissible paths $p \in \Gamma$ in (D.3), we take $p_\varepsilon(t) = p_n(t) - \varepsilon\alpha(t)v(t)$, for $\varepsilon > 0$ sufficiently small. We also observe that $d(p_n, p_\varepsilon) \leq \varepsilon$. Our choice of α guarantees that $p_\varepsilon(0) = p_n(0) = 0$ and $p_\varepsilon(1) = p_n(1) = v_0$; hence, $p_\varepsilon \in \Gamma$. Let $t_\varepsilon \in [0,1]$ such that

$$J_\varepsilon(t_\varepsilon) = \max_{t\in[0,1]} J(p_\varepsilon(t)).$$

Setting $p = p_\varepsilon$ in (D.3), we obtain

$$J\left(p_n(t_\varepsilon) - \varepsilon\alpha(t_\varepsilon)v(t_\varepsilon)\right) - \max_{t\in[0,1]} J(p_n(t)) + \frac{\varepsilon}{n} \geq 0. \tag{D.5}$$

On the other hand, we have the asymptotic estimate

$$J\left(p_n(t_\varepsilon) - \varepsilon\alpha(t_\varepsilon)v(t_\varepsilon)\right) = J\left(p_n(t_\varepsilon)\right) - \varepsilon\langle J'(p_n(t_\varepsilon)), \alpha(t_\varepsilon)v(t_\varepsilon)\rangle + o(\varepsilon), \tag{D.6}$$

as $\varepsilon \to 0$. Relations (D.5) and (D.6), combined with the observation that

$$J(p_n(t_\varepsilon)) \leq \max_{t\in[0,1]} J(p_n(t)),$$

yield

$$\langle J'\left(p_n(t_\varepsilon)\right), \alpha(t_\varepsilon)v(t_\varepsilon)\rangle \leq \frac{1}{n} + o(1) \qquad \text{as } \varepsilon \to 0.$$

But $\alpha(t_\varepsilon) = 1$, provided $\varepsilon > 0$ is small enough. Hence,

$$\langle J'\left(p_n(t_\varepsilon)\right), v(t_\varepsilon)\rangle \leq \frac{1}{n} + o(1) \qquad \text{as } \varepsilon \to 0.$$

Using now (D.4), we deduce that

$$\|J'(p_n(t_\varepsilon))\|_{E^\star} \leq \frac{2}{n} + o(1) \qquad \text{as } \varepsilon \to 0. \tag{D.7}$$

We can assume, passing eventually at a subsequence, that $t_\varepsilon \to t_n$ as $\varepsilon \to 0$. Because $t_n \in M_n$, relation (D.7) yields $\|J'(p_n(t_n))\|_{E^\star} \leq 2/n$. Now, it is enough to take $u_n := p_n(t_n)$ and we obtain a sequence of "almost critical points" satisfying the conclusion. In the case where J satisfies the Palais–Smale condition, a standard compactness argument implies that c is a critical value of J corresponding to a nontrivial critical point. □

We point out that the conclusion in the mountain pass theorem remains true if assumption (D.1) in the Palais–Smale condition is replaced with

$$\sup_{n\geq 1} |J(u_n)| < \infty \quad \text{and} \quad \lim_{n\to\infty} (1 + \|u_n\|)\, \|J'(u_n)\|_{E^\star} = 0\,.$$

The previous weaker hypothesis corresponds to the Cerami compactness condition. More generally, Theorem D.1.3 is still valid if assumption (D.1) becomes

$$\sup_{n\geq 1} |J(u_n)| < \infty \quad \text{and} \quad \lim_{n\to\infty} g(\|u_n\|)\, \|J'(u_n)\|_{E^\star} = 0\,,$$

provided $g : [0, \infty) \to [1, \infty)$ is a continuous function such that $\int_1^\infty dt/g(t) = \infty$. We left this property as an exercise to the reader.

D.2 Application to the Emden–Fowler equation

A direct consequence of the mountain pass theorem is related to the existence of solutions for the subcritical Emden–Fowler boundary value problem

$$\begin{cases} -\Delta u = u^p & \text{in } \Omega, \\ u > 0 & \text{in } \Omega, \\ u = 0 & \text{on } \partial\Omega, \end{cases} \tag{D.8}$$

where $\Omega \subset \mathbb{R}^N$ is a bounded domain with smooth boundary and $1 < p < (N+2)/(N-2)$ if $N \geq 3$ or $1 < p < \infty$ if $N \in \{1, 2\}$. This equation was introduced by Emden [76] and Fowler [77]. Existence results for problem (D.8) are related not only to the values of p, but also to the geometry of Ω. For instance, problem (D.8) has no solution if $p \geq (N+2)/(N-2)$ (if $N \geq 3$) and if Ω is starshaped with respect to some point (say, with respect to the origin). This property was observed by Rellich [176] in 1940 and refound by Pohozaev [163] in 1965, and its proof relies on the Rellich–Pohozaev identity. More precisely, after multiplication by $x \cdot \nabla u$ in equation (D.8) and integration by parts we find

$$\int_\Omega \left(\frac{N}{p+1}\, u^{p+1} - \frac{N-2}{2}\, u^{p+1} \right) dx = \frac{1}{2} \int_{\partial\Omega} \left(\frac{\partial u}{\partial n} \right)^2 (x \cdot n) d\sigma(x). \tag{D.9}$$

Because Ω is starshaped with respect to the origin, then the right-hand side of (D.9) is positive; hence, problem (D.8) has no solution if $N/(p+1) - (N-2)/2 \leq 0$, which is equivalent to $p \geq (N+2)/(N-2)$. We point out that a very general variational identity was discovered in 1986 by Pucci and Serrin [168].

The situation is different if we are looking for *entire solutions* (that is, solutions on the whole space) of the Emden–Fowler equation either in the critical or in the supercritical case. For instance, the equation

$$-\Delta u = u^{(N+2)/(N-2)} \quad \text{in } \mathbb{R}^N,\ (N \geq 3),$$

admits the family of solutions

$$u_\lambda(x) = \frac{[CN(N-2)]^{(N-2)/4}}{(C^2+|x|^2)^{(N-2)/2}}\,,$$

for any $\lambda > 0$, where $C = \lambda^{2/(N-2)}N^{-2}(N-2)^{-2}$.

If Ω is not starshaped, Kazdan and Warner [114] showed in 1975 that problem (D.8) has a solution for any $p > 1$, provided Ω is an annulus in $\mathbb{R}^N$. We also point out that if $\Omega \subset \mathbb{R}^N$ is an *arbitrary* bounded domain with smooth boundary, then the perturbed Emden–Fowler problem

$$\begin{cases} -\Delta u = |u|^{p-1}u + \lambda u & \text{in } \Omega, \\ u \not\equiv 0 & \text{in } \Omega, \\ u = 0 & \text{on } \partial\Omega \end{cases}$$

has a solution, provided λ is an *arbitrary* real number and $1 < p < (N+2)/(N-2)$ if $N \geq 3$ or $1 < p < \infty$ if $N \in \{1, 2\}$. The proof combines the mountain pass theorem (if $\lambda < \lambda_1$) and the *dual variational method* (if $\lambda \geq \lambda_1$). As usually, λ_1 denotes the first eigenvalue of the Laplace operator $(-\Delta)$ in $H_0^1(\Omega)$.

D.3 Mountains of zero altitude

The limiting case $c = \max\{J(0), J(v_0)\}$ in the statement of the mountain pass theorem corresponds to *mountains of zero altitude*. This degenerate situation was studied by Pucci and Serrin [167]. They considered a functional $J : E \to \mathbb{R}$ of class C^1 that satisfies the following geometric conditions:

(a) There exist real numbers α, r, R such that $0 < r < R$ and $J(v) \geq \alpha$ for every $v \in E$ with $r < \|v\| < R$.

(b) $J(0) \leq \alpha$ and $J(v_0) \leq \alpha$ for some $v_0 \in E$ with $\|v_0\| > R$.

Under these hypotheses, combined with the standard Palais–Smale compactness condition, Pucci and Serrin established the existence of a critical point $u \in E \setminus \{0, v_0\}$ of J with corresponding critical value $c \geq \alpha$. Moreover, if $c = \alpha$, then the critical point can be chosen with $r < \|u\| < R$. Roughly speaking, the mountain pass theorem continues to hold for a mountain of zero altitude, provided it also has nonzero thickness. In addition, if $c = \alpha$, then the pass itself occurs precisely on the mountain, in the sense that it satisfies $r < \|u\| < R$.

REFERENCES

[1] Aftalion, A. and Reichel, W. (1997). Existence of two boundary blow-up solutions for semilinear elliptic equations. *J. Differential Equations*, **141**, 400–421.

[2] Agarwal, R. and O'Reagan, D. (2001). Existence theory for single and multiple solutions to singular positone boundary value problems. *J. Differential Equations*, **175**, 393–414.

[3] Alaa, N. E. and Pierre, M. (1993). Weak solutions of some quasilinear elliptic equations with data measures. *SIAM J. Math. Anal.*, **24**, 23–35.

[4] Alama, S. and Tarantello, G. (1996). On the solvability of a semilinear elliptic equation via an associated eigenvalue problem. *Math. Z.*, **221**, 467–493.

[5] Amann, H. (1971). On the existence of positive solutions of nonlinear elliptic boundary value problems. *Indiana Univ. Math. J.*, **21**, 125–146.

[6] Amann, H. (1976). Existence and multiplicity for semilinear elliptic boundary value problems. *Math. Z.*, **150**, 567–597.

[7] Ambrosetti, A. and Gámez, J. L. (1997). Branches of positive solutions for some semilinear Schrödinger equations. *Math. Z.*, **224**, 347–362.

[8] Ambrosetti, A. and Rabinowitz, P. H. (1973). Dual variational methods in critical point theory and applications. *J. Funct. Anal.* **14**, 349–381.

[9] Aris, R. (1975). *The Mathematical Theory of Diffusion and Reaction in Permeable Catalysts*. Clarendon Press, Oxford.

[10] Badiale, M. and Tarantello, G. (1997). Existence and multiplicity results for elliptic problems with critical growth and discontinuous nonlinearities. *Nonlinear Anal., T.M.A.*, **29**, 639–677.

[11] Ball, J. M. (1977). Convexity conditions and existence theorems in nonlinear elasticity. *Arch. Rational Mech. Anal.*, **63**, 337–403.

[12] Ball, J. M. (1998). The calculus of variations and materials science. Current and future challenges in the applications of mathematics. *Quart. Appl. Math.*, **56**, 719–740.

[13] Bandle, C., Diaz, G., and Diaz, I. J. (1994). Solutions d'équations de réaction–diffusion non linéaires explosant au bord parabolique. *C. R. Acad. Sci. Paris, Ser. I Math.*, **318**, 455–460.

[14] Bandle, C., Greco, A., and Porru, G. (1997). Large solutions of quasilinear elliptic equations: Existence and qualitative properties. *Boll. Unione Matematica Italiana*, **7**, 227–252.

[15] Bandle, C. and Marcus, M. (1992). 'Large' solutions of semilinear elliptic equations: Existence, uniqueness and asymptotic behavior. *J. Anal. Math.*, **58**, 9–24.

[16] Bandle, C. and Porru, G. (1994). Asymptotic behavior and convexity of

large solutions to nonlinear equations. *World Sc. Series in Appl. Anal.*, **3**, 59–71.

[17] Bandle, C., Pozio, A., and Tesei, A. (1987). The asymptotic behavior of the solutions of degenerate parabolic equations. *Trans. Amer. Math. Soc.*, **303**, 487–501.

[18] Bandle, C. and Stakgold, I. (1984). The formation of the dead core in parabolic reaction–diffusion problems. *Trans. Amer. Math. Soc.*, **286**, 275–293.

[19] Barbu, V. (1998). *Partial Differential Equations and Boundary Value Problems.* Kluwer Academic Publishers, Dordrecht.

[20] Benci, V., Fortunato, D., and Pisani, L. (1998). Solitons like solutions of a Lorentz invariant equation in dimension 3. *Rev. Math. Phys.*, **10**, 315–344.

[21] Benguria, R., Brezis, H., and Lieb, E. H. (1981). The Thomas–Fermi–von Weizsäcker theory of atoms and molecules. *Comm. Math. Phys.*, **79**, 167–180.

[22] Beznea, L. and Boboc N. (2004). *Potential Theory and Right Processes.* Kluwer Academic Publishers, Dordrecht.

[23] Bénilan, P., Brezis, H., and Crandall, M. (1975). A semilinear equation in $L^1(\mathbb{R}^N)$. *Ann. Scuola Norm. Sup. Pisa*, **4**, 523–555.

[24] Berestycki, H., Nirenberg, L., and Varadhan, S. R. S. (1994). The principal eigenvalue and maximum principle for second order elliptic operators in general domains. *Comm. Pure Appl. Math.*, **47**, 47–92.

[25] Bertsch, M. and Rostamian, R. (1985). The principle of linearized stability for a class of degenerate diffusion equations. *J. Differential Equations*, **57**, 373–405.

[26] Bieberbach, L. (1916). $\Delta u = e^u$ und die automorphen Funktionen. *Math. Ann.*, **77**, 173–212.

[27] Blanchard, P. and Brüning, E. (2003). *Mathematical Methods in Physics. Distributions, Hilbert Space Operators, and Variational Methods.* Progress in Mathematical Physics, Birkhäuser, Boston.

[28] Bocher, M. (1914). The smallest characteristic numbers in a certain exceptional case. *Bull. Amer. Math. Soc.*, **21**, 6–9.

[29] Bratu, G. (1914). Sur les équations intégrales non linéaires. *Bull. Soc. Math. France*, **42**, 113–142.

[30] Brezis, H. (1992). *Analyse Fonctionnelle. Théorie, Méthodes et Applications.* Masson, Paris.

[31] Brezis, H. and Browder, F. (1998). Partial differential equations in the 20th century. *Adv. Math.*, **135**, 76–144.

[32] Brezis, H. and Kamin, S. (1992). Sublinear elliptic equations in $\mathbb{R}^N$. *Manuscripta Math.*, **74**, 87–106.

[33] Brezis, H. and Nirenberg, L. (1983). Positive solutions of nonlinear elliptic equations involving critical Sobolev exponents. *Comm. Pure Appl. Math.*, **36**, 437–477.

[34] Brezis, H. and Nirenberg, L. (1989). A minimization problem with critical

exponent and nonzero data. *Symmetry in Nature, (a volume dedicated in honour of L. Radicati)*, vol. I, 129–140, Scuola Normale Superiore di Pisa.

[35] Brezis, H. and Nirenberg, L. (1991). Remarks on finding critical points. *Comm. Pure Appl. Math.*, **44**, 939–964.

[36] Brezis, H. and Nirenberg, L. (1993). H^1 versus C^1 local minimizers. *C. R. Acad. Sci. Paris, Ser. I*, **317**, 465–472.

[37] Brezis, H. and Oswald, L. (1986). Remarks on sublinear elliptic equations. *Nonlinear Anal., T.M.A.*, **10**, 55–64.

[38] Callegari, A. and Nachman, A. (1980). A nonlinear singular boundary value problem in the theory of pseudoplastic fluids. *SIAM J. Appl. Math.*, **38**, 275–281.

[39] Chayes, J. T., Osher, S. J., and Ralston, J. V. (1993). On singular diffusion equations with applications to self-organized criticality. *Comm. Pure Appl. Math.*, **46**, 1363–1377.

[40] Chen, H. (1997). On a singular nonlinear elliptic equation. *Nonlinear Anal., T.M.A.*, **29**, 337–345.

[41] Cheng, K. S. and Ni, W. M. (1992). On the structure of the conformal scalar curvature equation on $\mathbb{R}^N$. *Indiana Univ. Math. J.*, **41**, 261–278.

[42] Choi, Y. S., Lazer, A. C., and McKenna, P. J. (1998). Some remarks on a singular elliptic boundary value problem. *Nonlinear Anal., T.M.A.*, **3**, 305–314.

[43] Choi, Y. S. and McKenna, P. J. (2000). A singular Gierer–Meinhardt system of elliptic equations. *Ann. Inst. H. Poincaré, Anal. Non Linéaire*, **17**, 503–522.

[44] Choi, Y. S. and McKenna, P. J. (2003). A singular Gierer–Meinhardt system of elliptic equations: The classical case. *Nonlinear Anal.*, **55**, 521–541.

[45] Choquet–Bruhat, Y. and Leray, J. (1972). Sur le problème de Dirichlet quasilinéaire d'ordre 2. *C. R. Acad. Sci. Paris, Ser. A*, **274**, 81–85.

[46] Ciarlet, P. G. (2002). *The Finite Element Method for Elliptic Problems*. SIAM Classics in Applied Mathematics, SIAM, Philadelphia.

[47] Ciarlet, P. G. (2005). *An Introduction to Differential Geometry with Applications to Elasticity*. Springer-Verlag, Heidelberg.

[48] Cîrstea, F., Ghergu, M., and Rădulescu, V. (2005). Combined effects of asymptotically linear and singular nonlinearities in bifurcation problems of Lane–Emden–Fowler type. *J. Math. Pures Appl.*, **84**, 493–508.

[49] Cîrstea, F. and Rădulescu, V. (2002). Blow-up boundary solutions of semilinear elliptic problems. *Nonlinear Anal., T.M.A.*, **48**, 521–534.

[50] Cîrstea, F. and Rădulescu, V. (2002). Existence and uniqueness of blow-up solutions for a class of logistic equations. *Commun. Contemp. Math.*, **4**, 559–586.

[51] Cîrstea, F. and Rădulescu, V. (2002). Uniqueness of the boundary blow-up solution of logistic equation with absorbtion. *C. R. Acad. Sci. Paris, Ser. I*, **335**, 447–452.

[52] Cîrstea, F. and Rădulescu, V. (2002). Entire solutions blowing up at infinity for semilinear elliptic systems. *J. Math. Pures Appl.*, **81**, 827–846.

[53] Cîrstea, F. and Rădulescu, V. (2003). Solutions with boundary blow-up for a class of nonlinear elliptic problems. *Houston J. Math.*, **29**, 821–829.

[54] Clément, P. and Peletier, L. A. (1979). An anti-maximum principle for second order elliptic operators. *J. Differential Equations*, **34**, 218–229.

[55] Coclite, M. M. and Palmieri, G. (1989). On a singular nonlinear Dirichlet problem. *Comm. Partial Differential Equations*, **14**, 1315–1327.

[56] Cohen, D. S. and Keller, H. B. (1967). Some positive problems suggested by nonlinear heat generators. *J. Math. Mech.*, **16**, 1361–1376.

[57] Crandall, M. G., Rabinowitz, P. H., and Tartar, L. (1977). On a Dirichlet problem with a singular nonlinearity. *Comm. Partial Differential Equations*, **2**, 193–222.

[58] Dancer, E. N. (1996). Some remarks on classical problems and fine properties of Sobolev spaces. *Differential Integral Equations*, **9**, 437–446.

[59] Dávila, J. and Montenegro, M. (2003). Positive versus free boundary solutions to a singular equation. *J. Anal. Math.*, **90**, 303–335.

[60] de Gennes, P. G. (1985). Wetting: Statics and dynamics. *Rev. of Mod. Physics*, **57**, 827–863.

[61] del Pino, M. (1992). A global estimate for the gradient in a singular elliptic boundary value problem. *Proc. R. Soc. Edinburgh Sect. A*, **122**, 341–352.

[62] del Pino, M. (1994). Positive solutions of a semilinear elliptic equation on a compact manifold. *Nonlinear Anal., T.M.A.*, **22**, 1423–1430.

[63] del Pino, M., Kowalczyk, M., and Chen, X. (2001). The Gierer–Meinhardt system: The breaking of homoclinics and multi-bump ground states. *Commun. Contemp. Math.*, **3**, 419–439.

[64] del Pino, M., Kowalczyk, M., and Wei, J. (2003). Multi-bump ground states of the Gierer–Meinhardt system in $\mathbb{R}^2$. *Ann. Inst. H. Poincaré, Anal. Non Linéaire*, **20**, 53–85.

[65] Diaz, G. and Letelier, R. (1993). Explosive solutions of quasilinear elliptic equations: Existence and uniqueness. *Nonlinear Anal., T.M.A.*, **20**, 97–125.

[66] Diaz, J. I. (1986). *Nonlinear Partial Differential Equations and Free Boundaries. Elliptic Equations.* Pitman Adv. Publ., Boston.

[67] Diaz, J. I., Morel, J. M., and Oswald, L. (1987). An elliptic equation with singular nonlinearity. *Comm. Partial Differential Equations*, **12**, 1333–1344.

[68] Donsker, M. and Varadhan, S. R. S. (1976). On the principal eigenvalue of second-order elliptic differential operators. *Comm. Pure Appl. Math.*, **29**, 595–621.

[69] Doob, J. L. (2001). *Classical Potential Theory and Its Probabilistic Counterpart.* Springer-Verlag, Berlin.

[70] Du, Y. and Huang, Q. (1999). Blow-up solutions for a class of semilinear elliptic and parabolic equations. *SIAM J. Math. Anal.*, **31**, 1–18.

[71] Dumont, S., Dupaigne, L., Goubet, O., and Rădulescu, V. (2007). Back

to the Keller–Osserman condition for boundary blow-up solutions. *Adv. Nonlinear Studies*, **7**, 271–298.

[72] Dupaigne, L., Ghergu, M., and Rădulescu, V. (2007). Lane–Emden–Fowler equations with convection and singular potential. *J. Math. Pures Appl.*, **87**, 563–581.

[73] Dynkin, E. B. (2002). *Diffusions, Superdiffusions, and Partial Differential Equations*. Colloquium Publications, vol. 50. American Mathematical Society, Providence, RI.

[74] Edelson, A. (1989). Entire solutions of singular elliptic equations. *J. Math. Anal. Appl.*, **139**, 523–532.

[75] Ekeland, I. (1974). On the variational principle. *J. Math. Anal. Appl.*, **47**, 324–353.

[76] Emden, R. (1907). *Die Gaskügeln*. Teubner, Leipzig.

[77] Fowler, R. H. (1931). Further studies of Emden and similar differential equations. *Quart. J. Math.*, **2**, 259–288.

[78] Fowler, A. C. (1997). *Mathematical Models in the Applied Sciences*. Cambridge University Press, Cambridge.

[79] Fraile, J. M., Koch Medina, P., López–Gómez, J., and Merino, S. (1996). Elliptic eigenvalue problems and unbounded continua of positive solutions of a semilinear elliptic equation. *J. Differential Equations*, **127**, 295–319.

[80] Frank-Kamenetskii, D. A. (1955). *Diffusion and Heat Exchange in Chemical Kinetics*. Princeton University Press, Princeton.

[81] Fulks, W. and Maybee, J. (1960). A singular nonlinear equation. *Osaka J. Math.*, **12**, 1–19.

[82] Gelfand, I. M. (1963). Some problems in the theory of quasilinear equations. *Amer. Math. Soc. Transl.*, **29**, 295–381.

[83] Ghergu, M. and Rădulescu, V. (2003). Sublinear singular elliptic problems with two parameters. *J. Differential Equations*, **195**, 520–536.

[84] Ghergu, M. and Rădulescu, V. (2003). Bifurcation and asymptotics for the Lane–Emden–Fowler equation. *C. R. Acad. Sci. Paris, Ser. I*, **337**, 259–264.

[85] Ghergu, M. and Rădulescu, V. (2004). Bifurcation for a class of singular elliptic problems with quadratic convection term. *C. R. Acad. Sci. Paris, Ser. I*, **338**, 831–836.

[86] Ghergu, M. and Rădulescu, V. (2005). Multiparameter bifurcation and asymptotics for the singular Lane–Emden–Fowler equation with convection term. *Proc. R. Soc. Edinburgh Sect. A*, **135**, 61–84.

[87] Ghergu, M. and Rădulescu, V. (2005). On a class of sublinear singular elliptic problems with convection term. *J. Math. Anal. Appl.*, **311**, 635–646.

[88] Ghergu, M. and Rădulescu, V. (2006). Singular elliptic problems with convection term in anisotropic media. In C. Niculescu and V. Rădulescu (Eds.), *Mathematical Analysis and Applications*, 74–89, Amer. Inst. Phys., Melville, NY.

[89] Ghergu, M. and Rădulescu, V. (2007). On the influence of a subquadratic convection term in singular elliptic problems. In O. Cârjă and I. Vrabie (Eds.), *Applied Analysis and Differential Equations*, 127–138, World Scientific, Singapore.

[90] Ghergu, M. and Rădulescu, V. (2007). Ground state solutions for the singular Lane-Emden-Fowler equation with sublinear convection term. *J. Math. Anal. Appl.*, **333**, 265–273.

[91] Ghergu, M. and Rădulescu, V. (2007). On a class of singular Gierer–Meinhardt systems arising in morphogenesis. *C. R. Acad. Sci. Paris, Ser. I*, **344**, 163–168.

[92] Ghergu, M. and Rădulescu, V. (2008). A singular Gierer–Meinhardt system with different source terms. *Proc. R. Soc. Edinburgh Sect. A*, in press.

[93] Gidas, B., Ni, W. M., and Nirenberg, L. (1979). Symmetry and related properties via the maximum principle. *Comm. Math. Phys.*, **68**, 209–243.

[94] Gierer, A. and Meinhardt, H. (1972). A theory of biological pattern formation. *Kybernetik*, **12**, 30–39.

[95] Gilbarg, D. and Trudinger, N. S. (1983). *Elliptic Partial Differential Equations of Second Order*. Springer-Verlag, Berlin.

[96] Gomes, S. M. (1986). On a singular nonlinear elliptic problem. *SIAM J. Math. Anal.* **17**, 1359–1369.

[97] Gossez, J. P. and Omari, P. (1994). Non-ordered lower and upper solutions in semilinear elliptic problems. *Comm. Partial Differential Equations*, **19**, 1163–1184.

[98] Granas, A. and Dugundji, J. (2003). *Fixed Point Theory*. Springer-Verlag, New York.

[99] Gui, C. and Lin, F. H. (1993). Regularity of an elliptic problem with a singular nonlinearity. *Proc. R. Soc. Edinburgh Sect. A*, **123**, 1021–1029.

[100] Habets, P. and Omari, P. (1996). Existence and localization of solutions of second order elliptic problems using lower and upper solutions in the reversed order. *Topol. Methods Nonlinear Anal.*, **8**, 25–56.

[101] Haitao, Y. (2003). Multiplicity and asymptotic behavior of positive solutions for a singular semilinear elliptic problem. *J. Differential Equations*, **189**, 487–512.

[102] Hernández, J. and Mancebo, F. J. (2006). Singular elliptic and parabolic problems. In M. Chipot and P. Quittner (Eds.), *Handbook of Differential Equations: Stationary Partial Differential Equations*, 317–400, Elsevier, North-Holland.

[103] Hernández, J., Mancebo, F. J., and Vega, J. M. (2005). Nonlinear singular elliptic problems: Recent results and open problems. In H. Brezis et al. (Eds.), *Progress in Nonlinear Differential Equations and Their Applications*, 227–242, Birkhäuser, Basel.

[104] Hernández, J., Mancebo, F. J., and Vega, J. M. (2002). On the linearization of some singular elliptic problems. *Ann. Inst. H. Poincaré, Anal. Non Linéaire*, **19**, 777–813.

[105] Hess, P. (1981). An anti-maximum principle for linear elliptic equations with an indefinite weight function. *J. Differential Equations*, **41**, 369–374.

[106] Hess, P. (1991). *Periodic–Parabolic Boundary Value Problems and Positivity.* Longman Scientific and Technical, Harlow, UK.

[107] Hess, P. and Kato, T. (1980). On some linear and nonlinear eigenvalue problems with indefinite weight function. *Comm. Partial Differential Equations*, **5**, 999–1030.

[108] Hopf, E. (1927). Elementare Bemerkungen über die Lösungen partieller Differentialgleichungen zweiter Ordnung vom elliptischen Typus. *Sitz. Ber. Preuß. Akad. Wissensch. Berlin*, **19**, 147–152.

[109] Huiqiang, J. (2006). Global existence of solutions of an activator–inhibitor system. *Discrete Contin. Dyn. Syst.*, **14**, 737–751.

[110] Iscoe, I. (1998). On the supports of measure-valued critical branching Brownian motions, *Ann. Probab.*, **16**, 200–221.

[111] Jacobsen, J. and Schmitt, K. (2002). The Liouville–Bratu–Gelfand problem for radial operators. *J. Differential Equations*, **184**, 283–298.

[112] Joseph, D. D. and Lundgren, T. S. (1973). Quasilinear Dirichlet problems driven by positive sources. *Arch. Rational Mech. Anal.*, **49**, 241–269.

[113] Karamata J.(1933). Sur un mode de croissance régulière de fonctions. Théorèmes fondamentaux. *Bull. Soc. Math. France*, **61**, 55–62.

[114] Kazdan, J. L. and Warner, F. W. (1975). Remarks on some quasilinear elliptic equations. *Comm. Pure Appl. Math.*, **28**, 567–597.

[115] Kazdan, J. L. and Warner, F. W. (1975). Scalar curvature and conformal deformation of Riemannian structure. *J. Differential Geometry*, **10**, 113–134.

[116] Keener, J. P. (1978). Activators and inhibitors in pattern formation. *Stud. Appl. Math.*, **59**, 1–23.

[117] Keller, J. B. (1956). Electrohydrodynamics. I. The equilibrium of a charged gas in a container. *J. Rational Mech. Anal.*, **5**, 715–724.

[118] Keller, J. B. (1957). On solution of $\Delta u = f(u)$. *Comm. Pure Appl. Math.*, **10**, 503–510.

[119] Kim, E. H. (2004). A class of singular Gierer–Meinhardt systems of elliptic boundary value problems. *Nonlinear Anal.*, **59**, 305–318.

[120] Kim, E. H. (2005). Singular Gierer–Meinhardt systems of elliptic boundary value problems. *J. Math. Anal. Appl.*, **308**, 1–10.

[121] Krantz, S. G. and Parks, H. P. (2002). *The Implicit Function Theorem. History, Theory, and Applications*. Birkhäuser, Boston.

[122] Kusano, T. and Swanson, C. A. (1985). Entire positive solutions of singular semilinear elliptic equations. *Japan J. Math.*, **11**, 145–155.

[123] Ladyzhenskaya, O. A. and Ural'tseva, N. N. (1968). *Linear and Quasilinear Elliptic Equations*. Academic Press, New York.

[124] Lair, A. V. and Shaker, A. W. (1996). Entire solution of a singular semilinear elliptic problem. *J. Math. Anal. Appl.*, **200**, 498–505.

[125] Lair, A. V. and Shaker, A. W. (1997). Classical and weak solutions of a singular semilinear elliptic problem. *J. Math. Anal. Appl.*, **211**, 371–385.

[126] Lair, A. V. and Shaker, A. W. (2000). Existence of entire large positive solutions of semilinear elliptic systems. *J. Differential Equations*, **164**, 380–394.

[127] Lazer, A. C. and McKenna, P. J. (1991). On a singular nonlinear elliptic boundary value problem. *Proc. Amer. Math. Soc.*, **3**, 721–730.

[128] Lazer, A. C. and McKenna, P. J. (1993). On a problem of Bieberbach and Rademacher. *Nonlinear Anal., T.M.A.*, **21**, 327–335.

[129] Lazer, A. C. and McKenna, P. J. (1994). Asymptotic behavior of solutions of boundary blow-up problems. *Differential and Integral Equations*, **7**, 1001–1020.

[130] Le Gall, J. F. (1997). A probabilistic Poisson representation for positive solutions $\Delta u = u^2$ in a planar domain. *Comm. Pure Appl. Math.*, **50**, 69–103.

[131] Leray, J. and Schauder, J. P. (1934). Topologie et équations fonctionnelles. *Ann. Sci. Ec. Norm. Sup.*, **51**, 43–78.

[132] Liouville, J. (1853). Sur l'équation aux dérivées partielles $\partial^2 \log \lambda / \partial u \partial \pm 2a^2\lambda = 0$. *J. Math. Pures Appl.*, **18**, 71–72.

[133] Loewner, C. and Nirenberg, L. (1974). Partial differential equations invariant under conformal or projective transformations. In L.V. Ahlfors et al. (Eds.), *Contributions to Analysis*, 245–272, Academic Press, New York.

[134] Luning, C. D. and Perry, W. L. (1984). An interactive method for solution of a boundary value problem in non-Newtonian fluid flow. *J. Non-Newtonian Fluid Mech.*, **15**, 145–154.

[135] Marcus, M. (1992). On solutions with blow-up at the boundary for a class of semilinear elliptic equations. In G. Buttazzo et al. (Eds.), *Developments in Partial Differential Equations and Applications to Mathematical Physics*, 65–77, Plenum Press, New York.

[136] Marcus, M. and Véron, L. (1997). Uniqueness and asymptotic behavior of solutions with boundary blow-up for a class of nonlinear elliptic equations. *Ann. Inst. H. Poincaré Anal. Non Linéaire*, **14**, 237–274.

[137] McKenna, P. J., and Reichel, W. (2001). Sign-changing solutions to singular second-order boundary value problems. *Adv. Differential Equations*, **6**, 441–460.

[138] Meadows, A. (2004). Stable and singular solutions of the equation $\Delta u = 1/u$. *Indiana Univ. Math. J.*, **53**, 1681–1703.

[139] Meinhardt, H. (1976). Morphogenesis of lines and nets. *Differentiation*, **6**, 117–123.

[140] Meinhardt, H. and Gierer, A. (1980). Generation and regeneration of sequence of structures during morphogenesis. *J. Theoret. Biol.*, **85**, 429–450.

[141] Minakshisundaran, S. and Pleijel, A. (1949). Some properties of the eigenfunctions of the Laplace operator on Riemann manifolds. *Can. J. Math.*, **1**, 242–256.

[142] Mironescu, P. and Rădulescu, V. (1993). A bifurcation problem associated to a convex, asymptotically linear function. *C. R. Acad. Sci. Paris, Ser. I*, **316**, 667–672.

[143] Mironescu, P. and Rădulescu, V. (1996). The study of a bifurcation problem associated to an asymptotically linear function. *Nonlinear Anal.*, **26**, 857–875.

[144] Miyamoto, Y. (2006). An instability criterion for activator–inhibitor systems in a two-dimensional ball. *J. Differential Equations*, **229**, 494–508.

[145] Murray, J. D. (2003). *Mathematical Biology II: Spatial Models and Biomedical Applications.* Springer, New York.

[146] Nagumo, M. (1937). Ueber die Differentialgleichung $y'' = f(x, y, y')$, *Proc. Phys. Math. Soc. Japan*, **19**, 861–866.

[147] Nash, J. (1956). The imbedding problem for Riemannian manifolds. *Ann. Math.* **63**, 20–63.

[148] Ni, W. M. (1998). Difussion, cross-difusion, and their spike-layer steady states. *Notices Amer. Math. Soc.*, **45**, 9–18.

[149] Ni, W. M. (2004). Diffusion and cross-diffusion in pattern formation. *Atti Accad. Naz. Lincei Cl. Sci. Fis. Mat. Natur. Rend. Lincei Mat. Appl.*, **15**, 197–214.

[150] Ni, W. M., Suzuki, K., and Takagi, I. (2006). The dynamics of a kynetics activato-inhibitor system. *J. Differential Equations*, **229**, 426–465.

[151] Ni, W. M. and Wei, J. (2006). On positive solutions concentrating on spheres for the Gierer–Meinhardt system. *J. Differential Equations*, **221**, 158–189.

[152] O'Reagan, D. (1997). *Existence Theory for Nonlinear Ordinary Differential Equations.* Kluwer Academic Publishers, Dordrecht.

[153] O'Regan, D. and Precup, R. (2001). *Theorems of Leray–Schauder Type and Applications.* Gordon and Breach Science Publishers, Amsterdam.

[154] Osserman, R. (1957). On the inequality $\Delta u \geq f(u)$. *Pacific J. Math.*, **7**, 1641–1647.

[155] Ouyang, T. (1992). On the positive solutions of semilinear equations $\Delta u + \lambda u - hu^p = 0$ on the compact manifolds. *Trans. Amer. Math. Soc.*, **331**, 503–527.

[156] Ouyang, T., Shi J., and Yao, M. Exact multiplicity of positive solutions for a singular equation in unit ball, preprint.

[157] Palais, R. (1966). Ljusternik–Schnirelmann theory on Banach manifolds. *Topology*, **5**, 115–132.

[158] Palais, R. and Smale, S. (1964). A generalized Morse theory. *Bull. Amer. Math. Soc.*, **70**, 165–171.

[159] Perron, O. (1923). Eine neue Behandlung der Randwertaufgabve für $\Delta u = 0$. *Math. Z.*, **18**, 42–54.

[160] Perry, W. L. (1984). A monotone iterative technique for solution of pth order ($p < 0$) reaction–diffusion problems in permeable catalysis. *J. Comput. Chemistry*, **5**, 353–357.

[161] Phelps, R. R. (1974). Support cones in Banach spaces and their applications. *Adv. Math.*, **13**, 1–19.

[162] Pleijel, A. (1950). On the eigenvalues and eigenfunctions of elastic plates. *Comm. Pure Appl. Math.*, **3**, 1–10.

[163] Pohozaev, S. (1965). On the eigenfunctions of the equation $\Delta u + \lambda f(u) = 0$. *Dokl. Akad. Nauk SSSR*, **165**, 36–39.

[164] Poincaré, H. (1890). Sur les équations aux dérivées partielles de la physique mathématique. *Amer. J. Math.*, **12**, 211–294.

[165] Poincaré, H. (1894). Sur les équations de la physique mathématique. *Rend. Circ. Mat. Palermo*, **8**, 57–155.

[166] Poincaré, H. (1898). Les fonctions fuchsiennes et l'équation $\Delta u = e^u$. *J. Math. Pures Appl.*, **4**, 137–230.

[167] Pucci, P. and Serrin, J. (1984). Extensions of the mountain pass theorem. *J. Funct. Anal.*, **59**, 185–210.

[168] Pucci, P. and Serrin, J. (1986). A general variational identity. *Indiana Univ. Math. J.*, **35**, 681–703.

[169] Pucci, P. and Serrin, J. (2000). A note on the strong maximum principle for singular elliptic inequalities. *J. Math. Pures Appl.*, **79**, 57–71.

[170] Pucci, P. and Serrin, J. (2004). The strong maximum principle revisited. *J. Differential Equations*, **196**, 1–66.

[171] Pucci, P. and Serrin, J. (2006). Dead cores and bursts for quasilinear singular elliptic equations. *SIAM J. Math. Anal.*, **38**, 259–278.

[172] Pucci, P. and Serrin, J. (2007). Maximum principles for elliptic partial differential equations. In M. Chipot (Ed.), *Handbook of Differential Equations: Stationary Partial Differential Equations*, 355–482, Elsevier, North-Holland.

[173] Rademacher, H. (1943). Einige besondere problem partieller Differentialgleichungen. *Die differential und Integralgleichungen der Mechanik und Physik I*, 838–845, Rosenberg, New York.

[174] Rădulescu, V. (2005). Bifurcation and asymptotics for elliptic problems with singular nonlinearity. In C. Bandle et al. (Eds.), *Studies in Nonlinear Partial Differential Equations: In Honor of Haim Brezis, Fifth European Conference on Elliptic and Parabolic Problems: A special tribute to the work of Haim Brezis*, 349–362, Birkhäuser, Basel.

[175] Rădulescu, V. (2007). Singular phenomena in nonlinear elliptic problems. From blow-up boundary solutions to equations with singular nonlinearities. In M. Chipot (Ed.), *Handbook of Differential Equations: Stationary Partial Differential Equations*, 483–591, Elsevier, North-Holland.

[176] Rellich, F. (1940). Darstellung der Eigenwerte von $\Delta u + \lambda u = 0$ durch ein Randintegral. *Math. Z.*, **46**, 635–636.

[177] Riesz, F. (1926) Sur les fonctions subharmoniques et leur rapport à la théorie du potentiel. *Acta Math.*, **48**, 329–343.

[178] Sattinger, D. H. (1972). Monotone methods in nonlinear elliptic and parabolic boundary value problems. *Indiana Univ. Math. J.*, **21**, 979–1000.

[179] Scorza Dragoni, G. (1931). Il problema dei valori ai limiti studiato in grande per gli integrali di una equazione differenziale del secondo ordine. *Giornale di Mat. (Battaglini)*, **69**, 77–112.

[180] Seneta, E. (1976). *Regularly Varying Functions*. Springer-Verlag, Berlin.

[181] Serrin, J. (1964). Local behavior of solutions of quasi-linear equations. *Acta Math.*, **111**, 247–302.

[182] Shi, J. and Yao, M. (1998). On a singular nonlinear semilinear elliptic problem. *Proc. R. Soc. Edinburgh Sect. A*, **128**, 1389–1401.

[183] Stampacchia, G. (1966). *Équations elliptiques du second ordre à coefficients discontinus.* Presses de l'Université de Montréal, Montréal.

[184] Struwe, M. (1990). *Variational Methods. Applications to Nonlinear Partial Differential Equations and Hamiltonian Systems*. Springer-Verlag, Berlin.

[185] Sullivan, F. (1981). A characterization of complete metric spaces. *Proc. Amer. Math. Soc.*, **83**, 345–346.

[186] Taliaferro, S. D. (1979). A nonlinear singular boundary value problem. *Nonlinear Anal., T.M.A.*, **3**, 897–904.

[187] Tarantello, G. (1992). On nonhomogeneous elliptic equations involving critical Sobolev exponent. *Ann. Inst. H. Poincaré, Anal. Non Linéaire*, **9**, 281–304.

[188] Trembley, A. (1744). *Mémoires pour servir à l'histoire d'un genre de polype d'eau douce, à bras en forme de corne.* Verbeek, Leiden, Netherland.

[189] Turing, A. M. (1952). The chemical basis of morphogenesis. *Philos. Trans. R. Soc. London B*, **237**, 37–72.

[190] Vázquez, J. L. (1984). A strong maximum principle for some quasilinear elliptic equations. *Appl. Math. Optim.*, **12**, 191–202.

[191] Véron, L. (1992). Semilinear elliptic equations with uniform blow-up on the boundary. *J. Anal. Math.*, **59**, 231–250.

[192] Wei, J. (1998). On the interior spike layer solutions for some singular perturbation problems. *Proc. R. Soc. Edinburgh Sect. A*, **128**, 849–874.

[193] Wei, J. and Winter, M. (2002). Spikes for the Gierer–Meinhardt system in two dimensions: The strong coupling case. *J. Differential Equations*, **178**, 478–518.

[194] Wei, J. and Winter, M. (2004). Existence and stability analysis of asymmetric patterns for the Gierer–Meinhardt system. *J. Math. Pures Appl.*, **83**, 433–476.

[195] Widman, K. O. (1967). Inequalities for the Green function and boundary continuity of the gradient of solutions of elliptic differential equations. *Math. Scand.*, **21**, 17–37.

[196] Willgoose, G., Bras, R. L., and Rodriguez–Iturbe, I. (1991). A coupled channel network growth and hillslope evolution model. *Water Resourc. Res.*, **27**, 1671–1684.

[197] Wong, J. S. W. (1975). On the generalized Emden–Fowler equation. *SIAM Rev.*, **17**, 339–360.

[198] Yanagida, E. (2002). Mini-maximizers for reaction–diffusion systems with skew-gradient structure. *J. Differential Equations*, **179**, 311–335.

[199] Yijing, W. and Shujie, L. (2003). Structure of ground state solutions of singular semilinear elliptic equations. *Nonlinear Anal.*, **55**, 399–417.

[200] Zhang, Z. (1996). A remark on the existence of entire solutions of a singular semilinear elliptic problem. *J. Math. Anal. Appl.*, **215**, 579–582.

[201] Zhang, Z. (2006). The existence and asymptotic behaviour of the unique solution near the boundary to a singular Dirichlet problem with a convection term. *Proc. R. Soc. Edinburgh Sect. A*, **136**, 209–222.

[202] Zhang, Z. and Cheng, J. (2004). Existence and optimal estimates of solutions for singular nonlinear Dirichlet problems. *Nonlinear Anal.*, **57**, 473–484.

INDEX